The Fisheries Co-management Experience

Fish and Fisheries Series

VOLUME 26

The Kluwer Fish and Fisheries Series begins with the new millennium. The volumes in this series will cover topics ranging from the biology of individual species or groups of fishes, to broader concepts in fisheries science, conservation and management. The Series is directed to professionals and researchers in fish biology, senior undergraduate and postgraduate students, and those concerned with commercial production or harvest of fishes.

It is difficult to know if this is the best of times or the worst of times for fish and fisheries. For example, many of the historically important marine fisheries are at or very near harvest capacity, according to the best scientific data and predictions. Many of the changes in commercial harvests fit a predictable, depressing pattern. We tend to produce simplified communities, harvested by ever more efficient technology, at increasing rates of exploitation. Some would suggest that nothing can stop the apparently inevitable destruction of all commercially harvested fishes. Fish habitats seem to be increasingly degraded, and the deliberate or accidental introductions of exotic species threaten endemic native fishes. We always do things to the limit of our technology.

However, in contrast, we can point to a number of very favourable examples of current success and future promise in fish and fisheries. Our knowledge of the basic biology of fishes continues to expand. We disseminate that knowledge with ever increasing speed to libraries and personal computers around the world. Many fishes are increasingly recognized as fundamentally important subjects for basic research. Studies of the zebrafish, *Brachydanio rerio*, have produced a veritable explosion of fundamental scientific information at major research institutions around the world. Fishes as diverse as arctic charr, *Salvelinus alpinus*, and stickleback, *Gasterosteus aculeatus*, are providing insights and new understanding of the fundamental processes of natural selection and speciation. Science and technology give us a better understanding of the implications of long term climate change for fish populations. We continue to see fundamental breakthroughs in our understanding of development, genetics and evolution of fishes on almost a daily basis. Production of fishes through our increasingly sophisticated and efficient aquaculture rivals or exceeds the harvest of wild fishes in many places. Our knowledge and understanding continue to develop to the limits of our science and technology.

Science and the promise it holds for us to deal with our questions and concerns about fish and fisheries, is the basis for this Series. The future is certainly not what it used to be. This and forthcoming volumes in the Kluwer Fish and Fisheries Series will define the scientific basis for our future interactions with fishes. It is truly an exciting time.

Dr. David L. G. Noakes
Series Editor, Fish and Fisheries Series
Professor of Zoology, University of Guelph
Guelph, Canada

The Fisheries Co-management Experience

Accomplishments, Challenges and Prospects

Edited by

Douglas Clyde Wilson
Jesper Raakjaer Nielsen
and
Poul Degnbol

The Institute for Fisheries Management
and Coastal Community Development,
Hirtshals, Denmark

KLUWER ACADEMIC PUBLISHERS
DORDRECHT / BOSTON / LONDON

A C.I.P. Catalogue record for this book is available from the Library of Congress.

ISBN 1-4020-1427-9

Published by Kluwer Academic Publishers,
P.O. Box 17, 3300 AA Dordrecht, The Netherlands.

Sold and distributed in North, Central and South America
by Kluwer Academic Publishers,
101 Philip Drive, Norwell, MA 02061, U.S.A.

In all other countries, sold and distributed
by Kluwer Academic Publishers,
P.O. Box 322, 3300 AH Dordrecht, The Netherlands.

Printed on acid-free paper

Printed in the Netherlands.

Table of Contents

Co-management: bringing it together

This is a time of uncertainty and change for fisheries management; the only thing certain is uncertainty. It is a time of unprecedented change, of shifts in concepts and priorities. The global climate is changing, perhaps as a decadal shift. Native species are becoming endangered at what appear to be unprecedented rates. Exotic species are invading with potentially catastrophic consequences for individual native species and communities. Harvests of wild fish stocks are declining. The demands on fisheries and aquatic sciences are ever increasing, from an ever-increasing number of individuals, organizations, agencies and nations. Fresh water itself is increasingly seen as a major resource and is the basis for conflicting management demands. Aquaculture is the fastest growing sector of captive animal production, but is perceived by some as yet another conflicting demand on water and fisheries production.

In this context, this volume is notable and timely for a number of reasons. The authors are remarkably diverse in their interests, their experiences and their geographic locations. They range from graduate students to senior scientists, from anthropologists to ecologists. Their experiences, and their chapters, include examples of fisheries management from around the globe.

This volume is forward looking. It not only looks towards the future, it also presents a view of that future and a blueprint for the future. It offers new insights for conflict resolution. It is unusual within the field of fisheries in that it combines theory and reality. It includes, and is directed towards, both academics and those responsible for management.

This volume is particularly timely for anyone concerned with research or management of fishes. I am located in the midst of a very large freshwater system, the Laurentian Great Lakes. My interactions with fish and fisheries in the Great Lakes include First Nations (with aboriginal treaty rights), federal government ministries and agencies in both Canada and the U. S. A., provincial ministries and state departments with responsibilities for fisheries management in both countries, the Great Lakes Fishery Commission (an agency formed by treaty between Canada and the U. S. A.), various groups with interests in recreational fishing, private corporations licensed to operate nuclear and hydroelectric generating stations, groups representing the interests of commercial fishing other than aboriginal treaties, groups with interests in commercial aquaculture production, several organizations with interests in conservation and restoration of endangered species, among others.

Each of these entities has its particular interests, priorities and objectives. All operate within a maze of federal, provincial and international legislation, agreements and treaties. History, economics, politics and biology are obvious concerns of most of these parties, but there is often strong disagreement as to the importance of these disciplines and the necessity for including other disciplines. Some parties claim special jurisdiction or regulatory responsibilities. Some have strongly contradictory demands and priorities. Many feel that they have a right, or responsibility, to be involved in fisheries management or co-management.

The history of fisheries management in the Laurentian Great Lakes is well documented. The invasion of parasitic sea lamprey, Petromyzon marinus, into the upper Great Lakes is a classic textbook example. The sequential impacts of invading species, changing land use practices, urbanization and industrialization, and conflicting demands of growing human populations have all been clearly documented in the Great Lakes over the past 500 years. At various times the lakes have been declared dying or even dead. We have variously treated them as inexhaustible reservoirs of water and fishes, sewers for domestic and industrial wastes, pathways for commerce and exploration, and inspiration for legends and songs.

It is remarkable that we, and the Lakes, have survived as well and as long as we have. Their continued stability and resilience are probably a tribute in large part to their sheer size. We sometimes seem to have stumbled from one management crisis to another. I welcome this volume, and the new insights and advice it has to offer. We can all learn a great deal from it.

Dr. David L.G. Noakes
Series Editor, Fish and Fisheries Series
Professor of Zoology, University of Guelph
Guelph, Canada

Contributors

Juan-Luis Alegret studied social anthropology at UAB in Barcelona. He was enrolled in the Ph.D. programme on fisheries issues at the Université Laval (Quebec). He worked as a teacher in Nicaragua (UCA-Managua) and Barcelona (UAB). He now teaches at UdG in Girona (Spain). His research work deals mainly with the management, heritage and history of fisheries in Catalonia (Spain).
Universitat de Girona, Plaça de Sant Doménec 3, 17071 Girona, Spain. juan.alegret@udg.es

Alpina Begossi, Ph.D. in Ecology (University of California, Davis, 1989). Has been studying Amazonian and Atlantic Forest fisheries as one of her main research lines, among other human ecological studies. She has about 80 published works, that include articles published in periodicals, such as Environmental Development and Sustainability, Fisheries Research, Human Ecology, Journal of Ethnobiology, and Maritime Anthropological studies, among others. She is currently Associate Coordinator of the Center of Environmental Studies and Research (NEPAM) at the State University of Campinas (UNICAMP), Brazil, and she teaches Human Ecology at the Graduate Group in Ecology, at UNICAMP.
Center of Environmental Studies and Research (NEPAM), State University of Campinas (UNICAMP), CP 6166, Campinas SP 13081-970, Brazil. alpina@nepam.unicamp.br

David Brown is Sociologist with the CARICOM (Caribbean Community) Fisheries Unit in Belize; a regional institution for promoting the sustainable development and management of the fisheries resources of 12 CARICOM states. He coordinates activities for the empowerment of small-scale fishers through institution and capacity building and strengthening programmes for resource co-management. He holds a Ph.D. in Sociology of Development from McMaster University, Canada. His current areas of interest include social impact assessment, fisheries socio-economics, and community-based resource co-management, with special reference to the English-speaking Caribbean.
CARICOM Fisheries Unit, P.O. Box 642, Princess Margaret Drive, Belize City, Belize, C.A. dbrown@caricom-fisheries.com

Poul Degnbol has worked as fisheries biologist with FAO and the Danish Institute for Fisheries Research, where he was research director until 1999, when he took up the position as director at IFM. His experience includes development and implementation of fisheries management in Europe and several countries in southeast Asia and Africa. From 2003 he chairs the Advisory Committee of Fisheries Management of the International Council for the Exploration of the Sea.
Institute for Fisheries Management and Coastal Community Development, P.O. Box 104, DK-9850 Hirtshals, Denmark. pd@ifm.dk

Jay J.C. Ginter oversees the Regulatory Operations Branch at the Alaska Regional office of the National Marine Fisheries Service in Juneau, Alaska. He has worked for this government agency in Alaska since 1985. His academic training is in fishery management policy at the University of Washington, Seattle, marine environmental studies at the State University of New York at Stony Brook.
National Marine Fisheries Service, Alaska Region, P.O. Box 21668, Juneau, Alaska 99802-1668. Jay.Ginter@noaa.gov

Susan Hanna is professor of marine economics at Oregon State University. Her research and publications are in the areas of fishery economics and management, fishery policy and property rights. She has served as a scientific adviser to the Pacific Fishery Management Council, National Marine Fisheries Service, and National Oceanic and Atmospheric Administration.
Department of Agricultural and Resource Economics, Oregon State University, 229 Ballard Extension Hall, Corvallis, Oregon 97331-3601, USA. susan.hanna@orst.edu

Mafaniso Hara is a Senior Researcher at the Programme for Land and Agrarian Studies at the University of Western Cape, South Africa. He obtained his Ph.D. in 2001. His research interests are in the area of Community Based Management and Co-management of Fisheries and Coastal resources and Indigenous Fisher Knowledge.
Programme for Land and Agrarian Studies (PLAAS), School of Government, University of the Western Cape, Private Bag X17, Modderdam Road, Bellville, ZA-7535 Bellville, Cape Town, South Africa. mhara@uwc.ac.za

Michael Harte is currently Economic Adviser to the Falkland Island Government. He is working with the local fishing industry on a series of major legislative and institutional reforms. Prior to the Falkland Islands, Dr. Harte worked for the New Zealand Seafood Industry Council where he was a strong advocate for co-management.
Secretariat, Falkland Islands Government, Falkland Islands, South Atlantic. MHarte@sec.gov.fk

Hans-Kristian Hernes is Associate Professor of Political Science at the University of Tromsø where he teaches political theory and public policy. He has done research on relations between the state and interest groups in Norwegian fisheries, and is currently engaged in research on globalization and democracy and on the 'greening' of marine resource management.
Faculty of Social Science, University of Tromsø, N-9037 Tromsø, Norway. hansh@sv.uit.no

Svein Jentoft is Professor of Sociology at the Department of Planning and Community Studies at the University of Tromsø, Norway. He has worked on fisheries related management throughout his whole career and has published books and articles on topics such as fisheries management, aquaculture, community development and industrial organisation.
Department of Planning and Community Studies, University of Tromsø, N-9037 Tromsø, Norway. sveinj@sv.uit.no

Duncan Leadbitter joined the Marine Stewardship Council as International Fisheries Director and is currently the Regional Director, Asia Pacific region. Previously, he was the Executive Director of Ocean Watch Australia. He has also worked for the New South Wales, Australia fisheries agency and was the Deputy Director of the Australian Seafood Industry Council.
PO Box 305, Stanwell Park, NSW 250, Australia. paradise45@bigpond.co

Laura Loucks is a Ph.D. candidate at the School for Resource and Environmental Management at Simon Fraser University, in Burnaby British Columbia. Her thesis research focuses on sustainable fisheries governance, particularly the interlinkages between socio-economic decision making and marine conservation. In addition, Laura is co-president of eco-Planning Consulting and works on the West Coast of Vancouver Island designing sustainable community-based resource tenure systems.
Simon Fraser University, 8888 University Drive, Burnaby, B.C., Canada V5A 1S6. laloucks@sfu.ca

Bonnie McCay is Board of Governors Distinguished Professor of Anthropology and Ecology at Rutgers University in New Jersey, USA. She remains involved in research on fishing communities in Newfoundland, Nova Scotia, and New Jersey - all 'new' places - and their relationship to fisheries management.
Department of Human Ecology, Cook College, Rutgers the State University, 210 Cook Office Building, 55 Dudley Road, New Brunswick, New Jersey, 08901, USA. mccay@AESOP.RUTGERS.EDU

Rebecca Metzner is currently a Fishery Officer - Fishing Capacity in the Fishery Policy and Planning Division at the Food and Agriculture Organization of the United Nations in Rome, Italy. Prior to FAO, she held the positions of Principal Policy Officer, Fisheries Western Australia and Principal Economist, Australian National fisheries Adjustment Scheme project.
Via Aventina 3A, int. 11, 00153 Rome, Italy. rmetzner@alumni.princeton.edu

Knut H. Mikalsen is Professor of Political Science at the University of Tromsø. He is currently engaged in research on issues of stakeholder management in European fisheries.
Department of Political Science SVF, University of Tromsø, N-9037 Tromsø, Norway. knutm@sv.uit.no

Jesper Raakjær Nielsen, Research Professor/Principal researcher at IFM, Ph.D. in Institutional Economics. Main research interest is governance in fisheries with emphasis on the role of management institutions and conditions under which they work effectively. Substantial research experience in Europe, Southern Africa and Southeast Asia and a frequently used consultant for international development organisations. Has extensive practical experience in the fishing industry.
Institute for Fisheries Management and Coastal Community Development, P.O. Box 104, DK-9850 Hirtshals, Denmark. jrn@ifm.dk

Evelyn Pinkerton, maritime anthropologist, School of Resource and Environmental Management, Simon Fraser University, has authored some 30 peer-reviewed publications on co-management of fisheries and related resources over the last 15 years. She is currently researching the impact of co-management arrangements on government agencies, and other issues surrounding implementation.
School of Resource and Environmental Management, Simon Fraser University, Burnaby, B.C., Canada V5A 1S6. epinkert@sfu.ca

Caroline Pomeroy is a Research Scientist and Lecturer at the University of California, Santa Cruz. Her work has included studies of the social and economic organization of the California squid/wetfish fishery, the socio-economic impacts of pinniped interactions on California salmon trollers, and the human dimensions of marine reserves.
Institute of Marine Sciences, Earth and Marine Sciences Bldg. A 316, UC Santa Cruz, Santa Cruz, CA 95064, USA. cpomeroy@cats.ucsc.edu

Robert S. Pomeroy is currently an Associate Professor in Agricultural and Resource Economics at the University of Connecticut. He has worked at the World Resources Institute in Washington DC and the International Center for Living Aquatic Resources Management. He has worked on co-management research and development projects in Asia, Africa, Caribbean and Latin America.
380 Marine Science Building, 1080 Shennecossett Road, University of Connecticut - Avery Point, Groton, CT 06340, USA. robert.pomeroy@uconn.edu

Pekka Salmi is a sociologist working in the Finnish Game and Fisheries Research Institute. His main research themes include conflicts, governance systems based on private water ownership and social and institutional sustainability in fisheries. Recently he has studied sustainability of aquaculture and coastal fishing in a European multidisciplinary project.
Saimaa Fisheries Research and Aquaculture, Laasalantie 9, FIN-58175 Enonkoski, Finland. pekka.salmi@rktl.fi

Nathalie Steins was trained as a rural sociologist at Wageningen Agricultural University (Netherlands), where she also obtained her doctorate (1998). From 1996 to 1998 she worked at the University of Portsmouth (UK) and was involved in various projects on inshore fisheries management. In October 1998 she was employed as a policy officer by the Dutch Fish Product Board and is now deputy head of its fisheries section.
Dutch Fish Product Board, P.O. Box 72, 2280 AB, Rijswijk, The Netherlands. nsteins@pvis.nl

David Symes. Recently retired as Reader Emeritus in geography at the University of Hull where he taught for over 40 years and published widely in the fields of European agricultural development and fisheries management. Currently, working on a regionalised, ecosystem based approach to fisheries management in Europe.
The University of Hull, Department of Geography, Faculty of Science, Hull HU6 7RX, United Kingdom. b.simons@geo.hull.ac.uk

Riku Varjopuro is a cultural anthropologist whose research has dealt with fisheries conflicts, fisheries management and environmental regulation of fish farming. Interactions between fisheries and nature conservation is currently his main research theme. He is working in the Finnish Environment Institute (Research Programme for Environmental Policy).
Finnish Environment Institute, Research Programme for Environmental Policy, P.O. Box 140, FIN-00251 Helsinki, Finland. riku.varjopuro@ymparisto.fi

Kuperan Viswanathan, a Resource Economist, Research Scientist within the Policy Research and Impact Assessment Programme of the WorldFish Center (formerly known as ICLARM), Penang, Malaysia. He leads a global project on Fisheries Co-management that examines the use of co-management as a strategy for improving fisheries management in developing countries in Asia and Africa.
WorldFish Center, PO Box 500, GPO 10670, Penang, Malaysia. k.viswanathan@cgiar.org

James A. Wilson, an Economist, is Professor of Marine Sciences and Resource Economics and the Director of the Marine Policy Program at the University of Maine. He specializes in the regulation of renewable resources, especially fisheries, and market institutions. He has performed research and published in the areas of: Fisheries regulation, property rights and resource use, ecosystem management, predictability and the management of chaotic systems, transactions costs and the organization of markets, scientific uncertainty and institutional structure. He has served as: Chair of the Scientific and Statistical Committee of the New England Fisheries Mgt. Council; Chair of the committee that established the Maine lobster zones; and, Chair of the Socio-economic subcommittee of the lobster technical committee of the Atlantic States Marine Fisheries Commission.
School of Marine Sciences, University of Maine, Orono, ME 04469-5782, USA. Jwilson@Maine.edu.

Douglas Clyde Wilson is a Senior Researcher at the Institute for Fisheries Management and Coastal Community Development in Hirtshals, Denmark. His primary research interests lie in the sociology of fisheries management in both the North and the South. The main foci of his work are 1) the effective participation of stakeholders in fisheries management, 2) the social processes involved in the creation of a scientific knowledge base for fisheries management, and 3) the tensions between stakeholder participation and construction and using valid science in management.
Institute for Fisheries Management and Coastal Community Development, P.O. Box 104, DK-9850 Hirtshals, Denmark. dw@ifm.dk

Preface

This book came into being for a number of reasons. The most important, perhaps, was a feeling among the IFM staff that, while for two decades many scholars around the globe had been looking at and talking about fisheries co-management, there was little sense of progress in our understanding of it. The reason was not that the progress was not there. Rather there was no place where what we had learned had been examined as a whole. Such a resource, we felt, was needed by the management and academic communities focussed on fisheries and natural resources, and perhaps most importantly, by students who needed a concise, general picture of fisheries co-management as a field.

Initial attempts at a general examination of co-management had been done in the first phase of the Fisheries Co-management Research Project, in which IFM is a partner. Through this project, some general lessons about co-management work in parts of Asia and Africa had been collated and published. In the second phase of the project, we felt, we needed to expand those attempts by drawing in people who had been examining co-management in other contexts.

This book is the result. It is, in a sense, a hybrid. It is an edited volume with contributions reflecting the individual perspectives of many different researchers. It is also a single work in the sense that these authors were invited to write on topics designed around three basic dimensions.

The first dimension runs from the local to the global. The myriad different co-management programmes discussed in these pages take place on local scales. The heart of the book, however, the reports on experiences with fisheries co-management in six continent-sized regions, places these local efforts within larger and even global trends.

The second dimension runs from the empirical to the theoretical. All the chapters, and particularly the 'experiences' section, discuss a substantial number of specific co-management programmes. The second half of the book seeks to build general lessons based on these experiences disciplinary perspectives from, among others, ecology, sociology, anthropology, and economics.

The third dimension runs from the past to the future. The first section emphasizes places where the underlying ideas of co-management emerged. It culminates in Evelyn Pinkerton's wonderful discussion of what it means for a co-management programme to have become 'fully developed'. The second and third sections focus on the present through reports on experiences both geographically and from the perspective of different stakeholder groups. The last section looks to the future by examining what the authors see as the 'intellectual edges' in the fisheries co-management field.

Many people contributed to this effort. All of the authors deserve the warmest thanks for both their creative efforts and infinite patience. Svein Jentoft and Susan Hanna deserve special thanks for not only contributing chapters but doing the major work in pulling together the 'big picture' through the introduction and conclusion. Dorte Holmgaard Jensen volunteered both extra hours over many days and a keen critical and creative eye to the transformation of a collection of papers into an actual book. Finally, the DANIDA-funded WorldFish Center / IFM / NARS Fisheries Co-management Research Project, under the leadership of Kuperan Viswanathan, made this book possible through both financial support and by providing the single largest source of current information about co-management.

Douglas Clyde Wilson
Hirtshals, January 2003

CO-MANAGEMENT - THE WAY FORWARD

SVEIN JENTOFT
Department of Planning and Community Studies, University of Tromsø, Norway

1. INTRODUCTION

Co-management as a concept has a relatively short history in fisheries, not more than a quarter of a century. As a practice, however, co-management has been with us for a much longer time period. Co-management systems have existed in some fisheries in many parts of the world for decades, in some instances for centuries (cf Pomeroy and Viswanathan, Chapter 6). If this is the case, why has it become such a 'hot issue' in fisheries in recent years? Why are there now so many hopes and expectations connected with this concept? Why is the idea spreading globally - as this book clearly demonstrates? I believe the reasons are several:

History does not speak for itself but must be interpreted. Sometimes, history must be rediscovered, and for this we need a new perspective that allows us to see something that is otherwise concealed. The concept of co-management has brought attention to management practices that would perhaps otherwise have been neglected. Institutions become part of the reality we take for granted. Unavoidably they eventually assume a naturalness and obviousness that make them invisible to us. Fisheries management institutions are victims of the same process. The longer they have been with us, the less we see their significance or can imagine how life would have been without them. Once co-management was launched as a term, fisheries management systems that had existed for a long time, suddenly made sense in a new way. This happened for instance to the Spanish Cofradías, the French Prud'homie, the Lofoten management system, the Polish Mazoperias, the Indian Panchayat system, the Indonesian Sasi, and the Japanese community co-operatives, to name a few of the examples that have drawn much interest from social scientists in recent years.

Fisheries management, as we think of it today, is a recent thing. Fish was perceived as an inexhaustible resource into the 1960s. As Thomas H. Huxley claimed in his Inaugural Address to the London International Fisheries Exhibition of 1883: 'I believe that the great cod fishery, the herring fishery, the pilchard fishery, and probably all great fisheries are inexhaustible; that is to say that nothing we do seriously affects the number of fish'. (Quoted by Knauss, 1994). We know now that Huxley was mistaken. With the modern harvesting technology in use on the world's fishing grounds, we are fully capable of destroying the resource base. We are still struggling with how to take this new understanding into account. If fish stocks are indeed exhaustible, what measures do we to take to avoid it? How do we secure the carrying capacity of marine ecosystems? The debate on how to manage fisheries is still in flux. Co-management has been part of this discourse, and it took some time before it obtained the recognition it deserves.

Some major initiatives have been taken to curb over-fishing, most notably the Exclusive Economic Zones of 200 nautical miles that came as a result of the United Nations Conference of the Law of the Sea in the late 1970s. The new regime provided national governments with extended regulatory authority and a mandate in fisheries. However, ambitious states did not always recognise the history of fisheries management, even within their own territories. The lessons embedded in customary resource regimes which in many instances were targeting other goals than resource conservation were not brought up or regarded as pertinent.

We know that the government top-down, command and control approach to fisheries management, does not work well. Partly, as a consequence of this insensitivity to the diversity of customary fisheries management practices and local ecological knowledge, government management initiatives have often met mixed results, if not dismally failed. Crises in the fishery are still rampant and the legitimacy of state fisheries management has been eroded. New ideas are in constant demand, and co-management has been proposed as another vehicle with untapped potentials.

Fisheries management research used to be the domain of marine biologists. Increasingly, it has also become an area of interest to social scientists as well. Successful co-management requires a renewed examination of political, social and institutional matters. These are certainly among the areas where social scientists have expertise, so an increasing number of social researchers believe they have a mission in fisheries. During the last twenty years or so we have seen an impressive increase of contributions to the fisheries management discourse from social scientists, and their voices are being heard by managers and stakeholders more so than they used to be.

Co-management is a response to many concerns and those who support it do so for very different reasons. Some perceive co-management as a way of rectifying basic flaws in the perspectives underpinning current management systems, where the role of civil society is not part of the equation (cf Jentoft and McCay, Chapter 17). For some, co-management is a way of recognizing and formalizing what is already occurring informally at the local level or between government and fisheries organizations. Some are drawn into co-management for ideological reasons. They see co-management as the realization of the classic Rochdale co-operative principles into the sphere of fisheries management. Fisheries managers usually take a more pragmatic attitude. They also regard co-management as a way of enhancing legitimacy, and, hence, compliance to management rules and regulations (cf Pomeroy, Chapter 12). This is also an expectation that is shared by economists, whose important principal concern is transactions costs (cf Hanna, Chapter 3). They see co-management as a possible approach to increasing the cost-effectiveness of fisheries management through lowering the 'ex post' transaction costs, ie, the costs involved in implementation and enforcement.

Loucks, Wilson and Ginter (Chapter 9) point out that the concept of co-management in fact originates from the idea 'of social equality by vesting power of government in the people being governed'. Co-management could thus be regarded as a continuation of a process that occurs in other spheres of society not only in the public but also in the private sector, for instance workplace democracy. Co-management, then, is a way of living up to some basic social values, something to be supported out of principle rather than some instrumental purpose, for instance compliance. It is, of course, possible to be supportive of co-management for all these reasons. Together they make a strong case. However, an important explanation for co-management's increasing appeal is that it is possible to support it for some or just one of these reasons. This is why co-management has become an issue of multi-disciplinary academic interest. This is also why co-management holds so much promise.

2. CO-MANAGEMENT DEFINED

Co-management can mean different things in different settings. As Pomeroy points out in Chapter 14, there no 'blueprint formula' that can be applied everywhere. This is partly because the concept is broad. It contains some basic characteristics that may assume different organizational forms, as will be clear from this book. Co-management often reflects distinct national styles of governance and the specific ecological, social and cultural context within which it set to operate. The chapter by Loucks and co-authors (Chapter 9) comparing the US and Canadian system of fisheries governance provides a good illustration of this. Although principles such as democracy, transparency, accountability, and sustainability are key defining attributes of co-management, the way they are converted into concrete management institutions may vary from one country to another and from one fishery to another. The context into which co-management is introduced must always be taken into account. This means that co-management as a concept is, and has to be, sufficiently flexible to be generally useful.

I believe the chapter authors would agree on the following definition: co-management is a collaborative and participatory process of regulatory decision-making between representatives of user-groups, government agencies, research institutions, and other stakeholders. Power sharing and partnership are an essential part of this definition. Admittedly, this is a broad characterization, perhaps even too broad, since it frequently leads to questions of what co-management really is. Therefore it seems easier to state what co-management is not than what it really is. There are, however, limitations to how precise definitions can be. To avoid circularity, a definition must employ terms that do not simply restate that which they define, which may in themselves need clarification, like collaboration, participation, representation, sharing and so on (cf Metzner, Harte and Leadbitter, Chapter 10). Defining these terms, however, would again bring in new concepts that ideally should be defined, and the problem continues. Therefore, precision can never be complete.

However, if 'everything' becomes co-management, the concept loses its edge. Given its positive connotations, managers would be tempted to label any of their initiatives co-management. As Metzner and co-authors point out in Chapter 10, definitions can become 'intensively subjective' making stakeholders free to define them according their particular interest. In New Zealand, these authors hold, both fishers and managers have deliberately avoided the term because they regard it as synonymous with state abdication. This is not what co-management is. Co-management is partnership and power sharing. If 'anything' is co-management, people may draw whatever lesson they prefer from co-management experiences, regardless of whether the experience they are drawing their inferences from are 'true' co-management or not. Thus, the broader the concept, the greater the risk of false learning. Evelyn Pinkerton discusses this issue at length in Chapter 4. She argues that over the years the concept has become so broad that 'it risks losing important aspects of its original thrust', and that there is now time to assign it with a more specific meaning. Whether we can arrive at some consensus of what these specifics are is, however, another issue. Her own list of seven criteria is a good start.

All forms of user-participation do not qualify as co-management. In my own perception, the 'co' in co-management stands for cooperative and not consultative management. In the consultation mode of management, government agencies ask user-groups for advice before management decisions are made, but they have no obligation to follow the advice they get, or even listen. Co-management is, of course, more than a symbolic gesture aimed at relieving the political pressure from user-groups. Co-management is real, and to be real, power sharing is a must. As Hildebrand (1997): points out: 'Genuine participation is only achieved when power is shared' (p. 2). Wilson (Chapter 1) stresses that co-management developed out of a notion of 'authentic' participation. Co-management cannot just aim at relieving 'government of some of its burdens but none of its power'. Hara and Nielsen (Chapter 5) hold: 'Unless users are genuinely allowed and empowered to participate in the setting of management objectives on equal terms with government, co-management cannot really be considered as a serious institutional innovation'. Evelyn Pinkerton (Chapter 4) argues: 'Co-management is misnamed unless it involves the right to participate in making decisions about how, when, where, and how much fishing will occur'.

3. SUBSIDIARITY

Co-management requires that management functions be delegated to user-organizations that make autonomous decisions. These organizations would then assume both rights and responsibilities. Users, or their elected representatives, must have a hand on the pen when rules are crafted, and for some rules they may also be the single author. Co-management should involve all kinds of rules where users have a stake, be they constitutional, collective, or operational (cf Schlager and Ostrom, 1992). But the degree of autonomy, delegation and decentralization may vary between levels. Co-management is formal; it has a charter, it specifies mandates, membership and procedures for election, representation, and provision of scientific advice. Co-management also means that there exist rules for deliberation, conflict resolution, voting, administration, reporting, monitoring, appeal and so on.

We have argued elsewhere (McCay and Jentoft, 1996) that fisheries management should adopt the principle of subsidiarity. This principle states that management authority should be vested at the lowest possible organization. For some management functions, the lowest level would be the vessel or firm, for some the community or some other representative unit of organization at the village level. It some situations decision-making authority and responsibility must rest with the regional, national or international level. Co-management may occur at different levels of authority. Large scale complicates but does not make participation impossible, although it requires other procedures for representation and deliberation (cf Jentoft, Mikalsen and Hernes, Chapter 16).

The subsidiarity principle also states that the 'burden of proof' for centralization rests with the higher organization. In other words, the lower level does not need to justify decentralization. The higher organization also has an obligation to help facilitating decentralization. In some instances, the higher authority must help resolve a institutional 'deficit' at the lower level, for instance assist in forming and equipping organizations capable of handling management responsibilities. As Pomeroy and Viswanathan argue in Chapter 6, successful co-management and meaningful partnerships can only occur when the community is empowered and organized. The subsidiarity principle makes government co-responsible for building these community institutions. Indeed, as Begossi and Brown demonstrate in Chapter 8, government can play a constructive role in this respect. Thus, the 'lowest possible organization' is not something that is given once and for all.

A third criterion of the classic subsidiarity principle pertains to local autonomy: Local-level institutions should not be acting as mere agents of decisions that are made by higher authorities. This is why it is important to distinguish between delegation and decentralization in fisheries co-management. The former concept implies endogenous management of user-groups, while the latter may or may not do so. Co-management is endogenous management but within an enabling institutional framework established by a higher authority such as a national assembly of legislators (cf Pomeroy, Chapter 14).

In many countries, if the subsidiarity principle were used as a yardstick for current management systems, it would suggest the need for instant radical reform. The European Union, which has formally adopted this principle, would be no exception here (cf Symes, Steins and Alegret, Chapter 7). But reform should, most wisely, be a gradual process. As Metzner, Harte and Leadbitter argue (Chapter 10), co-management should first be introduced in those settings where it is most likely to succeed until it has 'become part of the cultural landscape of fisheries management'.

4. CONFLICT AND POWER

Given both the widespread optimism and scepticism, any co-management failure can lead to people immediately starting to draw negative conclusions about its viability in fisheries. As all management reforms do, co-management inevitably produces winners and losers, and for the latter group failure may, of course, be most welcome. Bureaucrats are sometimes the most adamant and vociferous sceptics. Cicin-Sain and Knecht (1998) point out:

> 'It is well known that agencies jealously guard their missions and the responsibilities and resources that accompany them. Indeed, survival of the agency depends on its keeping the mission and resources intact (or better yet, expanding them): Anything that threatens the mission or the resource base tends to be resisted with great vigour and tenacity' (p. 217).

There is no doubt that co-management puts government under duress. New demands would easily be met with scepticism and even active resistance within management agencies. As Hara and Raakjær Nielsen (Chapter 5) report from Africa, the reorientation of departments has proven difficult, especially when it meant reorganization. Co-management requires a new management philosophy, new skills, and new ways of interaction with fishing communities. It may even imply the possible threat of jobs for existing staff. No wonder, therefore, when co-management sometimes is met with opposition.

Sometimes the hindrances do not lie with government, but with user-groups. Co-management does not only demand cooperation among competitors in the fisheries commons. It would also require cooperative links to the government where lack of trust is sometimes mutual and has a deep history. Building trust would therefore be a necessary ingredient of any attempt at co-management. As Pomeroy points out (Chapter 14), the state may easily operate in ways that is contrary to the interest of the community. The state is sometimes heavy handed. It has no fingers, only thumbs, as the economist Charles Lindblom once remarked. A distant bureaucracy is handicapped in relating effectively to local circumstances, for instance due to a lack of local knowledge. Frequently in fisheries, there is simply too much diversity for central government to handle adequately. Interestingly, as Begossi and Brown explain based on the Jamaican experience (Chapter 8), there is no necessary conflict between invoking the state and supporting the local community's self-reliance in fisheries management. 'This co-management arrangement enhances community democracy and empowerment, while acknowledging the duty of the state to be involved in the process' as a facilitator and capacity builder of local institutions.

Co-management is not free from conflict. As Wilson (Chapter 11) holds, conflict is part of every fishery, and can be a creative as well as constructive force for management. Consensus is rare and compromise is the best one can hope for. Fisheries are too complex, diverse and dynamic to permit permanent, once and for all, settlement of conflicting interests and world views. Therefore, co-management does not by itself eliminate conflict, but should ensure good procedures for conflict resolution. If not, co-management may become a costly exercise (cf Loucks *et al.* Chapter 9). It improves the legitimacy of the management decisions (Pomeroy Chapter 12) if the ways through which these decisions are reached are considered fair: As Bo Rothstein (1998) maintains: 'Just institutions matter'. But even though consensus is outside reach, it is still something to strive for. Therefore, co-management must also allow communication and deliberation among involved stakeholders.

Importantly, as Wilson argues (Chapter 11) co-management cannot rely on shared norms. Stakeholders have backgrounds, interests, world views, agendas and interests that are too different for them to feel committed to the same norms. Therefore, shared understandings are more important. For this to happen, rich and varied communication among involved stakeholders is essential. In negotiations the parties may have to give and take, but the process must be structured so that they arrive at some mutual understanding of what stakes are involved. Shared norms may well come out of such a process, but in most situations that is simply too much to hope for.

5. PROPERTY-RIGHTS

The zero-sum nature of fisheries management is no less evident when we consider access- and property rights. The key attribute of property is not the resource itself but the social relation it involves - the fact that property legalises the right to reserve the resource flow for oneself while excluding others (cf Hann, 1998; Taylor; 1987). A rights-based fishery does exactly that. For this reason, it also sometimes triggers non-compliance. The 'have-nots' would not willingly abide with rules that benefit only the 'haves'. If co-management involves property rights to resources in addition to rights to manage, this may cause problems.

Co-management is not synonymous with a particular property rights system. It may work within state, private, communal and collective group property. This fact is demonstrated throughout the chapters of this book. Co-management based on private property defines and limit who the participants are, who can be held responsible and who should be blamed if management does not work. If the resource right is fully privatised, as within an Individual Transferable Quota (ITQ), then the co-management system would have fewer sanctions at its disposal if rules are violated. For instance, the co-management board would be in no position to withdraw the resource right from violators. In this situation, a third party, such as a government agency, would be needed to monitor and enforce rules.

A community based property rights system on the other hand, would induce mutual control and sanction. According to Symes and co-authors in their portrayal of the Dutch co-management system (Chapter 7): 'In a scenario where offences by one fisherman have immediate consequences for his colleagues, the offender does not make himself popular'. But a system of mutual enforcement may be unfair; indeed it could be cruel (cf van der Schans, 2001). Property-rights are also an incentive for members to pull up the ladder, guarding their privileges by excluding newcomers, who in many instances are the hungry poor, as Hara and Raakjær Nielsen report in Chapter 5. A co-management system only for the privileged few runs counter to its basic ideals - despite the fact that it may be an effective tool in conserving the resource base from over-exploitation. This is the well-known ethical dilemma of the lifeboat. Wilson (Chapter 11) points out that we have long understood that communities are rarely homogenous. Neither was the Titanic, where people on first class were allowed space on the lifeboats while people on lower classes perished. Therefore, as Loucks and co-authors (Chapter 9) maintain: 'we must continuously ask the question 'who benefits from co-management? ... particularly as various rights and responsibilities are exchanged'.

Ideally, property invokes interest in, and responsibility for, keeping the resource benefits flowing. As Aristotle proclaimed: 'Men pay most attention to what is their own: they care less for what is in common'. Scott Gordon (1954) alluded to this insight when he argued that 'nobody's property is nobody's responsibility'. But there is no guarantee that property owners - be they private or communal - will always care for, and invest in, their property, as any owner of a house or a car would know. As Clark (1974) demonstrated, also a sole owner may overfish if the immediate value of catching all now is higher than the present value of all future resource rent.

Importantly, new management institutions must relate to existing institutions. Co-management is never created in an institutional vacuum. Sometimes co-management will reduce or replace existing management arrangements. In other instances, co-management will have to be adapted to an existing institutional environment that will remain dominant. Sometimes, the existing institutional structures are such that they can be put to use for co-management. A good example is the Producers' Organizations within the European Union, which were originally formed for marketing purposes rather than fisheries co-management. In some EU countries, most notably in Great Britain, their mandate has been broadened to include fisheries management functions (cf Symes and co-authors, Chapter 7).

Therefore, co-managers often find themselves in the situation where they cannot simply choose the most preferable property rights system. As Varjopuro and Salmi point out (Chapter 13): 'In discussion of potential of creating or developing a co-management system the existing regimes and institutions must be taken into account'. Hara and Raakjær Nielsen (Chapter 5) provide a vivid illustration in the African report when they discuss how co-management challenges traditional authority structures within the community. In Finland, the present management system is based on private property to seawater territories. 'It would be difficult to arrange it in any other way, since private owners are protected by the constitutional law'. Investing effort in changing the constitution is hardly the place to start if one wants to install co-management in fisheries.

I have made a similar argument regarding the introduction of co-management for the Saami fishing communities in Norway. Legally, fish in Norway is 'nobody's property'. This is a principle that is largely taken as given by the majority of fishers in Norway. Changing that principle, for instance by instituting a Saami indigenous collective ownership of fishing territories, would be a cumbersome process that would be met with heavy resistance from ethnic Norwegians. But this should not exclude the Saami people from having a co-management system where they are considerably empowered as compared to their current situation (Jentoft, 2000a). Notably, the co-management system in the Lofoten region of Norway is based on the 'nobody's property' institution (Jentoft and Kristoffersen, 1989).

6. REPRESENTATION AND KNOWLEDGE

If the democratic principle that those affected by a decision should have a say in the decision-making process (cf Dahl, 1989) applies to fisheries co-management, then more than just fishermen must be involved. The number of stakeholders is in some instances considerable. For instance, Mikalsen and Jentoft (2001) identified eighteen groups with interests at stake in the Norwegian fisheries management, with only a few of them represented. Who all the stakeholders are in a particular fishery is not always clear and, obviously, some have more at stake than others do. How this should be accounted for when co-management boards are put together is an issue and this is related to the question of power. Those whose lives are most dependent on the fishery and whose concerns are most urgent are not always the most powerful of stakeholders. A truly democratic co-management system must be designed so that it corrects for such inequities.

The greater the number and diversity of voices in the management system, the greater the challenge on the communicative process. Varjopuro and Salmi (Chapter 13) illustrate some of these complexities in the Finnish fishery. The different stakeholders form what they call a 'user-community', but the community does not work as an integrated whole. The user-community consists of those who fish for leisure, for subsistence, and for commercial purposes. Some of the users are residents of rural communities and have private ownership to water areas. Other users live in urban centres, and their only connection to a particular geographical area is that they sometimes fish there or have a weekend cottage nearby. Only a few are well organized and thus politically vigorous in the management process. How to make up for these imbalances in the co-management system is therefore no easy task. Rather, co-management risks entrenching inequities that are already prevalent in the user-community (cf also Davis and Bailey, 1996). But as other democratic institutions, co-management has formal rules for representation and decision-making that may well compensate for such inequities, for instance the one- person- one-vote principle. Jentoft, Mikalsen and Hernes (Chapter 16) provide insights into this matter in their discussion of the issue of representation in co-management.

As Hanna points out in Chapter 3, involving the largest number of stakeholders possible in management decision-making increases complexity and makes the management process cumbersome, time-consuming, and costly. This is the classic dilemma between internal democracy and external effectiveness (cf Dahl and Tufte, 1973). However, co-management may also bring more rationality since it broadens the knowledge base on which decisions are made. As Pomeroy points out (Chapter 12), 'diverse participants in the process can bring particular knowledge and concerns to the management process, and assist in the clear definition of the problem in ecological, social and economic terms'. Participants do not only bring their stakes to the negotiations and deliberations, they also bring their experience-based knowledge (cf Wilson, Chapter 15).

Neither should one underestimate the interactive learning process that occurs when people participate and communicate in democratic settings (cf Pateman, 1970). Thus, Degnbol (Chapter 2) regards co-management as a vehicle for bridging 'the gap' between two discourses: one that occurs among researchers and managers and is typically addressing 'averaged', large scale issues such as stock abundance and recruitment, and another discourse that takes place in the user community. The latter is characteristically derived from 'high resolution' knowledge of spatial variation of fish abundance in time and space, i e. knowledge that is essential for fishers when pursuing their fishery. Efforts to reduce the gap should aim at creating a 'joint scientific culture' built on mutual trust (cf Wilson, Chapter 15).

This may, however, be easier said than done. One problem is that some of this knowledge is tacit, and will only be invoked if challenged. Wilson (Chapter 15) argues: 'How well fishers' knowledge can be articulated in management debates has important implications for co-management both from the perspectives of mobilizing fishers' knowledge for rational management and from the perspective of equitable control over the knowledge base'. We must assume that this point is valid for other stakeholders as well. We should also expect that when stakeholders draw on their knowledge in the management discourses, they do so selectively and in accordance with their particular interests and needs. This is, of course, legitimate, but may well hamper the process of reaching rational management decisions. 'Opportunism' - as 'the selective disclosure or distortion of the data to which each party uniquely has access' (cf Williamson, 1975: 32) - is something that does not only occurs in markets, but also in organization albeit to a lesser degree. Clearly, this is also a factor to reckon with in fisheries co-management.

Much would be gained, though, if users, managers, researchers and other stakeholders could agree on some eco-system indicators, as Degnbol (Chapter 2) argues. These indicators must be generally observable, understandable, acceptable, cost-efficient, and relevant to management. In many quarters internationally there is now work going on in order to identify what these indicators might be. For this, Degnbol argues, interdisciplinary research is needed, where biologists and social scientists cooperate.

7. THE COMMUNITY

Co-management systems must work at different scale levels depending on the nature of ecosystem and social system characteristics within which they must operate. Therefore, regional approaches to co-management must be applied when the scale of the resource is larger than local (cf Wilson, Chapter 11). Therefore, to address scale issues, co-management may form federative systems. Examples are the Spanish Cofradías (Symes and co-authors, Chapter 7) and the Belize fisheries co-operative system described by Begossi and Brown (Chapter 8). This, of course, does not exclude local representation in regional co-management boards. Co-management must also relate to framing institutions that impact and put demands on fisheries in general and fisheries management in particular. Generally, this is the notion of 'nested systems' (Ostrom, 1990). Thus, co-management systems are never designed in an institutional vacuum but are generally part of a larger network on institutions. Co-management, as does other social institutions (cf Scott, 1995), also reflects the deeper cultural and social values that are nested in human communities. In fact, if co-management institutions fail to consider and integrate these values, their legitimacy will be questioned and their decisions opposed. Conflicts triggered by management are in many instances cultural rather than interest driven. In several chapters authors express beliefs in the role that local communities can play in fisheries co-management. Wilson (Chapter 1) regards co-management as having been heavily influenced by the tradition of community development. As he points out, the question of the commons is fundamentally a question of community. Similarly, the Tragedy of Commons in Garrett Hardin's rendering, is depicted on the assumption of a lack of such (McCay and Jentoft, 1998). Current management systems that are inspired by Hardin and by neo-liberal economic theory, for instance those that involve exchangeable resource rights (cf Jentoft and McCay Chapter 17), typically neglect the existence or community - be it of place or interest - and its potential contribution to fisheries management. Unfortunately, this negligence leads to a process that becomes self-fulfilling: fisheries management impacts negatively on the social cohesion and solidarity of communities, and communities therefore become less and less capable of handling management responsibilities. As a consequence, they become increasingly dependent on the state to perform management tasks, which again is eroding the qualities of communities essential to management, such as trust in their own capabilities. Co-management should therefore aim at turning the vicious circle into a process that is positively self-reinforcing. For this to happen, government has a supporting role to play. As Pomeroy and Viswanathan (Chapter 6) point out, governments must provide the necessary 'enabling legislation that foster the partnerships needed between communities and governments...'. If local communities and their organizations are allowed to assume co-management roles and rights, they will be strengthened in social and political terms. No less important is that they may then also be able to generate large revenues for community well being and growth, as Loucks and co-authors (Chapter 9) demonstrate in the case of the community development quota programme in the case of Alaska.

8. THE WAY FORWARD?

We should take a pragmatic attitude to co-management. Co-management may not work in all settings. There are also risks involved that may jeopardize well-intentioned co-management systems, some of which are mentioned above and in several chapters such as Hanna in Chapter 3. We are not living in an ideal world, and we should not expect that co-management would bring us there. Neither should co-management be criticized for this. Undoubtedly, co-management holds promise when compared to other management systems. Just as democracy has its shortcomings, so does co-management, but democracy is still the best form of governance that we can think of. As can democracy, co-management systems can be improved by addressing its concrete problems in real situations. In many instances, these problems are not inherent to the co-management model but are caused by its context specific designs that can also be altered and improved. Co-management is the way forward, despite its risks and problems.

We should avoid all dogmatism when it comes to the particular design of co-management. Co-management must assume different organizational forms in different social, cultural and ecological settings. To avoid grave mistakes, co-management reforms should preferably be tried out in small scale before implemented in large scale. We should avoid throwing dice, as when Caesar crossed Rubicon. Learning is essential, and we should not exclude the possibility that we might change our minds. Therefore, I have argued elsewhere, that ludism, the playful, experimental attitude, should guide our efforts (Jentoft, 2000b). As James March (1976) argues, playfulness is rational when conditions are complex and unstable.

Hara and Raakjær Nielsen (Chapter 5) are convinced that in the case of Africa, co-management has to be a learning process. For this region 'and for each specific fishery, unfortunately the specific design needs to be tailor-made'. Their experience applies to other regions as well. Social research should be integral to such a learning process. If the decision-making process is going to be less top-down and more participatory, we need to think through what this means, what the main issues are, and what problems need to be assessed and addressed. There are also things we need to now about how to overcome barriers of communication among stakeholders, particularly how to bridge the gap between local and scientific knowledge through interactive learning. Another focus should also be on how interest and power may play out in these settings. We need to know more about contextual factors and how they impact on co-management. We know that social and cultural conditions are conducive to transactions, cooperation and communication. Jentoft and McCay (Chapter 17) are pleading for a much stronger emphasis on a working civil society as a precondition for successful fisheries co-management. They propose two research themes; the 'colonization thesis' launched by Jürgen Habermas (cf also Wilson Chapter 11), claiming that the market and the state are eroding the capacities of civil society, included communities, to play an effective role in fisheries management. Co-management may be regarded as a countervailing force in this respect. The other is the 'embeddedness thesis' (drawing on Mark Granovetter), which argues that economic action is situated in, and is nourished by, social networks. In other words, we need to better understand how the social and cultural underpinnings of co-management determine its success or failure.

The best way to get a closer grip on these issues is to carefully study the co-management schemes that are currently working in fisheries. This book clearly demonstrates that there are many experiences internationally to tap from and that there are lessons to be gained by comparing them. These are lessons that should also inform the new co-management initiatives that are now underway in many countries. As social researchers, Jentoft and McCay (Chapter 17) argue, we should not shy away from advocacy of the co-management model, as long as long as it is firmly rooted in empirical research. Importantly, we should se ourselves not as experts with a curing medicine, but as contributors to the co-management learning process - a process which in the spirit of co-management must be bottom up.

9. BOOK OUTLINE

This book came about in a somewhat unusual way, falling somewhere between an edited collection that collates papers addressed to a broad topic from authors' individual perspectives and a book by a small group of authors that develops a single, internally coherent perspective on the topic. For this book the editors created an outline of topics related to fisheries co-management and invited specific people to contribute papers on a topic. The result is that the chapters represent authors' individual perceptions, but they are individual perceptions of particular selected issues within fisheries co-management, rather than individual perceptions of co-management as a whole. This result, hopefully, reflects some of the strengths of both the single author and edited collection approach. The following briefly describes the outline the book.

Section One has two purposes. The first is to give readers, particularly students, a historical perspective on fisheries co-management. As discussed above, co-management, like any powerful idea, has a history of both of antecedent ideas and of programmatic successes and failures. Understanding what co-management is today means understanding where it came from. The second purpose is to introduce the reader to the conceptual framework from which fisheries co-management takes its meaning. Co-management does not mean anything divorced from other concepts such as fisheries management, fishing community, the tragedy of the commons, property rights, transaction costs, top-down management, participatory democracy, and action research.

Section One begins with an overview by Douglas Clyde Wilson of the traditions of community development and community-based management of natural resources from which fisheries co-management emerged. Then three other chapters take disciplinary perspectives and discuss the development of the co-management idea from their science's point of view. In Chapter 2, Poul Degnbol discusses co-management through the eyes of fisheries science, which has always held the central place in the scientific knowledge base of management. Then in Chapter 3, Susan Hanna gives the fisheries economics perspective emphasizing the importance of incentives and transaction costs and the role that co-management has to play in balancing them. Then Evelyn Pinkerton discusses the various meanings that social science in general, and the common property research tradition in particular, have attached to the co-management idea. She ends the section with some suggestions about what 'complete' co-management would entail.

In Section Two, teams of fisheries management researchers, who have a long history of working in particular areas, review the current fisheries co-management situation in six world regions. Each of these chapters covers the trends and future prospects for programmatic fisheries co-management activities and the legal and policy issues faced by co-management programmes in the region. While all the chapters address these broad issues, they do so in quite different ways. The reader will understand the appropriateness of these different approaches while reading about the very different forms that co-management is taking, and the very different reactions it is getting, in these regions.

In Chapter 5, Mafaniso Hara and Jesper Raakjær Nielsen, both of whom have been researching African co-management for a decade as part of the ICLARM/IFM Co-management Research Project, summarize the situation on that continent. In Chapter 6, Robert Pomeroy and Kuperan Viswanathan, who worked on the Asian team of that same ten-year research project, report on the situation in Southeast Asia and Bangladesh. Chapter 7 turns to Europe, where the intense current debate around the revision of European Union fisheries management system sets the context for David Symes, Nathalie Steins & Juan-Luis Alegret's discussion of the status and role of co-management within that system. Next, Alpina Begossi and David Brown review the situation in Latin America and the Caribbean, with an emphasis on how co-management has evolved out of different forms of local management into a multiplicity of different kinds of programmes. Then Chapter 9 on North America, written by Laura Loucks, James Wilson and Jay Ginter, also demonstrates the rich contrasts that can be found within continents as it compares both Canadian and US cases and cases of co-management operating on very different scales. Finally, in Chapter 10 on Australia and New Zealand, Rebecca Metzner, Michael Harte and Duncan Leadbitter report on a very different policy environment where convincing people of the value of co-management has been a particular burden, but where some exciting experiments have emerged.

Section Three examines issues around involving multiple stakeholders in co-management, some of the demands they are placing on fisheries management in general, and the role that co-management can play in addressing them. In Chapter 11, Douglas Clyde Wilson begins with a theoretical discussion of conflict in fisheries management. He argues that conflict is deeply related to the scale over which the co-management programme is operating, and that these two realities of conflict and scale converge to give communities of multiple stakeholders an inescapably central role in management. Then the section turns to some specific issues in multiple stakeholder approaches. For instance, in Chapter 12 on marine protected areas, Caroline Pomeroy focuses on what has emerged as a top priority for environmental groups concerned with fishing. Then in Chapter 13, on recreational fishing, Riku Varjopuro and Pekka Salmi review a trend toward non-commercial fishing that is having profound implications for fisheries management in Europe, North America and Oceania. Finally, Robert Pomeroy takes a closer look at the government's role in fisheries co-management, looking at the government as a unique kind of stakeholder.

The final section looks to the future by focussing on a selection of edge intellectual issues that have emerged through research on fisheries co-management. Chapter 15, written by Douglas Clyde Wilson, examines the relationship between co-management and fisheries science, asking the question of how co-management can simultaneously increase the democratic basis of management while leaving the scientific validity of management decisions fully intact. Svein Jentoft, Knut H. Mikalsen, and Hans-Kristian Hernes then turn to the question of democracy and representation and how they might make possible the benefits of co-managements on greater than local scales. The section concludes in Chapter 17 with a suggestion by Svein Jentoft and Bonnie McCay about the future directions that research in support of fisheries co-management might take.

In conclusion, Susan Hanna takes a step back from specific issues of co-management. She pulls together many of the various strands found within the chapters by asking the question of what the future role of co-management approaches can be within the much broader arena of fisheries management in general.

REFERENCES

Cicin-Sain, B. and Knecht, R. W. (1998) Integrated Coastal Zone and Ocean Management: Concepts and Practices. Washington D. C. : Island Press.

Clark. C. W. (1974) The Economics of Over-exploitation. *Science*, **181**, 630-634.

Dahl, R. A. and Tufte, E. R. (1973) Size and Democracy. Stanford: Stanford University Press.

Dahl, R. A. (1989) Democracy and Its Critics. New Haven: Yale University Press.

Davis, A. and Bailey, C. (1996) Common in Custom, Uncommon in Advantage: Common Property, Local Elites, and Alternative Approaches to Fisheries Management. *Society and Natural Resources*, **9**(3), 251- 266.

Gordon, S. (1954) The Economic Theory of a Common Property Regime: The Fishery. *Journal of Political Economy*. **62**, 124-42.

Hann, C. N. (ed.) (1998) Property Relations: Renewing the Anthropological Tradition. Cambridge: Cambridge University Press.

Hildebrand, L. P. (1997) Introduction to Special Issue on Community-based Coastal Management. *Ocean & Coastal Management*, **36** (1-3).

Jentoft, S. and Kristoffersen, T.I. (1989) Fisheries Co-management: The Case of the Lofoten Fishery. *Human Organization*, **48**(4), 355-365.

Jentoft, S. (2000a) Rights to Nature - A Natural Right? Fisheries Management from a Saami Perspective. In D. Symes (ed.): Fisheries Dependent Regions. Oxford: Fishing News Books, Blackwell.

Jentoft, S. (2000b) Co-managing the Coastal Zone: Is the Task too Complex? *Ocean & Coastal Management*, **43**, 527-535.

Knauss, J. A. (1994) The State of the World's Marine Resources. In C. W. Voigtlander (ed.): The State of the World's Fisheries Resources. Proceedings of the World Fisheries Congress Plenary Sessions. New Dehli: Oxford & IBH Publishing.

March, J. G. (1976) The Technology of Foolishness. In J. G. March and J. P. Olsen (eds): Ambiguity and Choice in Organizations. Oslo: Universitetsforlaget, 69-81.

McCay, B. J. and Jentoft, S. (1996) From the Bottom Up: Participatory Issues in Fisheries Management. *Society and Natural Resources*, **9**, 237-250.

McCay, B. J. and Jentoft S. (1998) Market or Community Failure? Critical Perspectives on Common Property Research. *Human Organization*, **57**(1), 21-29.

Mikalsen, K. H. and Jentoft S. (2001) From User-groups to Stakeholders?: The Public interest in Fisheries Management. *Marine Policy*, **25**, 281-292.
Ostrom, Elinor (1990) Governing the Commons. The Evolution of Institutions for Collective Action. Cambridge: Cambridge University Press.
Pateman, C. (1970) Participation and Democratic Theory. Cambridge: Cambridge University Press.
Rothstein, B. (1998) Just Institutions Matter: The Moral and Political Logic of the Universal Welfare State. Cambridge: Cambridge University Press.
Schans, J. W. van der (2001) Governance of Marine Resources: Conceptual Clarifications and Two Case Studies. Delft: Eburon.
Schlager, E. and Ostrom E. (1992) Property-rights Regimes and Natural Resources: A Conceptual Analysis. *Land Economics*, **68**(3), 249-262.
Scott, W. R. (1995) Institutions and Organizations. London: Sage Publications.
Taylor, M. (1987) The Possibility of Cooperation. Cambridge: Cambridge University Press.
Williamson, O. E. (1975) Markets and Hierarchies: Analysis and Antitrust Implications. New York: The Free Press.

Section One

THE FISHERIES CO-MANAGEMENT IDEA

Chapter 1

THE COMMUNITY DEVELOPMENT TRADITION AND FISHERIES CO-MANAGEMENT

DOUGLAS CLYDE WILSON
Institute for Fisheries Management and Coastal Community Development
Hirtshals, Denmark

1. INTRODUCTION

Fisheries co-management, like any powerful idea, has a history of both antecedent ideas and of practical successes and failures. Appreciating co-management means understanding its family history, where it came from and who its cousins are. This section of the book traces how co-management emerged from the main branches of its family tree: the participatory community development efforts that are its closest programmatic cousins and the disciplines of fisheries science, economics and social science that make up the knowledge base of fisheries management as a whole. This chapter examines participatory community development programmes and how ideas that significantly influence fisheries co-management today developed through the history of these efforts.

People have always participated in building their communities and managing the natural resources. Programmes that have self-consciously sought to create community cooperation to achieve development or resource management goals have existed at least since the turn of the 20th Century. Approaches to creating this community cooperation have had different outcomes in different places and a great deal has been learned in these attempts. In keeping with this section's focus on understanding the antecedents of the co-management idea, the present chapter traces this history up until the 1980s when the idea of fisheries co-management began to appear frequently in fisheries management discussions.

2. BASIC COMMUNITY DEVELOPMENT CONCEPTS

2.1. Community Development

Sanders (1958) was an early and insightful observer of community development. He offers a complementary way to understand what people mean by the term. His, now classic, outline of four different ways that people look at community development reveals the complex dimensions of the concept. Sanders (1958) said that people see community development as:

> A **process**. 'CD as a process moves by stages from one condition or state to the next....it is a neutral, scientific term, subject to fairly precise definition and measurement expressed chiefly in social relations...Emphasis is upon what happens to *people*'.
> A **method**. 'CD is a means to an end; a way of working so that some goal is attained.. Emphasis is upon some *end*'.
> A **programme**. 'The method is stated as a set of procedures and the content as a list of activities. Emphasis is upon *activities*'.
> A **movement**. 'CD is a crusade, a cause to which people become committed. It is dedicated to progress as a philosophic and not scientific concept. ..it stress the *idea* of community development as interpreted by its devotees' (Sanders, 1958 p 5).

A generation later, Christenson and Robinson (1980) noted that community development programmes could be classified into three basic types. The first is community self-help. This approach works toward slow and sustainable change, often in middle class communities, and sees the role of the outside change agent as that of a facilitator or educator. The second approach is technical assistance where the change agent is seen as an advisor or consultant who works mainly with community leaders and administrators. The third approach is the conflict approach where the change agent is seen as an organizer or advocate and the goal is a fundamental shift in community power and control of resources. The conflict approach is most commonly taken poor and minority communities. The most famous model for the conflict approach to community development is Saul Alinski, the Chicago community activists in the 1960s who advocated polarizing communities and empowering their most vulnerable groups.

While the ways these things are expressed have changed in the last 40 years, the realities these scholars were pointing at can still be seen in both discussions of community development and fisheries co-management. We would now call Sander's dimensions 'discourses' and would note that these same four discourses appear in discussions of fisheries co-management today. To some extent they represent disciplinary biases with managers and fisheries scientists tending to think of co-management as a method, social scientists as a process and local organizers as a programme and a movement. Current fisheries management programmes can also be helpfully classified using Christenson and Robinson categories. The degree to which co-management should be thought of as a bottom-up self-help approach to management and the degree to which is should be a reflection of outside fisheries agents 'teaching' the community how to manage the resource is an important debate in most co-management programmes. Co-management as an method for empowering poor communities can also be seen in the approaches to management, for example, of the International Collective in Support of Fishworkers.

2.2. Participation

Participation has been a watchword in community development for a long time for both ethical and practical reasons. The ethical reason is that people simply have a right to a say in decisions which will affect their lives. The practical reason is that participation in making both decisions and investments strengthens people's commitment to outcomes.

People use the word 'participation' in association with development to mean many things. How participation is defined is not an academic issue because the real definition emerges from who does what when within practical activities and decision making processes. Hence, it is more revealing to speak of different types of practical participation than it is to try to give the word a definition beyond the Oxford Dictionary's 'the action of taking part in something' (Hornby, 1995). Contributors to the international development literature have underscored the questions of who participates and what the effect of that group's participation is. Cohen and Uphoff (1977) suggest a framework for analysis. They emphasize looking separately at local residents, local elites, government personnel and agency personnel in terms of their level of participation in decision making, programme implementation, the distribution of benefits and the evaluation of the programme's effectiveness.

One helpful way of distinguishing types of participation is to oppose 'instrumental' to 'developmental' goals of participation. The instrumental approach emphasizes using participation to achieve the goals of an outside agent. The outside agent is usually the government or an NGO. Instrumental participation means encouraging people to 'get on board' in order to more efficiently meet the agents goals.

The developmental approach to community participation places its primary emphasis on the building of a community's capacity for ongoing development. Littrell and Hobbs (1989) offer a philosophical basis for the developmental approach:

> 'Helping communities achieve a capacity for self-help is fundamental to both the theory and practice of community development. If a spirit of self-help doesn't exist within a community... then, from the perspective of community development or empowerment, a capacity for self-help may be instigated with the assistance of an outside community development practitioner or organization. (49)'.

Two aspects of this quote are particularly helpful. The first is the use of the word 'spirit'. Spirit implies a general attitude in the community toward its own development. Every developmental process, decision, or action can either contribute to or detract from the self-help spirit. The second is the claim that people from outside the community can help to catalyse and/or strengthen this spirit. In the developmental approach to community development this is the primary role of any outside force. One crucial aspect of the developmental approach is the creation of a participatory decision making process. Capacity building creates a process through which a community can continue to do its own development, even after the outside agent has left.

Arguments for participation in fisheries management reflect these same two basic rationales of more effective and more ethical management. Participation in fisheries is also defined in many different ways depending on who is doing the defining and what their objectives are for participation (Wilson and McCay, 1998). As discussed by Jentoft and colleagues in Chapter 16, the dynamics described above as 'instrumental' and 'developmental' participation is very much alive in debates about the design of co-management institutions.

2.3. Action Research

Who participates in something, and how they participate, is closely related to, indeed inseparable from, the question of who is defining and describing the problems that the participatory action is meant to address. Action research (AR) is a tradition that brings participation together with the knowledge on which decisions are based. The seminal work on AR is Freire's *The Pedagogy of the Oppressed* (1970). Its main emphasis is on conscientization, a process of building awareness about social and political realities. In Freire's vision the poor carry out their own research on the question of how they came to be poor and what can be done about it. The inspiration is essentially Marxist.

More mainstream approaches to community development, however, cannot avoid the question of who defines the problems and generates the knowledge used to address them. Vandenberg and Fear (1983) make a distinction between a utilitarian approach and a radical approach to AR. The radical approach, with Freire as the main exemplar, seeks social transformation as the ultimate goal of AR. This approach rejects standard scientific methods such as surveys as 'alienating dominating or oppressive in character' (Hall, 1979). The radical approach is basically educative in that it is focussed on changing local understandings of the situation rather than gathering research information (Vandenberg and Fear, 1983). The utilitarian approach uses standard research methods but insists that they be site specific. It is basically instrumental rather than educative and involves data gathering that yields information for decision making and may even contribute to theoretical knowledge (Vandenberg and Fear, 1983). Utilitarian AR has evolved into a very mainstream approach with programmatic foci, nearly every development project or programme in world today begins, at least nominally, with 'participatory rural appraisal' or a similarly named process.

This issue of defining and describing the problems to be addressed by the community programme is absolutely central to debates about fisheries co-management. Chapters 2 and 11 in this book focus on two different aspects of the issue of fisheries research and fisheries co-management. Long term and well structured cooperation between fishers and scientists, very often in the form of cooperative research, is central to the most successful current co-management programmes (Wilson, 1999). These programmes are almost entirely what Vandenberg and Fear (1983) would describe as utilitarian approaches, however Chapter 11 argues that a critical aspect of co-management is the mobilization of fishers' tacit knowledge as a resource for management, a process that is very much about the empowerment of the fishing community.

3. EARLY EXPERIENCES WITH COMMUNITY DEVELOPMENT

Community development has been around as long as there have been people, but perhaps is was only in the 20th Century that outside groups began to try to do community-level organization to bring about economic and social development. Community development began in the early parts of the century in both the North and South.

In the US, President Theodore Roosevelt's Country Life Commission in 1908 identified a lack of community organization as the major problem in rural development. At the same time the university-based cooperative extension movement was beginning to experiment with community development approaches, which led to Congress writing a mandate for community development work into the legislation that created the Cooperative Extension Service in 1914 (Phifer, 1980). The New Deal in the 1930s was perhaps the first very large scale use of community development, in this case focussed on community services and municipal planning (Holdcroft, 1982). In 1935, sixteen states had rural community development programmes in place (Booth and Fear, 1985).

The British Colonial Offices' first official use of the term 'community development' was in 1948, but they had been carrying out various types of community mobilizations in the colonies since about 1920 (Holdcroft, 1982). Much of the Colonial Office's focus, particularly in Africa, was on resource conservation, especially preventing soil erosion, and the 'community development' aspects were, if fact, often coercive. The programmes consisted of mobilizing local labour for terracing and other heavy conservation work, accompanied by unpopular measures such as reductions in number of livestock (Korir-Koech, 1991).

Support for community development spread rapidly in the late 1940s to both external donor agencies and national governments. Holdcroft (1982) argues that the inspiration for the global community development movement that emerged in the 1950s had both religious and secular aspects, and that three early sources were particularly important. The first was the rural reconstruction movement in India in the 1930s that was animated and lead by such figures as Ghandi and Tagore. Policy makers in both the United States and the United Nations were influenced by their understanding of this movement. The second source was the experiences of both secular volunteers and missionaries engaged with poor communities throughout the world. The Ford Foundation played an important early role in the supporting these efforts. The third major source of the inspiration was the experience, coming out of the Depression, with adult education, community services and social welfare programmes in both the United States and Europe.

Throughout the 1950s, community development was a prominent official aspect of the nascent international development mission that Western Governments were creating as a new way to relate to the areas that were rapidly becoming ex-colonial dominions. The first national community development campaign was launched in India in 1952 with support from the Ford Foundation and the United States government. Soon afterwards, national programmes were established in the Philippines, Indonesia, Iran and Pakistan. At the height of the community development wave the United States alone supported community development programmes in 30 countries, employed 105 community development experts, providing approximately 50 million dollars while the UN provided a somewhat smaller amount. Community development newsletters and conferences proliferated (Holdcroft, 1982). While this was not a lot of money compared with military and infrastructure development, it was still a substantial amount for programmes that provided more 'software' than 'hardware'.

The community development philosophy that practitioners articulated at the time was not very different from what is heard today. The emphasis was on catalysing self-initiative and organization in communities, doing adequate research and evaluation, ensuring the co-ordination of community development with technical services, and working with village councils. In US and British influenced approaches, the common method was to train secondary school graduates in community organizational methods and send them into rural areas where they would work with the existing village leadership (Holdcroft, 1982). The *animation rurale* approach in the former French colonies was slightly different. Here the trained development agents were recruited, trained, and then sent back to work in their own village. *Animation rurale* programmes were more rigidly designed than the US and British efforts with more of an emphasis on teaching technical skills to the development agents in addition to organizational skills (Gow and Vansant, 1983).

The early 1960s saw a rapid decline of support for community development on the part of both donors and developing country governments. The major reason was that the anticipated benefits of community development for rural economies had not materialized. Measured programme performance was, in fact, very disappointing. While defenders of community development claimed that the politicians did not understand how long such processes should take, the programmes were often not accepted by the people and failed to reach the very poor (Holdcroft, 1982). While the programmes were operating at a local village level, the participation by the local people was in most cases acting as recipients and inexpensive labour. In the end, the advocates of community development within governments and donor agencies lost out to the advocates of more specialized and technical approaches to development. These people had real, practical proposals, while community development advocates were advocating abstractions (Holdcroft, 1982).

4. PARTICIPATORY COMMUNITY DEVELOPMENT AS A CRITICAL PERSPECTIVE

The collapse of community development as a central theme of large scale development efforts in the middle 1960s was hardly the end of the story. Community development's move to a more marginal place within the development landscape was a sort of reverse co-optation. Community development could begin to be elaborated into a community-based, critical alternative to mainstream development. The formation of the Community Development Society and the founding of its Journal in 1969 symbolized and gave form to this new status.

By the end of the 1960s, community development retained three major institutional homes. One was in the universities. Community development was still an important part of agricultural extension programmes in Western countries. Beyond this, by 1976, 48 institutions in the United States were offering degree programmes in community development, as were many other institutions throughout the world (Cary, 1976). The second main institutional home was in non-governmental and private, voluntary development organizations. During the 1960s, these organizations began to move away from a 'relief and welfare' idea of aiding the rural poor towards a more community-development, local capacity building strategy that would finally evolve into the regional approach discussed below (Korten, D., 1986). The third major institutional location for community development was in offices within line ministries in developing countries where some community development workers continued to operate but took more of a back seat, supporting role than they had in the 1950s.

The community-based, critical alternative to mainstream development that began to develop was critical of previous local development efforts in a number of ways.

One of the most commonly heard criticisms of the community development efforts was that they treated rural communities as homogeneous. The main issue, particularly among early critics, was class. Village level community development agents worked through village elites for several reasons. The rural elite class includes the recognized community leadership who play a role in facilitating the entry to communities for any outsider, particularly for official functions or programmes. As secondary school graduates, community development agents came from and were biassed toward the elite class. Working through the existing community leadership and other prominent citizens was simply easier for programme implementation and meant quicker results (Gow and Vansant, 1983; Holdcroft, 1982). The programme designs themselves were at best class neutral, as in programmes that were based on creating village consensus, or at worst they would benefit elites directly without aiding the poorer people at all. A common example of the later were programmes that relied on distributing new technology to 'progressive farmers' under the assumption that these technologies would then be adopted by the 'laggards'. Holdcroft (1982) argues that the major failing of the 1950s efforts were that the programmes were not intended to address structural barriers to development. Basic conflicts over resources and opportunities within the villages were simply too deep to resolve through persuasion and consensus methods. Government run community development projects, in fact, could do little about the main causes of rural poverty, for example land distribution, because their influence on the social and political relations at the local level was negligible (Gow and Vansant, 1983).

The second major problem in the assumption of community homogeneity was systematic gender differences. A series of studies beginning in the 1970s debunked the assumption that gender differences in communities (McCall, 1987) and households (Guyer, 1981) could be ignored by development efforts. Studies found that for women participation meant that the burden of development labour was heavier, while difficulties continued in gaining full access to its fruits (Palmer, 1979). In Rogers' (1990) review of the literature on development and households she identified four critical causes of differential costs and benefits of development efforts between men and women. The first was differences in the time available to different household members, many times allocation studies have found that women in rural households work much longer hours than men to. This fact led McCall (1987) to suggest that the dominance of 'male heads-of-households' in decision making in Tanzanian villagization programmes might be related to the fact that women's work loads left them little time to spend in endless village debates. Rogers' (1990) second point related to the allocation of household tasks. A common finding was that men tended to control cash crops while women controlled food crops. This and other household patterns lead to the third issue, which was differential control over household income. Rogers' (1990) last point was that men and women have unequal access to goods for both production and consumption. One critical good was access to information from agricultural extension services that assumed that men were making the agricultural decisions so targeted them for the training and benefits (eg Potash, 1985).

Community development was also criticised for being a vehicle for the promotion of government programmes rather than a way to promote local goals. Gow and Vansant (1993) argue that this approach not only undercut village initiatives but led the community development workers to be pressured by many different state agencies and become, in essence, the sales representatives of line ministries. The relationship between community development efforts and the state became very problematic. Community development was seen by the villagers as a demand by the state for their time and labour to achieve the state's goals (Hirsch, 1990; Holmquist, 1984). At the same time, Holmquist (1984) argues, community development was leading to mobilized villagers making increased demands on the state for resources, which quickly undercut the state's support for the programmes.

The 'project approach' to community development led to a tendency to treat villages as isolated entities divorced from larger geographical, economic, and social realities (Gow and Vansant, 1983). Rural development requires a spatially balanced stimulation of the rural economy (Rondinelli, 1981). Early community development efforts, in contrast, were criticised for being concentrated in areas that were relatively prosperous (Holmquist, 1984) and easily accessible to outsiders (Chambers, 1983).

These and similar criticisms of community development came together in the late 1970s and early 1980s as a call for 'authentic' as opposed to 'pseudo' participation in development (Uphoff, 1984). Strong evidence was mounting that willing participation was the key to success in development efforts. Local people needed to be actors in development, not the recipients. A study by Development Alternatives (1975) looked at the effect of small holder participation in 36 rural development projects. They evaluated each project on: increased income, increased knowledge of agriculture, self help capacity and the probability that the benefits would be self sustaining. They found that local action by the farmers accounted for 49% of the variance of project success.

Some of the suggestions for making participation more 'authentic' seemed counter intuitive. One idea was that all participants, no matter how poor, should be required to invest their own resources in development efforts. Anyone will value something they have invested in more than something they have gotten without any commitment of their own resources (Gow and Vansant, 1983). Authentic participation also meant not trying to avoid or paper over differences but rather development efforts needed to recognize existing conflicts and be willingness to work with various parties in what we would call today a multiple stakeholder approach (Gow and Vansant, 1983).

Finally, the meaning of authentic participation that developed during these years included, but went well beyond, the considerations about which groups were carrying out which tasks, as in Cohen and Uphoff's (1977) framework described above. Authentic participation meant a willingness by government and other outside groups to allow local people the dignity of true partnership. Frances Korten (1986), looking at local resource management in Asia, argued that when a government views a local group as primarily an extension of their agency it fails to 'acknowledge the extent to which the development of local capacities depends on establishing commensurate rights and authorities. Local organizations can seldom command the respect of their own members if they cannot also command the respect of the agencies of the state' (1986, pp 286).

5. NEW APPROACHES TO PARTICIPATORY COMMUNITY DEVELOPMENT

Bringing about 'authentic' participation, its proponents argued, required several things. One was the reorientation of development bureaucracies to be willing to work cooperatively and responsively with local people (Uphoff, 1984). As anyone who has worked in local communities can tell you, success is dependent on meeting practical felt needs, local people are mainly interested in ways of increasing their production and wealth (Hirsch, 1990). Perhaps the most difficult requirement is ensuring that participation is comprehensive, that women and the poorer and less powerful community members also participate in decision making. This means, finally, that the participatory structures be democratic. Efforts that rely on non-democratic village or traditional leadership may end up benefiting only those groups (Ribot, 1998).

Symbolized by the 1987 publication of the Brundtland Commission's <u>Our Common Future</u>, during the 1980s the term 'sustainable development' was becoming increasing important. Some of the proposals for how to do 'sustainable development' represented a substantial break from previous practice, including a suggestion that development efforts move away from a 'project by project' approach and begin to consider regional, bottom up approaches to development. The idea was that the very idea of organizing development around 'projects' is inherently unsustainable. Projects are based on achieving a set of goals in a definite time period, usually because of funders needs, this means a bias toward visible and quantifiable outcomes rather than capacity building. Setting up development efforts in terms of projects can also lead to an inflexible use of knowledge and information. Projects enter an area with a predisposition towards particular solutions to problems based on the analysis of outside experts or rapid consultation with local people. Projects also continue patterns of uneven development by creating islands of intensified development within their own sectors that have no good way of spreading their benefits beyond their boundaries. Korten and Uphoff (1981) wrote:

> 'Within this framework (development projects) development is approached through a series of finite, discontinuous actions with discrete time-bounded outcomes and

which depend on special temporary injections of external funds. Following Western theory, planning is presumed to be separate from and preparatory to action. The implementor then follows this blueprint in the best tradition of policy-neutral administration. Institutional structures are regarded as largely fixed and are given little attention, as projects preoccupy both planners and implementors'.

In the 1970s community development efforts had emerged that were designed to catalyse authentic bottom up participation while operating on a regional basis. They focussed directly on the question of capacity building in development organizations and institutions. The most famous of them, the Bangladesh Rural Advancement Committee (BRAC), an effort which later evolved into the Grameen Bank, focussed on the provision of micro credit. BRAC was regional in that it focussed on organizing the rural poor across a wide area of Bangladesh. It was a great success and became an important model for other development efforts. The regional programmes that emerged at the time used community organization at the very local level as their strategy but focussed on regional, and even national level, linkages between villages. Some examples were sectoral, such as the Lampang Health Development Project in Thailand or the Indian National Dairy Development Board. Others did local institutional development such as the Sarvodaya Movement of Sri Lanka or the Institute of Cultural Affairs in Kenya, an organization with which the present author worked for many years.

These regional approaches focussed on catalysing authentic participation in local villages while linking them with other villages in the wider regions. The best of them brought about development participation by bringing village level leadership into the development effort at the regional level. Trained village leaders are often the best suited people to catalyse increases in development capacity in other villages. Regional approaches also represented a possible response to concerns about how traditional systems of natural resource use, which are locally ecologically adaptive, but limited in their contribution managing larger-scale resource, could play a role in regional-level resource management problems (Ruddle and Rondinelli, 1983).

Central to the new approaches to participatory community development that arose in the 1980s was the increased prominence of action research. Some of this was technical research focussed on applying the 'appropriate technology' that was a central technical theme in rural development at the time (Schumacher, 1973). But action research was much more than technical research, it was really about the relationship between knowledge and control over development decisions. A seminal expression of the relationship between community development and action research was David Korten's (1980) article 'Community Organization and Rural Development: A Learning Process Approach' which was very influential in arguing that successful development programmes involved an openness to new insights generated by a constant reevaluation of the programme from the perspective of the beneficiaries and a willingness to redirect the programme when necessary. Rondinelli (1983) took Korten's argument to the next logical step by arguing that development bureaucracies needed to incorporate such a learning process into their ongoing efforts. He also pointed out that the need of central bureaucracies and donor agencies for central control made this very difficult. He called for a bureaucratic reorientation which would bring the reward structure of government bureaucracies into line with the need for identifying and responding to problems.

The Farming System Research (FSR) approach that appeared in the late 1970s strongly reflected the thinking about development and action research emerging at the time. FSR is a cross-disciplinary approach that emphasizes holism by approaching agriculture as an integrated system (Conway, 1987). Shaner *et al.* (1982) argued that 'a clear picture of farmers and their environments' (19) is the key to a holistic approach to farming. FSR begins with the farmer and traces the physical and social systems that lead to farming decisions. Another emphasis of FSR is a focus on problem solving. Not only does a problem solving focus ensure a practical bent to the investigation, it helped developing participation and the training of new personnel (FAO, 1982). Another aspect of FSR emphasized by Shaner *et al.* (1982) among others is that the FSR approach is an iterative one that remains grounded by this problem solving orientation. The research team begins with partial information, increases its knowledge through studies and experiments, and then modifies its actions in response to the result. What it does not do is wait for laboratory studies and 'excessive precision' and then take the results onto the farm. FSR 'seeks to provide better solutions to farmers' conditions, but not necessarily the best solutions' (Shaner *et al.*, 1982, p 20, emphasis in the original).

Action research and the 'learning process approach' were held up as an alternative to the 'blueprint' model of the development project. Research components were to become part of all development efforts. Projects should be designed in iterative phases. First, a design phase based on a full flow of information between stakeholders, in which communities defined their own problems and interests. Second, an implementation phase in which all parties carried out activities. Third, an evaluation phase based on participatory evaluation research (Gow and Vansant, 1983; Korten, 1980; Uphoff, 1984).

In retrospect, the emergence in the late 1970s and early 1980s of concepts around the two themes of 'action research' and 'authentic participation' broke down important barriers and laid the foundations for what would emerge as a new community development orthodoxy. One barrier was that between conceptualization and action. A division between those who defined the problems, outlined the solutions and carried out the actions was no longer seen as tenable. People were beginning to be aware of and put into practice concepts that John Dewey (1929) had developed as philosophy a generation before. Dewey's arguments had centred on how knowledge and action came to be separated, not by their own nature, but by the practical constraints on humans to take the risks involved in linking them. Addressing the vast problem if rural poverty meant a willingness to take such risks.

The other barrier was that between development and conflict. Community development's earlier papering over real divisions through relying on community consensus had not worked because the consensus always reflected the interests of the community members whose needs were the least pressing. The 'radical' version of action research and the 'conflict' version of community development could no longer be ignored as outside the mainstream because real conflicts within communities had to be addressed if development efforts were to have any effect.

6. NEW APPROACHES FOR FISHERIES DEVELOPMENT

In parallel with the changes in community development that focussed on agricultural communities, the late 1970s and 1980s saw a renewed interest in the development of small-scale fisheries. In comparison to agriculture, if the relative contribution of the two sectors to nutrition is considered, fisheries development was badly neglected. Only the FAO paid much attention to small scale fisheries. Sommerville, (1986) in an analysis of the Sahel an area that includes the Niger valley, one of the most productive flood plain fisheries in Africa, found that between 1975 and 1983 fisheries received only 1.2% of overseas development aid. When aid did go to fisheries it tended to be misplaced, with too much going to large scale fishing. Robinson and Lawson (1986), in their analysis of the problems with fisheries development, argued that fisheries development should focus on semi-industrial fleets in order to achieve the full exploitation of the new 200 mile exclusive economic zones. They also argued that training and institutional capacity building for fisheries development was a major problem. Host governments expected that fisheries projects would pay off very quickly and this lead to neglecting training a local staff with long term capacity (Allsopp, 1985).

In keeping with the call for a development emphasis on small scale fishing, new 'appropriate technologies' began to be developed, especially for reducing the post-harvest losses which was seen as the most critical technical problem. The best known was a simple fish smoking method developed in Chokor, Ghana, which made small changes in the traditional method by adding wire trays, with greatly improved production and ease of use (Brownell and Lopez, 1985). Other developments included solar dryers, the use of fermentation as a preservation measure, and new cooling technologies (Diouf, 1985; N'jai, 1985). The importance that was given at the time to post-harvest losses meant that less attention was given to improving the actual fishing techniques of artisanal fishermen. Marketing and cooperatives was another area on which some experimentation was done (Toh, 1982).

In 1979 and 1980 the FAO held a major conference over two separate years on the development of research capacity in fisheries (FAO, 1981). The conference argued that fisheries research needs to be part of ongoing development efforts and develop research capacity in a practical way. One of the basic messages was that, like agriculture, fishery research has had a struggle shifting its northern, temperate biases to become effective in Third World development. This argument was made particularly in the context of shifting from single species research efforts to multiple species efforts more suited to tropical conditions. In spite of the enormity of the task research on fisheries received scant support (Vanderpool, 1989) and was too often designed to be a support system for developing an industrial fishing fleet (Ansa-Emmim, 1981).

It was not long before the emerging critique and new ideas in community development were taken up by people concerned specifically with fisheries. Ben-Yami and Anderson (1985) wrote a handbook for the FAO on fisheries development from a community development perspective. Their emphasis is on organizing fishing communities and the design and implementation of fisheries development and extension teams. The planning stage emphasizes training participation by local groups in 'joint planning groups' with extension workers. These plans were meant to cover infrastructure, training, boats, marketing and the financing of the development.

By the late 1980s, it was becoming clear that the central problem in fisheries was not development as much as management. Another contemporary concept was, to borrow the title of an important book of the time, the Question of the Commons (McCay and Acheson, 1987). The history of community development, and the changes that were going on in understandings of what community development meant, were clearly applicable to fisheries as well. The question of the commons was, and is, fundamentally a question about community, so it entirely appropriate that the community development tradition should be an important source in the search for answers. It was in this milieu that the term fisheries co-management began to appear.

7. LESSONS FOR FISHERIES CO-MANAGEMENT

The intent of this chapter is to describe the history of community development up to the point where the idea of fisheries co-management began to take root. Drawing some lessons for co-management from this history, however, requires a brief mention of how the community development ideas of the 1980s fared in the mean time. Both action research and the search for authentic participation have been to some degree institutionalized within small-scale development efforts around the world. They have also both given birth to new critical perspectives that are being brought to bear on current local management and development efforts.

Action research has mainly taken the form of a series of techniques used under names like 'participatory rural appraisal' (PRA) or 'participatory monitoring and evaluation'. PRA has been thoroughly institutionalized. Chambers (1999), who takes the word 'participatory' very seriously, lists 39 countries in which PRA *networks* exists. He also laments the vast amount badly done PRA that can now be found and is used to justify poor development work. Out of the action research tradition has emerged a very large, and heavily contested, literature on local (or indigenous or traditional) knowledge, a concept which extremely important in fisheries co-management and which figures heavily in Chapter 15.

The idea of authentic participation is now ubiquitous in the development discourse. Any small-scale or regional development project must address the question of how it will handle stakeholder participation. It has also been institutionalized in the enormous rise in the importance of small and medium scale Non-Governmental Organizations (NGOs) in the last twenty years. Development stakeholders are far more organized now than in the past, a fact that both enables and complicates participatory development. This institutionalization of participation has brought two critical questions to the forefront. The first is the question of scale and representation. What does participation actually mean when the development effort, or more pertinently, the natural resource is too large to base participation on stakeholders meeting under a baobab tree. This questions application to fisheries co-management is addressed directly in Chapter 16.

It is important in our thinking about fisheries co-management that we not reinvent the wheel. There are some important things that were learned in the community development field in the 1970s and 1980s that we need to keep in mind today. One is the discovery of the potential for regional approaches to participatory development that can be applied to the participatory management of larger than local scale resources. Another is that heterogeneity and conflict among stakeholders must be addressed, not hidden, in co-management efforts. Gender, class, ethnicity, etc. all make a difference. Conflict, as discussed at length in Chapter 11, is a part of every fishery and can be a creative as well as a destructive force for management. Another lesson is that government fisheries agencies must be willing to actually devolve some kinds of power to stakeholders if co-management is going to work. We cannot yield to the temptation to build co-management around using fishers to relieve government of some of its burdens but none of its power. At the same time, the growth of participation in the last 15 years has not happened because of extensive 'bureaucratic reorientation' (Uphoff, 1984). Government ministries have tasks to perform that are only tangentially concerned with local capacity building. Departments of Fisheries cannot 'reinvent' the demands on them that come from their mandate to manage national resources. Action research and participation in other arenas have taken their strongest form through the institutions of civil society, both local and international, working in concert with governments. Governments must be willing to work with, and even extend power to, other participating stakeholders, but they cannot themselves become a movement to bring about participatory resource management.

REFERENCES

Allsopp, W.H.I. (1985) Fishery Development Experiences. Farnham, Surrey: Fishing New Books Limited.

Ansa- Emmim, M. (1981) The Fishery Research Unit of Ghana:Past and Present. Annex 5 of *FAO Fisheries Report* # 251.

Ben-Yami, M. and Anderson, A. (1985) Community Fishery Centres: Guidelines for Establishment and Operation. *FAO Fisheries Technical Paper* # 264 Rome:FAO.

Booth, N. M. and Fear, F. A. (1985) Community Development: An Old Idea Comes of Age. *The Community Development Society*.

Brownell B. and Lopez, J. (1985) The Chokor Fish Smoking Method: A Truly Appropriate Technology. Proceedings of the FAO Expert Consultation on Fish Technology in Africa. *FAO Fisheries Report* #329 Rome:FAO.

Cary, L.J. (1976) Community Development Education: a overview of programmes throughout the world. *Journal of the Community Development Society* 7, 115-121.

Chambers, R. (1983) Rural development: Putting the last first. Longman, London.

Chambers, R. (1999) Relaxed and Participatory Appraisal: Notes on Practical Approaches and Methods. IDS mimeo available at http://www.ids.ac.uk.

Christenson, J.A. and Robinson, J.W. Jr. (1980) In search of community development. In J.A. Christenson and J.W. Robinson (Eds) Community Development in America Ames IA: Iowa State University Press, pp 3-17.

Cohen, J. and Uphoff, N. (1977) Rural Development Participation: Concepts and Measures for Project Design. Mimeo: Rural Development Committee, Cornell University.

Conway, G.R. (1987) The Properties of Agroecosystems. *Agricultural Administration* **22**, 95-117.

Development Alternatives Inc. (1975) Strategies for Small Farmer Development: An Empirical Study of Rural Development Projects. Washington: Development Alternatives Inc.

Dewey, J. (1929) The Quest for Certainty: a Study of the Relation Between Knowledge and Action. New York: Minton, Balch and Company.

Diouf, N. (1985) Etude de la Technologie Appropriée à la Conservation du Poisson Frais de Pêche Artisanale. Proceedings of the FAO Expert Consultation on Fish Technology in Africa. *FAO Fisheries Report* #329 Rome:FAO.

FAO (1981) Report of the Working Party on the Promotion of Resources Research in Developing Countries. *FAO Fisheries Report* #251 Rome:FAO.

FAO (1982) Proceedings of the FAO Expert Consultation on Fish Technology in Africa. *FAO Fisheries Report* #268 Rome:FAO.

Freire, P. (1970) Pedagogy of the Oppressed. New York: Herder and Herder.

Gow, David D. and Vansant, Jerry (1983) Beyond the Rhetoric of Rural Development Participation: How Can It Be Done. *World Development* **11**(5), 427-446.

Guyer, J.I. (1981) Household and Community in African Studies. *African Studies Review* **23**(2/3), 87-137.

Hall, B.L. (1979) Knowledge as a Commodity and Participatory Research. Prospects 9:4.

Hirsch, P. (1990) Development Dilemmas in Rural Thailand. Singapore: Oxford University Press.

Holdcroft, Lane, E. (1982) The Rise and Fall of Community Development in Developing Countries, 1950-1965: A Critical Analysis and Implications. In Jones, G. E. and M.J. Rolls (Eds) *Progress in Rural Extension and Community Development*. New York: John Wiley and Sons, pp 207-231.

Holmquist, F. (1984) Class Structure, Peasant Participation, and Rural Self-Help. Chapter 7 in Barkan J.D. (ed.) *Politics and Public Policy in Kenya and Tanzania*. Nairobi: Heinemann Kenya, pp. 171-197.

Horby, A.S. (1995) Oxford Advanced Learner's Dictionary of Current English Oxford: Oxford University Press.

Korir-Koech, Michael (1989) in Kiriro, A. and C. Juma (eds) *Gaining Ground: Institutional Innovations in Land-use Management in Kenya*. Nairobi: ACTS Press, pp. 21-34.

Korten, D.C. (1980) Community Organization and Rural Development: A Learning Process Approach. *Public Administration Review* **40**(5), 480-511.

Korten, D.C. and Uphoff, N. (1981) Bureaucratic Reorientation for Participatory Development. NASPAA Working Paper No. 1.

Korten, D. C. (1986) Micro-policy Reform - The Role of Private Voluntary Development Agencies. In Korten, D.C. (ed.) *Community Management: Asian Experience and Perspectives*. West Hartford: Kumarian Press, pp 309-317.

Korten, F.F. (1986) Creating and Enabling Setting. Chapter 20 in Korten, D.C. (ed.) *Community Management: Asian Experience and Perspectives*. West Hartford: Kumarian Press.

Littrell D.W. and Hobbs, D. (1989) The Self-Help Approach. Chapter 3 in Christenson J.A. and J.W. Robinson (eds) *Community Development in Perspective*. Ames IA: Iowa State University Press, pp 48-68.

McCall, M. (1987) Carrying heavier burdens but carrying less weight: some implications of villagization for women in Tanzania. In Momsen, J.H. and J.G.Townsend (eds) *Geography of Gender in the Third World*. Hutchinson: SUNY Press, pp 111-124.

McCay, B.J. and Acheson, J.M. (1987) Human Ecology and the Commons. In McCay, B.J. and J.M. Acheson (eds) *The Question of the Commons: The Culture and Ecology of Communal Resources*. Tucson: The University of Arizona Press, pp 1-34.

N'jai, A.E. (1985) Fermenting and Drying Fish in the Gambia. Proceedings of the FAO Expert Consultation on Fish Technology in Africa. *FAO Fisheries Report* #329 Rome:FAO.

Palmer, I. (1979) The Nenow Case. In Palmer, I. (ed.) *Case Studies of the Impact of Large Scale Development Projects on Women; a Series for Planners*. New York: The Population Council.

Phifer, B. M. (1980) History of community development in America. In Christensen, J.A. and Robinson, J.W. (eds) *Community Development in America*. Ames IA: Iowa State University Press, pp. 18-37 .

Potash, B. (1985) Female Farmers, Mothers-in-law and Extension Agents: Development Planning and a Rural Luo Community. Working Paper # 90 MSU WID.

Ribot, Jesse C. (1998) Theorizing Access: Forest Profits Along Senegal's Charcoal Commodity Chain. *Development and Change* **29**, 307-341.

Robinson, M. and Lawson, R. (1986) Some Reflections of Aid to Fisheries in West Africa. *Marine Policy* **10**, 101-110.

Rogers, B.L. (1990) The Internal Dynamics of Households: A Critical Factor in Development Policy. In Rogers, B.L. and N.P. Schlossman (eds) Intra-household Resource Allocation: Issues and Methods for Development Policy and Planning. Tokyo: United Nations University Press, pp. 1-19.

Rondinelli, D.A. (1981) Spatial Analysis for Regional Development: A Case Study of the Bicol River Basin of the Philippines. Resource Systems Theory and Methodology No. 2. Tokyo: United Nations University.

Ruddle, K. and Rondinelli, D.A. (1983) Transforming Natural Resources for Human Development: A Resource Systems Framework for Development Policy. Resource Systems Theory and Methodology Series No. 1. Tokyo: The United Nations University.

Sanders, I. T. (1958) Theories of Community Development. *Rural Sociology* **23**, 1-12.

SEATO Conference on Community Development (1960) Report Baguio, The Philippines.

Schumacher, E. F. (1973) Small is beautiful; economics as if people mattered. New York, Harper & Row.

Shaner W.W., Philipp, P.E. and Schmeil, W.R. (1982) Farming Systems Research and Development: Guidelines for Developing Countries. Boulder: Westview.

Sommerville, C. M. (1986) Drought and Aid in the Sahel: A Decade of Development Cooperation. Boulder: Westview.

Toh (1982) Collection and Distribution Programme for Fresh Fish in Small Scale Fisheries, Ondo State, Nigeria. Proceedings of the FAO Expert Consultation on Fish Technology in Africa. *FAO Fisheries Report* #286 Rome:FAO.

Uphoff, N. (1984) Participation in Development Initiatives; Fitting Projects to People. In Putting People First: Sociological Variables in Rural Development. Cernes, M. (ed.) Baltimore: Johns Hopkins University Press.

Vandenburg, L. and Fear, F. A. (1983) Participatory Research; An Analysis of Two Approaches. *Journal of Voluntary Action Research* **12**, 11-28.

Vanderpool, C. (1989) Food Security and Social Impact Assessment: Agricultural and Fisheries Production Systems in Senegal. Paper presented at the VIIth World Congress of Rural Sociology, Bologna.

Wilson, D.C. and McCay, B. J. (1998) How the Participants Talk About Participation in Mid-Atlantic Fisheries Management. *Ocean and Coastal Management* **41**, 41-61.

Wilson, D. C. (1999) Fisheries Science Collaborations: The Critical Role of the Community. *Institute for Fisheries Management Research Publication* No. 45.

Chapter 2

SCIENCE AND THE USER PERSPECTIVE: *The gap co-management must address*

POUL DEGNBOL
Institute for Fisheries Management and Coastal Community Development, Hirtshals, Denmark

1. KNOWLEDGE IN FISHERIES MANAGEMENT

The involvement of users in fisheries management is in many cases limited to consultation on implementation issues, but may also involve the development of shared understandings of objectives and the knowledge basis for management - the problems to be addressed by management and the characteristics and state of the resources and the fisheries. Jentoft (1993) distinguished between procedural legitimacy and content legitimacy. Procedural legitimacy comes from involvement in the specification of implementation modalities while content legitimacy comes from shared understandings of objectives and knowledge. In the public debate on fisheries management the issue of sharing knowledge is often translated into the need to disseminate research results to fishermen, with the underlying understanding that everybody in the management process share the same basic paradigm, and some actors just know better than others within this paradigm. However, the issue is much more complex than this, as the public debate also frequently indicates: fundamentally different understandings of the fish stocks are frequently presented and these differences cannot be reduced entirely to differences in interest. These differences must be understood as a first step to a shared understanding or - maybe more realistically - to mutual acceptance of differences.

The subject of this paper is the knowledge base for fisheries management decisions and specifically that part which relates to the functioning of the resource system. The paper will discuss how the mainstream discourse in fisheries science has developed over the last 100 years and how this development has led to a widening gap *vis a vis* the users perspective - a gap which co-management arrangements must address and bridge if they are to be truly inclusive.

Any technical or informal evaluation of the state of stocks and management options is based on explicit or implicit management objectives and will relate to a specific set of 'managers' who take note of the evaluation and implement management - whether this is a central government, a formal co-management committee or communities implementing access rules which may even not be understood as fisheries management by the communities in the first place. The character and relevance of biological knowledge for management is therefore constituted by the objectives for management and the identity of the 'managers'.

Modern fisheries biology has developed in close association with a management system characterised by both centralised decision making based on numerical control of input or output parameters through top-down control structures and by an explicit emphasis on resource conservation. Contemporary fisheries biology provides the cognitive basis for this system through stock assessments, which are basically predictions of short and long-term effects on stocks and yields given by various scenarios based on statistics. The development of this management system and its cognitive base is an example of broader developments in society's ideas about management. The modernization process has been one of continually incorporating purposive rationality into decision-making systems and should be analysed and understood within this historical and social context.

Within this process, fisheries research takes on the role of a regulatory science (Jasanoff, 1990). The research is carried out within specialised organizations where it produces formalised knowledge for use as a basis for management decisions and implementation by centralized bureaucracies interacting with representative democratic institutions. The management objectives in this model are in many cases not explicit, but the long term sustainability of the resource base has been the overriding objective whenever objectives are stated. The underlying rationality of this system is based on an assumption of predictability, ie an understanding that specific and predictable targets can be achieved by implementing specific regulatory measures such as catch or effort quotas or technical measures. A catch quota is within this rationality a means to regulate the fisheries such that the resulting pressure on the resource (as measured for instance by the fishing mortality) will be less than or equal to a reference pressure. The basic assumption is then that it is possible to predict the outcomes of a specific regulatory measure in terms of - in this case - the resulting fishing mortality. This normative and regulatory context has meant that the production of biological knowledge about stock dynamics and predictions of the response of stocks to fishing has been the dominating form of regulatory science within this model.

The specialised research organizations taking on this role were established in countries around the North Atlantic during the early part of the 20th century and are now an integral part of fisheries management systems in industrialised countries. In developing countries, development efforts based on the modern fisheries management model have emphasized the need to develop specialised research organizations that can produce this kind of knowledge. This has been done to the extent that this model for producing the cognitive base for management - including the encapsulation of cognitive validity within specific research institutions and the associated relevance criteria for knowledge - has been promoted by most national and multilateral developing agencies as an end in its own right, as something which is considered an essential component of any fisheries management system irrespective of normative, regulatory or social context.

This model for establishing a knowledge base for fisheries management has had limited success in both industrialised and developing countries so far in terms of achieving the stated objectives of management. As an example of the situation in industrialised countries, the European Commission (2001) concludes that 'as far as conservation is concerned, many stocks are at present outside safe biological limits. They are too heavily exploited or have low quantities of mature fish or both. The situation is particularly serious for demersal fish stocks such as cod, hake and whiting. If current trends continue, many stocks will collapse.

At the same time the available fishing capacity of the Community fleets far exceeds that required to harvest fish in a sustainable manner'. And 'Politically, the stakeholders do not feel sufficiently involved in the management of the policy and many believe that there is no level-playing field in terms of compliance and enforcement' (European Commission 2001). The situation in many developing countries is even more severe as coastal areas and their resources are under increasing pressure.

In relation to the knowledge base, scholars have argued that a decoupling of or even contradiction between the formalised research knowledge and the users' knowledge has contributed to the problem. This gap has been formulated variously as resulting from an inherent cultural contradiction (Finlayson, 1994), as a reflection of differing discourses and interests (Bailey and Yearley, 1999) or as a distortion resulting from the communicative properties of management institutions (Wilson and Degnbol, 2002; Wilson, 2002). This question is the main focus of Chapter 15 in this book.

The development of fisheries research around the North Atlantic in the early 20th century established the discourse which still forms the basis for mainstream international fisheries research. One of the major actors in the development of fisheries biology in the first half of the 20th century, Michael Graham, summarised some tendencies in fisheries research from 1900 onwards when he presented his Buckland lectures in 1939 (Graham, 1948). He noted that

> 'The underlying idea of the period of international research was that not enough was known about the life-histories of the food-fishes, about the causes of abundance and scarcity, the growth-rate, interchange of stocks, seasonal migrations, the proportion taken by fishermen, and other things that must be relevant to the problem of rational fishing. These things, naturally, had to be studied for each species separately, and consequently the work was first arranged according to species of fish. Then students of the several species became advocates of those particular measures that seemed best adapted for particular species that they studied, and the overfishing problem became, as it were, divided. The more diversity that was revealed, the less satisfactory did any simple action see; and this period lasted until 1935, when it became clear, as we shall see later, that it is possible to estimate the best possible course for all species of bottom-living fish taken together.'

He went on noting that

> 'the chief characteristic of the international period was that research discovered and adapted the sort of scale that was necessary for the solution of the overfishing problem. To give one example. In the early days men tried to find out *the* growth-rate of each species, whereas the international research showed that the aim was to find out the average growth-rate for any particular area and intensity of fishing, there being a wide variation according to season and grounds. This is a much more troublesome thing to discover. Statistics of various kinds had to bulk large in all the work.'

What Graham describes here are the changes in the perspective of fisheries biology which were associated with its transformation into a tool which could form the operational basis for practical measures within a management system based on international cooperation between governments and formalised research. The development of the International Council for the Exploration of the Sea (ICES) was at the core of this intellectual and institutional development (Mill, 1989; Rozwadowski, 2002) which according to Graham's account (Graham, 1948)

> 'has produced a new psychological phenomenon - the combined opinion of scientists and of chosen administrators, who mutually educate each other year by year at the meetings. This new kind of opinion, international and exceptionally well informed, is obviously a most powerful weapon for advancing a cause such as improvement of the fishery in the high seas.'

The discovery of 'the sort of scale that was necessary for the solution of the overfishing problem' went hand in hand with the emergence of an international community of managers and scientists, mainly working for governments, who shared norms and understandings regarding the fisheries. It may appear as a happy coincidence that 'the sort of scale' which was identified by fisheries biology happened to coincide with the scale needed by governments cooperating internationally to handle the political decision-making processes of fisheries management. This coincidence may, however, also reflect a sensitivity in the international community of fisheries science to the requirement that the science should be useful for management.

The scale of analysis in the research community may thus reflect the scale of the fisheries management institutions. This paper will investigate how changes in a specific aspect of the research discourse, the spatial and temporal scale of analysis, has reflected and contributed to the development of international management institutions and, in doing so, has removed itself from the perspective of users. The paper will also discuss the possibilities that the increasing awareness of the problems inherent in the present research discourse can lead to new approaches.

2. THE SCALE OF OBSERVATION AND THE INTERNATIONALISATION OF FISHERIES MANAGEMENT

Graham describes the development of management institutions and the transformation in research perspective as basically two sides of the same process - internationalization and formalization of the research base in management on one side and change of perspective from dealing with a range of spatial resolutions and a diversity of processes to an approach dealing with averages of a few key parameters over large scales on the other.

Graham's Buckland lecture was held at a time when this transformation was in its final stages, at least on the conceptual level.

This transformation is central to an understanding of the development of a research discourse, which is often considered remote from or even contradictory to fisher's perspective. Remoteness and contradiction are of course evident from the frequent accounts of disagreements in the fisheries press. The point to be made here is, however, that gaps between the perspectives are closely associated with the development of management institutions which required a specific type of scientific knowledge, namely knowledge based on large scale averages with low resolution in space and time and which could only be constructed on basis of sampling schemes and models which tried to overcome local variation rather than understanding it. Fishers' knowledge is generally described as having the opposite focus - knowledge on the local variation of fish abundance in time and space is essential if one is to be a successful fisherman.

The development of fisheries research began by addressing hypotheses that were based on observations of fishers, whalers and seafarers. The basic research question addressed by fisheries biologists into the 1920s was the reasons for variation in catches. The main working hypothesis was initially that variations were caused by fluctuations in migration patterns. This theory was already advanced in publications in the 18th century and the relation between migration and overfishing was discussed from the 1830s (Schwach, 2000). It was one of the two research problems which were placed in the foreground at the first meeting of ICES - the other being 'the problem of so-called overfishing' (Hoek, 1905). Studies of migrations required extensive observations of the local variations in time and space of fish abundance and the associated environment. Detailed studies were made which associated specific life history parameters such as growth rates with the local environment, and further with the local hydrography. An example of the output from such a study, demonstrating the high spatial and temporal resolution involved, is presented in Figure 1.

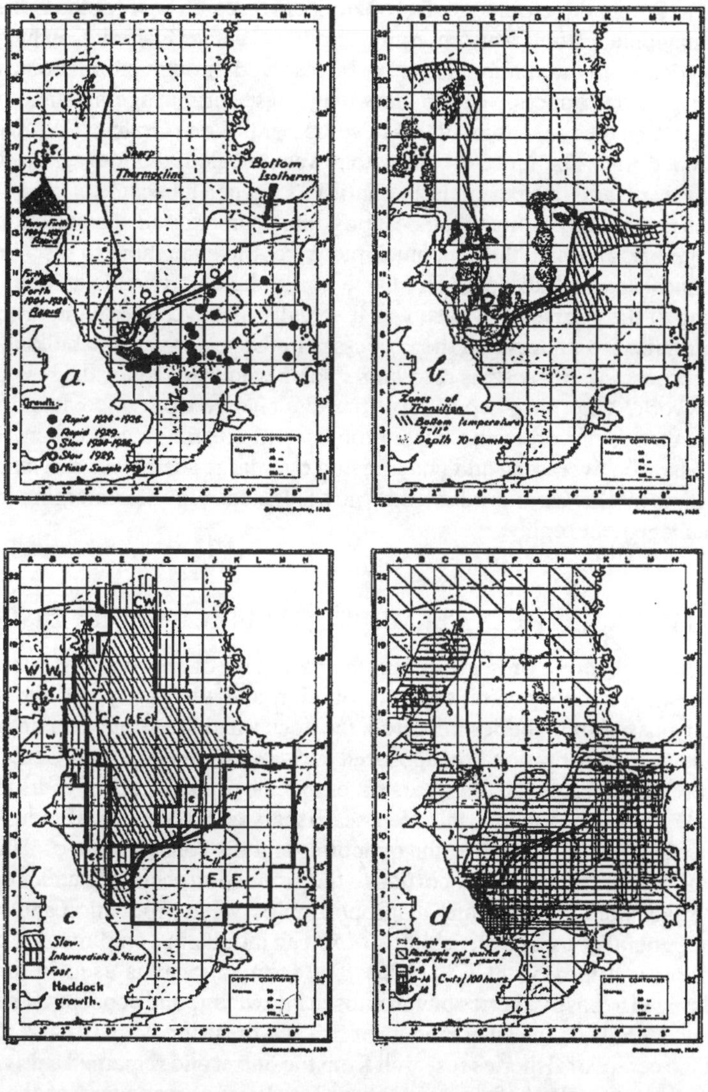

Figure 1. Association of rapid and slow growth of codling with environmental factors (Graham, 1934)

The issue of variations found a first closure after the studies following Johan Hjort's seminal paper on the Fluctuations in the great fisheries of northern Europe (Hjort, 1914) building on Heincke (1898), where he suggested that the research indicated that migration could not explain observed variations in catches and that variations in the success of year classes was a more likely explanation. This also represented a change in perspective. Even though the population concept was only used explicitly later, the 'year class' concept implies a basic population of fish which shares important aspects of their life history and which can be represented by parameters relating to the population rather than to individuals. The fish stock concept became increasingly a core concept on which theoretical developments and empirical studies was based.

The identification of the 'fish stock' as the central unit of analysis and management was fundamental to further development of an operational research base for the internationalised fisheries management that was emerging in the 1930s. Fisheries management was increasingly seen as an international issue to be resolved through international cooperation both in terms of the production of the knowledge base for management and management itself. ICES became the focal point of this development. When Graham (1948) in the quotes above concluded that 'The more diversity that was revealed, the less satisfactory did any simple action seem; and this period lasted until 1935, when it became clear, as we shall see later, that it is possible to estimate the best possible course for all species of bottom-living fish taken together' and that 'this is a much more troublesome thing to discover. Statistics of various kinds had to bulk large in all the work', he also reveals an ambiguity in his understanding of the course of events: was it actually discovered that the diversity did not need be represented to understand the processes in the sea or was the bulking of statistics a necessity because simple actions on a larger scale were needed for the new management approach to work? One may hypothesize that the change from understanding processes bottom-up at the resolution of the basic processes to creating a conceptual and research framework based on averaging and generalising over large scales was driven by this being a necessity if research was to produce the knowledge base for the emerging international, top-down management regime.

3. OPTIMALITY AND THE DETERMINISTIC PREDICTABILITY DISCOURSE

The change in perspective on scale and in the basic unit of analysis was accompanied by the development of a theory of 'rational exploitation' and, ultimately, on 'optimum fishing'. In the initial phases of fisheries research the focus had been on explaining variation. However, the concept of rationality appeared early on the agenda. The General Report of the Work of ICES covering the first years states that the principal endeavours included 'The solution of the problem, how far the deep-sea fishery as a commercial industry stands in general on a rational basis; whether the quantities and the consumption of fish, taken from the sea mentioned, are in proper proportion to the production occurring under the prevailing natural conditions, and whether any disproportion between production and consumption arises from a general or local overfishing, or from an injudicious employment of the fishing apparatus at present in use.' (ICES, 1905). The term 'rational' is used in this context to designate the need to base fisheries on formalised knowledge and concepts. This would also include knowledge about the whereabouts of fish and the technology to harvest them which had been at the centre of fisheries research from the outset and remained a driving force for research into the second half of the 20th century. In relation to management, the 'overfishing' concept was discussed extensively within the ICES community including early theories about what would now be called recruitment overfishing and destruction of habitat (Petersen, 1903) to production based considerations. The latter had by the 1930s developed to the concept of 'optimum catch' (see figure 2, Hjort *et al.*, 1933) which - as later extensions of the same approach such as MSY - remained a core concept in fisheries biology until the early 1990s and is still considered fundamental within some management regimes. The concept of rational fishing was expanded to include not just the need to base fisheries on formalised knowledge but also a requirement for *optimization*, specifically maximization of long-term yield.

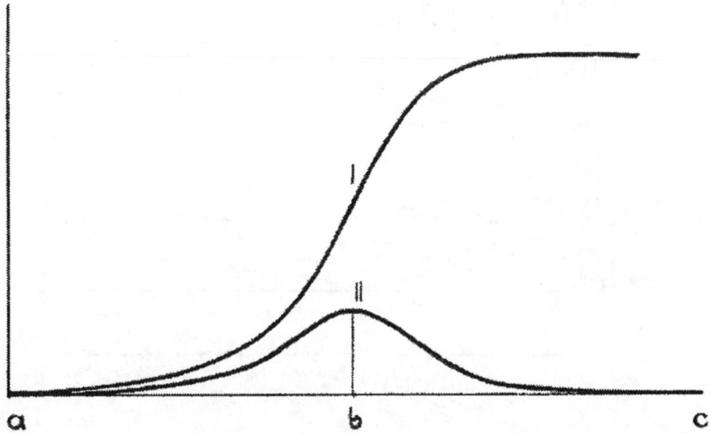

Fig. 67. Growth of a population of yeast cells. I. Growth curve. II. Curve representing the growth rate.

Figure 2. The 'optimum fishing' concept as illustrated by Hjort et al. *(1933). As the population increases its rate of increase will also increase until the population is roughly half its ultimate size after which the rate of increase is reduced.*

It is interesting to note that the concept of recruitment overfishing, that the parent stock may be too small to sustain recruitment, which was high on Petersen's (1903) list, had disappeared and was not to emerge again as an integral part of management advice till the 1980s. This may be seen as another indication of the level of generality that developed in the process, that even basic processes on the population level were disregarded if they could not be fitted into a simple conceptual framework.

The mathematical basis which was needed to operationalize this new concept of optimality was already developed by Baranov early in the century (Baranov, 1918), but this work was not known in the international research community until much later. It was not until the 1950s that the international breakthrough of a formalised base to operationalize optimality came fully about, initially by Beverton (1953) and culminating in the Principia Mathematica of fisheries biology, On the Dynamics of Exploited Fish Populations (Beverton and Holt, 1957). This represented the pinnacle in the abstract operationalization of fisheries management: fisheries can be optimised by adjusting two basic parameters, the overall fishing mortality and the lowest age at which fish are caught (Figure 3).

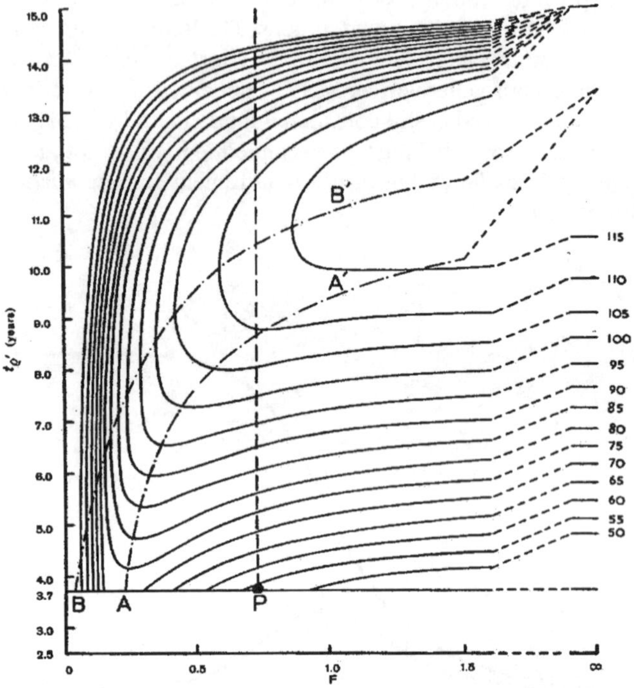

Figure 3. Yield isopleth diagram for plaice in the North Sea (Beverton, 1953). Maximum yield can be obtained by adjusting overall fishing mortality (the x-axis) and minimum landing size/mesh size (y-axis) in concert.

This approach was firmly based in the perspective that had developed during the first half of the century of internationalization of fisheries science and management. This perspective was built on the notion that the basic unit of fisheries and fisheries management was the stock which represents fish populations on large (100+ nautical miles) scales and that the dynamics of the stock and the impact of fisheries can be understood and managed by averaging life history parameters and stock abundances over the total stock area. It was also based on the implicit assumptions that the main source of variation is recruitment to the stock, that management can be implemented on a stock-by-stock basis and that the effects of specific management measures can be predicted whereby fisheries can be optimised in terms of maximizing long term yield. Each of these assumptions are contradicted by the perceptions of fishers, as will be discussed below.

One important task that remained to be done within this perspective was to develop methods to estimate the parameters of the model. From the publication of Beverton and Holt (1956) the estimation problem became the core of fisheries science. From the 1970s onwards, the main steps include the development of Virtual Population Analysis (Gulland, 1972) to estimate population sizes and fishing mortalities, the development of multispecies VPAs (Helgason and Gislason, 1979; Pope, 1979) to estimate natural mortalities, and the development of various 'tuning' methods (a range of ad hoc methods were developed and implemented by ICES during the late 1980s and early 1990s) and integrated statistical analysis models (such as Integrated Catch Analysis) to circumvent the overparameterization problem in the classical VPA.

In the period ca 1955 to ca 1990 the main research discourse can be described as rational fisheries with an optimization objective based on deterministic predictability. It can be characterised by an understanding that

- The basic unit of fisheries and fisheries management is the 'stock'.
- The stock represents fish populations on large (100+ nautical miles) scales.
- The dynamics of the stock and the impact of fisheries can be understood and managed by averaging life history parameters and stock abundances over the total stock area.
- These parameters can be estimated on basis of data sampling schemes and estimation models.
- The main non-explained source of variation is recruitment to the stock.
- But as far as management is concerned recruitment variation can be overcome by measuring the abundance of recruiting year classes before they enter the fishery.
- Management can be implemented on a stock-by-stock basis.
- The effects of specific management measures can be predicted.
- Whereby fisheries can be optimised in terms of maximizing long term yield.

4. PRECAUTIONARITY AND STOCHASTIC PREDICTABILITY

The scope of international fisheries management changed in the early 1990s when two new considerations entered the scene: the precautionary approach and the need to include considerations on the effects of the marine ecosystem at large into fisheries management. These additions were formalized in the Code of Conduct for Responsible Fisheries (FAO, 1995) and the United Nations Conference on Straddling Fish Stocks and Highly Migratory Fish Stocks (UN, 1995).

The precautionary approach implies a change in the role of knowledge. This change was first explicitly expressed in an international agreement text in the straddling fish stocks agreement which stated that 'States shall be more cautious when information is uncertain, unreliable or inadequate. The absence of adequate scientific information shall not be used as a reason for postponing or failing to take conservation and management measures' (UN, 1995). The precautionary principle changes the relationship between knowledge and exploitation. In an optimization scheme scientific knowledge is a useful and important but not mandatory guidance for management. Under the precautionary principle knowledge becomes a condition for exploitation in the first place and scientific uncertainty and allowable exploitation are coupled.

The requirement to include considerations on the effects of fisheries on ecosystems as expressed in the Code of Conduct for Responsible Fisheries implies a change in the scope of knowledge rather than a change in its basic role. The requirement that 'Management measures should not only ensure the conservation of target species but also of species belonging to the same ecosystem or associated with or dependent upon the target species' (article 6.2, FAO, 1995) expands the scope of knowledge required for management immensely. The combination of this requirement with the precautionary principle potentially implies either infinite demands on science or the closure of most fisheries.

These considerations are at their core a critique of the main fisheries research discourse on predictability - the precautionary approach is fundamentally about accepting the fact that uncertainty is an integral part of management. In spite of this, the precautionary approach as it emerged in the management debates in the 1990s was treated as a supplementary consideration, and regulatory fisheries research responded by internalizing uncertainty into the existing research discourse. Models were developed in which uncertainties were quantified and predictions were associated with probabilities of various outcomes. This approach may be described as *stochastic predictability* because the basic concept of predictability was maintained but the predictions of the effects of management measures were expanded to include an estimate of the associated uncertainty. Another adaptation to the new management discourse was maybe more fundamental although less noticed. The management discourse has implicitly changed its objectives from targeting production, with optimization being the core concept, to emphasizing conservation and risk management, with precautionarity being the core concept. The most important outcome to be predicted within the new stochastic predictability is, therefore, not catch but spawning stock biomass.

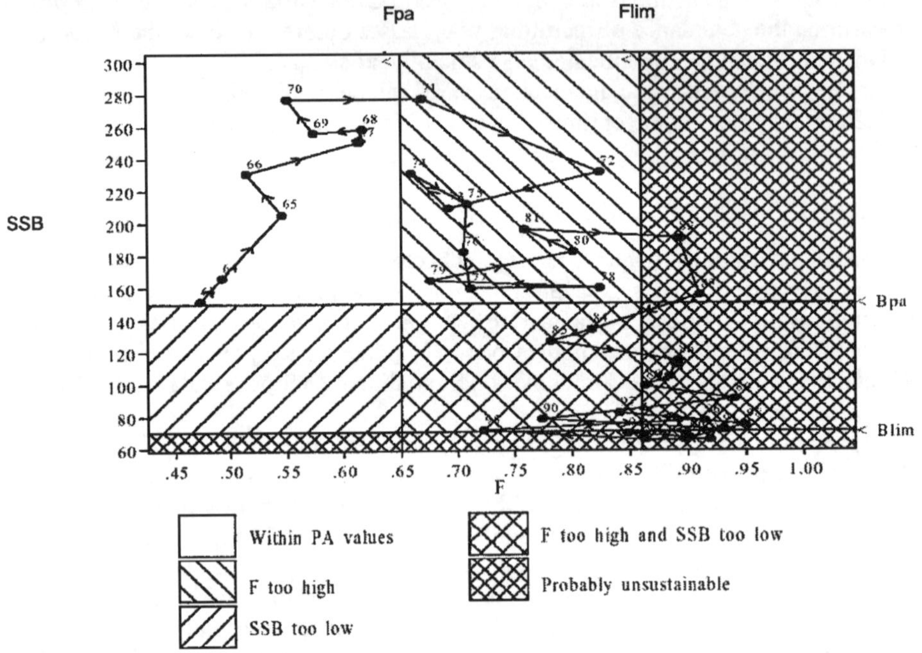

Figure 4. 'Precautionary approach plot', Cod in the North Sea as assessed by ICES in 2000 (ICES, 2001). The plot is a surface of two dimensions, the Spawning Stock Biomass (SSB), which is the state of the stock, and the fishing mortality (F) which is an expression of the pressure on the stock . For each dimension two reference points are identified, a 'limit' reference point, which should be avoided, and a 'pa' reference points which signals specific management action to be taken if the stock and the fishery bypasses these points, to prevent the stock and fishery to bypass the limit points. The shaded areas are danger zones within which specific responses or management measures should be taken to mitigate the situation, the darker the shade the higher the urgency of the situation.. The labels of individual points refer to the situation in specific years according to the assessment. The assessment thus indicates that the stock and the fisheries has developed from low-risk in the 1960s to high-risk in the 1990s as the spawning stock has dwindled and fishing pressure increased.

The changes can be illustrated by comparing the yield plots of Figure 3 with a 'precautionary approach plot' from the assessment and management advice on cod in the North Sea anno 2001 (Figure 4). The main parameter in optimization (Figure 3) is yield (a production outcome) and deterministic predictability implies that this parameter can be modelled as a single surface which is a function of the basic management instruments, in this case the fishing mortality and the age at first capture. Under the precautionary approach implemented as stochastic predictability the main parameter to watch is a risk outcome, in this case (Figure 4) the spawning stock biomass being above a certain critical minimum size which is considered the most critical parameter for the future sustainability of the stock. Another fundamental change relative to the deterministic optimization approach is that the main consideration is the probability of the risk outcome falling below critical levels. The surface to guide management is thus not a single surface with one optimal point (figure 3) to be aimed for in management but rather a map of risk zones (figure 4) which indicates the urgency and direction of action to be taken in a given situation.

From around 1990 the main research discourse can therefore be described as rational fisheries with an objective of risk avoidance in relation to stock conservation, based on stochastic predictability. Most of the basic assumptions and approaches of the optimization and deterministic predictability discourse have been maintained including notably that the basic unit of management still is the stock and that the relevant scale of relevant knowledge and management is still large (100+ nautical miles). The new components are that:

- These parameters can be estimated on basis of data sampling schemes and estimation models and *the estimates can be associated with uncertainty.*
- The main non-explained source of variation is recruitment to the stock, *but there is increased probability of low recruitment at low spawning stock sizes.*
- The effects of specific management measures can be predicted *with an associated uncertainty to the prediction.*
- Whereby *fisheries management measures can be devised which will be associated with a high probability of avoiding adverse situations.*
- *Adversity is defined as low spawning stock biomass.*

When comparing figure 3 and 4 two basic changes are apparent:

- There is no singular optimum state in the risk avoidance discourse as compared to the optimality discourse - there are danger zones of different intensity rather than a surface with a maximum.
- The basic parameters of the risk avoidance discourse do not include parameters referring to societal benefit such as yield.

There is thus an important change in the research discourse which reflects the changes in the management discourse. However, the basic approach has been maintained - to predict outcomes of management measures over large scales with the 'fish stock' as the basic unit.

Regulatory fisheries research has succeeded in embracing and operationalizing the precautionary approach by adjusting its existing discourse of predictability through internalization of uncertainty. These developments in management and fisheries biology are basically within the same paradigm - quantifiable objectives can be set and fisheries biology can provide quantitative models, which will quantify the regulatory parameters in relation to quantifiable objectives.

The contemporary management systems rely heavily on the predictability-based research paradigms. Most management systems rely on single stock TACs in one form or another. The requirement is real time knowledge of the state of the system and predictive models, with or without stochasticity. It is also a requirement that the unit of advice and thus of research is relevant to the scale of management, that is the stock concept defined on 100+ nautical miles scales must prevail.

The fisheries research discourse has thus developed though the 20th century in a response to emerging management issues. The emerging management issues have set specific research questions in the foreground and the fisheries research discourse has - since the concept of fish stock was made the basic concept - reacted by changing its scope to internalise these new issues within the same basic paradigm of quantifiable predictability over large scales. The last major break in discourse was when the perspective was changed from understanding processes on the local scale to large scale single stock descriptions that were compatible with the emerging management approach of rational fishing. The development of the fisheries management issues can again be related to the broader modernization process. This development is summarised below:

	Management issue	**Research issue**	**Research discourse**
1900	**Develop fisheries**	**Explaining variation** **Variation in migration**	**Understand the process bottom up at local scale**
1920	**Mitigate the 'overfishing problem'**	**Variation in year class strength**	**Average over large scales – 'fsih stock' basic unit of observation Hjort 1914)**
1935	**Rational fishing** **Optimization**	**Conceptualising optimization**	**Understand basic stock dynamics - Production dynamics (Hjort et al 1933)**
1956		**Operationalising optimization**	**Deterministic predictability 1: Formalised population processes (B&H 1956)**
1970			**Deterministic predictability 2: Estimation of parameters (VPA 1965, MSVPA 1979, tuning 1985+)**
1995	**Precautionarity**	**Operationalising precautionarity**	**Stochastic predictability : Quanitify Limit and 'pa' reference points**
2000	**Ecosystem considerations**	**Operationalising ecosystem approach**	**Still struggling (EcoQO's)**

5. LIMITS TO INTERNALISATION - THE END OF SHORT-TERM PROGNOSES?

The recent transformation from optimization to risk minimization represents an attempt to internalise a fundamental problem in the prevailing management system. The addition of stochasticity and ever more complex models in the transformation from optimization to risk minimization and in the inclusion of ever more complex goal functions does not represent a durable solution for two reasons: cost and chaos (Figure 5):

Cost: the marginal costs of adding another component to the models, another goal function, etc. are becoming prohibitive in terms of the data needed to support such models and model complexity.

Chaos: there are principal limits to the predictability of any natural system beyond which it is impossible to assemble sufficient detailed data and models to provide any reliability (Wilson *et al.,* 1994).

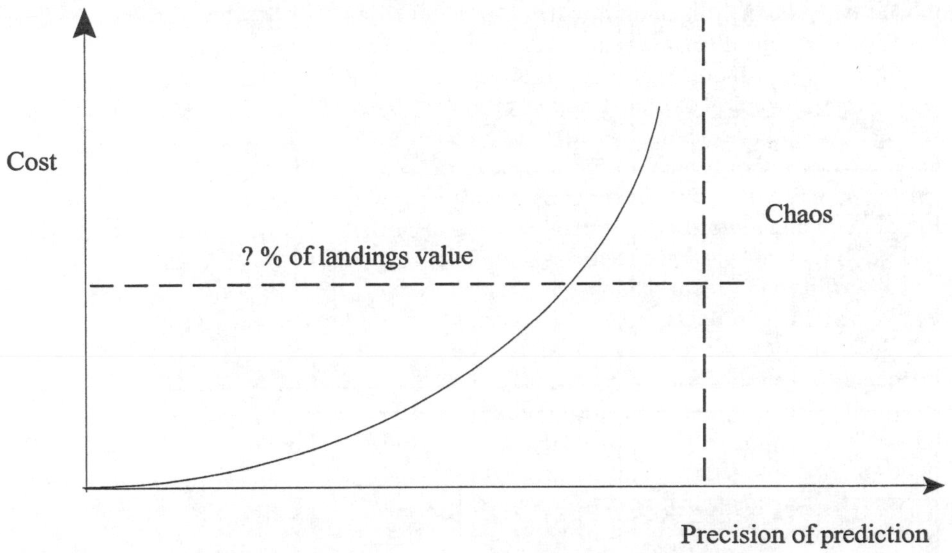

Figure 5. The cost-complexity trap of predictions. The precision of a prediction is associated with a price to produce the data required and to develop and implement the analytical and predictive model. There are absolute limits to the precision that can be obtained at any cost, given by the chaotic nature of the aquatic environment. The marginal returns in terms or precision gained by a given investment are expected to decrease, eventually to zero, when this limit is approached. Similarly, if predictions are considered instruments for management and not research in its own right, considerations of maximum acceptable costs to produce such predictions relative to the benefits to society will be relevant, for instance measured as a maximum fraction of the primary landings value of the fisheries in question.

These limitations relate to the costs and cognitive limitations of the production of research based knowledge for fisheries management. Another limitation relates to the acceptance of the research discourse among users. The list of basic understandings within the various predictability discourses listed above may be in fundamental conflict with the experiences of fishers. One of the basic problems is scale and the concept of average fish stock. The transformation within fisheries biology which took place in the early part of the 20th century when fisheries biology adapted itself to an emerging international, top-down management regime was, as explained above, also a transformation from observation and explanation on a scale of resolution that is similar to the resolution guiding the practices of the fisheries. The difference between the two approaches to scale is not so much range - fisheries operate over geographical scales that will routinely include several stock areas - but the scale of resolution. What significance does the local abundance of fish in association with specific bottom or hydrographic conditions have for the practices of a fisheries biologist and a fisher? To one, the local variation in abundance is a problem because it does not represent the stock mean, and this problem is to be overcome through an appropriate sampling design. To the other, the local variation represents opportunities or is even a condition for profitable harvest.

This may best be illustrated by comparing two sets of maps originating from the two fishing strategies - a sampling scheme to estimate mean abundance of plaice in the North Sea and a fishing operation to harvest flatfish (Figure 6).

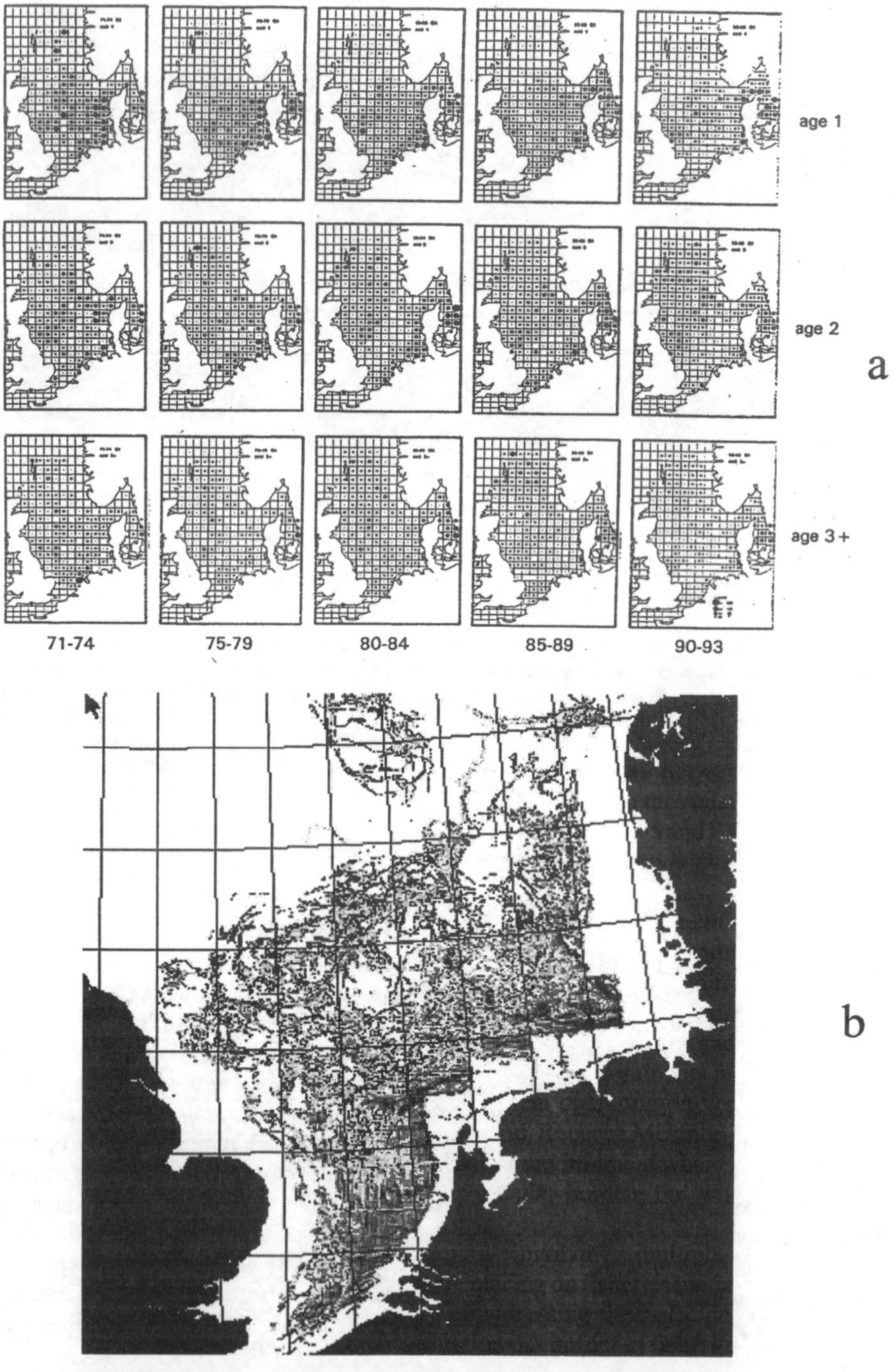

Figure 6. The abundance of plaice by year and age group over the entire North Sea as mapped through systematic sampling by research vessels (a, (ICES, 1994)) and the distribution of trawl tracks of Dutch beam trawlers in 1995 (b). (a) from ICES (1994), (b) courtesy A. Rijnsdorp.

The future adaptation of regulatory fisheries research to management requirements is therefore associated with two problems:

- Fisheries biology is approaching the limits of cost efficiency relative to the value of fisheries - and can still not deliver the goods in terms of numerical predictions.
- The models and concepts of fisheries biologists are becoming increasingly alien to stake holders. This gap is not just a question of lack of understanding or education on the side of fishers but is rather associated with the basic scales at which the resource basis for fisheries is observed and understood.

6. THE LIMITS TO KNOWLEDGE AND THE EMERGENCE OF INDICATOR BASED DISCOURSES

The present process of attempting to operationalize and internalise the requirement for ecosystem considerations may bring these problems more into the open but may also indicate new ways to address them.

It has proven considerably more difficult to operationalize ecosystem considerations than it was to operationalize the precautionary approach within the existing discourse. This must be expected considering that it is difficult to imagine a process of internalization that does not imply radical modification or even rejection of all the items in the list of basic assumptions and understandings of the existing discourse above. Considerable work has been done, both mandated directly by governments or management agencies (for instance National Marine Fisheries Service, 1999), in the primary literature (for instance Anon (2000a), reviews by Jennings and Kaiser (1998) and Hall (1999)) and within the international advisory bodies. Two different approaches are emerging:

- To internalise the issue in the same way as was done before when species interactions and uncertainty was internalised - that is by developing models with new layers of complexity which include all relevant processes and effects and thus enables ecosystem effects to be predicted within stochastic predictability.
- To develop a fundamentally new approach which does not pretend to understand or measure causal relationships and all relevant processes in detail but identifies specific measurable features that indicate the pressures on the system.

The first approach will add considerably to both the problem of costs versus predictability and to the alienation of users to the concepts and scales used.

The latter approach reflects a realization by some fisheries biologists that regulatory fisheries research has approached the point on the cost-precision curve where it is no longer tenable to try to solve the problem by demanding more resources to collect data and add complexity to models. This has created the basis for a discussion on *indicators for fisheries management* and *Ecological Quality Objectives* in relation to fisheries that is rapidly emerging.

7. INDICATOR DEVELOPMENT RESPONDING TO GLOBALISATION OR COST MINIMISATION

The concept of indicators in relation to fisheries sustainability has taken place within two different agendas.

One agenda is driven by globalization processes and is concerned with establishing indicators that can be used to govern policies in the international domain, in relation to international agreements on sustainable development/fisheries and in relation to market regulations and green labelling. This development is promoted by international organizations and NGOs and centres around the Indicators of Sustainable Development initiative of the UN Commission on Sustainable Development (CSD) (Commission on Sustainable Development, 2001), which is a body assigned to follow up on the UNCED agenda 21. OECD has likewise developed an indicator framework for environmental performance reviews (OECD, 1993). This agenda has been developed in relation to environmental sustainability in general but is also reflected in fisheries. An account of this development in relation to fisheries has been presented by Dahl (2000) and Garcia and Staples (2000), see also Anon (2000b). There has been a tendency for work within this agenda to add new layers of complexity to the knowledge base - by still requiring standard stock assessment procedures but adding ecological, social and economic dimensions and even developing specific methods to condense such diverse information into graphical or numerical metaindicators. Examples of such attempts are the RAPFISH approach (Pitcher and Preikshot, 2001) and the Sustainable Development Reference System approach of FAO (Garcia and Staples, 2000).

Another agenda is concerned with establishing a knowledge base to guide management while realising that regulatory fisheries research may have reached the cost and complexity limits. A response to the complexity wall has been explorations into an identification of proxies to the standard reference points of stock assessments and indicators that are assumed to capture the effects of fisheries pressures on the ecosystem. Reviews of this work were presented at the ICES/SCOR Symposium on Ecosystem Effects of Fishing (Anon, 2000a).

There has been some convergence between the two agendas and the international policy agenda should in principle be an extension of and build on the research agenda. The gap is, however, still very wide as indicated by the fact that the MSY concept, which now largely has been abandoned as a relevant and measurable reference point among fisheries biologists, is the only fisheries related indicator on the CSD list of indicator candidates (Commission on Sustainable Development, 2001). The US National Marine Fisheries Service has probably presented the most ambitious attempt to date to integrate scientific state of the art into a management framework in relation to ecosystem issues and the use of indicators. This was presented in a report to Congress 1999 (National Marine Fisheries Service, 1999).

8. INDICATORS AS MEANS TO ACCEPTANCE

The importance of the acceptance of indicators by stakeholders, and even their participation in identifying them, is alluded to within most of the literature. This is generally stated as an important issue without further consideration of the implications. Acceptance is dealt with as if it was a trivial add-on without implications for other parts of the management setup or the relevance of indicators.

However, acceptance is not a trivial issue, as discussed above in relation to the development of the knowledge base for fisheries management. Another approach to indicators, distinct from the globalization driven approach, is to try to identify scientifically valid indicators which reflect the perspective of users. Such indicators would serve to build agreement and acceptance between fishers, researchers, management authorities and other users. There are examples of such indicators having been implemented in fisheries management such as the fraction of juvenile fish in catches or the number of dead mussels in shellfish catches.

On the basis of the discussions above some of the key properties of such indicators would be that they are observable, make sense to both formal research and stakeholders, and are relevant to management. Indicators must be observable within the economic resources for research on a sustained basis, as well as by stakeholders, either directly or by transparency in the observation process. They should make sense in a research context and reflect features which correspond to stakeholders' understanding of the resource system. And they must be relevant to management by indicating direction of action and respond to management measures.

The scale gap between science and stakeholders will be one of the major obstacles to overcome in this context. Indicators which are observable and make sense across scales may be hard to come by.

The identification of indicators meeting these criteria and development of corresponding estimation methods and reference points is still in its infancy. This may be the real future challenge of fisheries biologists and social scientists working in concert.

9. CONCLUSIONS

Fisheries science and fishers observe and interpret the sea on different scales. Fisheries science is based on the 'stock' concept and interpretations are made on the scale of the stock while fishers are concerned with local abundance as required for successful fisheries operations. This scale gap is a challenge to co-management and must be addressed if co-management is to succeed. This is not an easy task as the scale gap is tied to the different practices and roles of fisheries science and fishers.

The large scale fisheries research discourse developed in response to the emergence of international fisheries management institutions and has developed through the 20th century in response to emerging fisheries management issues. The major break in this development was when the unit of observation changed from being processes on the local scale to the 'fish stock' averaged over large scales around 1920. This coincided with the internationalisation of fisheries management and 'rational fishing' emerging as the main management discourse, from an approach focussing on economic development and modernization.

The fisheries research discourse has adapted to emerging management issues through the remaining part of the 20th century including optimization and precautionarity, but these adaptations have been within the same basic paradigm of rational predictability of the outcomes of management on a 'fish stock' basis.

By adapting a large scale averaging approach fisheries research has at the same time alienated itself from the observations and understandings that are associated with commercial fishing activities, where predictability with a high resolution in space is required. This has led to loss of legitimacy for knowledge created through fisheries research among users. The fisheries research discourse will also approach cost limits as it is attempted to internalise more complex processes and systems (such as the emerging 'ecosystem approach' to fisheries management) and stochasticity within the same predictability paradigm. There are therefore both legitimacy and costs reasons to expect the present discourse to approach the end of its life as the ruling discourse for the knowledge base for fisheries management.

Emerging alternatives include the use of indicators for the fisheries pressures on stocks and ecosystems as guiding parameters for fisheries management. The international discussion on indicators is, however, ambiguous as to the basic rational for using indicators. A part of the discussion represents an attempt to internalise and accommodate the requirements for ecosystem considerations in fisheries management and another part is concerned with the requirements for international comparability and standardization following from international agreements on management principles and the global market for fish products. However, a discussion on the need to use indicators as a means to achieve legitimacy and cost effectiveness is also emerging. Within this rational for the use of indicators in management the requirements for useful indicators include that they must be observable by stakeholders and within reasonable economic means, acceptable to stake holders including both fishers and researchers and relevant as guidance to management action. When indicators meet these requirements they make possible a shared knowledge base on which effective co-management can be built. This need for shared, acceptable indicators also underlies the requirement, from the fisheries science viewpoint, to develop fisheries co-management institutions with the capicity to contribute to the knowledge base for management.

ACKNOWLEDGEMENTS

Thanks are due to Vera Schwach, Norwegian Institute for Research and Higher Education, for provision of valuable comments on historical aspects of this paper, to Doug Wilson, IFM, for constructive criticism and to Adriaan Rijnsdorp, RIVO, for allowing reproduction of figure 6b. Unfortunately we have not been able to contact the copyright owner for figure 2. For more information please contact Poul Degnbol, IFM, P.O. Box 104, DK-9850 Hirtshals, pd@ifm.dk

REFERENCES

Anon (2000a) Ecosystem Effects of Fishing. Proceedings of an ICES/SCOR Symposium. ICES *J.Mar.Sci.* **57**[3], 465-791.

Anon (2000b) Sustainability indicators in Marine Capture Fisheries: papers derived from a technical consultation organized by the Australian Department of Primary Industries in co-operation with FAO. *Mar.Freshwater Res.* 51.

Bailey, P.D. and Yeardley, S. (1999) Discourse in fisheries: constructing vessel monitoring systems and overfishing. ICES CM 1999/Q:9.

Baranov, T.I. (1918) On the question of the biological basis of fisheries. *Nauchn. Issledov. Ikhtiol. Inst. Izv.* **1**, 81-128 (in russian).

Bell, Simon and Morse, Stephen (1999) Sustainability indicators. Measuring the immeasurable. Earthscan Publishers. London.

Beverton, R. (1953) Some observations on the principles of fishery regulation. *Journal du Conseil*, XIX (1).

Beverton, R.J.H. and Holt, S.J. (1957) On the dynamics of exploited fish populations. Her Majesty's Stationary Office, London.

Commission on Sustainable Development (2001) Indicators of sustainable development. www.un.org/esa/sustdev/isd.htm.

Dahl, Arthur Lyon (2000) Using indicators to measure sustainability: recent methodological and conceptual developments. *Mar.Freshwater Res.* **51**, 427-433.

European Commission (2001) Green Paper on the Future of the Common Fisheries Policy. Com (2001) 135.

FAO (1995) Code of Conduct for Responsible Fisheries. *FAO Fisheries Technical Paper 350.* http://www.fao.org/WAICENT/FAOINFO/FISHERY/agreem/codecond/codecon.asp.

Finlayson, A.C. (1994) Fishing for truth. Institute of Social and Economic Research, Memorial University of Newfoundland.

Garcia, S. M and Staples, D.J. (2000) Sustainability reference systems and indicators for responsible marine capture fisheries : a review of concepts and elements for a set of guidelines. *Mar.Freshwater Res.* **51**, 385-426.

Graham, M. (1934) Report on the North Sea Cod. Min. Agric. and Fish. Invest., Ser. II, XIII, No 4, London, 1934.

Graham, M. (1948) Rational Fishing of the Cod of the North Sea. Being the Buckland Lectures for 1939. London: Edvard Arnold.

Gulland, J.A. (1965) Estimation of mortality rates. Annex to Arctic Fisheries working group report. ICES CM Doc 3.

Hall, S. J. (1999) The ecosystem effects of fishing. Chapman and Hall.

Heincke, F. (1898) Naturgeschichte des Herings I. Die Lokalformen und die Wanderungen des Herings in dem Europäischen Meeren, Abbandlungen der Deutschen Seefischereivereins, Vol. 2. Berlin: O.Salle.

Helgason, T. and Gislason, H. (1979) VPA-analysis with species interaction due to predation. ICES CM 1979/G:52.

Hjort, J. 1914. Fluctuations in the Great Fisheries of Northern Europe. Rapp. P.-V. Reun. XX, 1-13.

Hjort, J., Jahn, G. and Ottestad, P. (1933) The Optimum Catch. *Hvalrådets Skrifter*, **7**, 92-127.

Hoek, P.P.C. (1905) Introductory Review. App D-K, Rapp. P.-V. Reuin. III-e Edition Anglaise. Reprinted in Thomasson (ed.) 1981.

ICES (1905) General Report on the Work in the Period July 1902-July 1904. Rapp. P.-V. Reun. III-Edition Anglaise.

ICES (1994) Report on the Study Group on the North Sea Plaice Box. ICES CM 1994/Assess:14.

ICES (2001) Report of the ICES Advisory Committee on Fishery Management, 2001. ICES Cooperative Research Report 246.

Jasanoff, S. (1990). The fifth branch. Science advisers as policymakers. Harvard.

Jennings, S. J. and Kaiser, M.J. (1998) The effects of fishing on marine ecosystems. Advances in *Marine Biology* **34**, 201-252.

Jentoft, S. (1993) Fisheries co-management. *Marine Policy* **13**(2), 137-154.

National Marine Fisheries Service (1999) Ecosystem Based Fisheries Management. A Report to Congress. National Marine Fisheries Service. http://www.nmfs.noaa.gov/sfa/EPAPrpt.pdf.

Mill, E. (1989) Biological oceanography in early history 1870-1960. Cornell University Press, New York.

OECD (1993) OECD core set of indicators for environment performance reviews. OECD. http://www.oecd.org/env/docs/gd93179.pdf.

Petersen, C.G.J. (1903) What is overfishing? *Journal of the Marine Biological Association* **VI** (1900-1903), 587-594.

Pitcher, T. J. and Preikshot, D. (2001) RAPFISH: a rapid appraisal technique to evaluate the sustainability status of fisheries. *Fisheries Research* **49**, 255-270.

Pope, J.G. (1979) A modified cohort analysis in which constant natural mortality is replaced by estimates of predation levels. ICES CM 1979/H:16.

Rozwadowski, H. (2002) Without prejudice to the future: Marine science, North Atlantic fisheries, and one hundred years of International Council for the Exploration of the Sea (ICES). ICES. In press.

Schwach, V. (2000) Havet, fisken og vitenskapen. Fra fiskeriundersøgelser til havforskningsinstitut 1860 til 2000. Institute of Marine Research, Bergen, Norway. 405pp.

Smith, T. D. (1995) Scaling Fisheries. Cambridge University Press.

Thomasson, E.M. (Ed.) (1981) Study of the Sea. Fishing News Books.

UN (1995) United Nations Conference on Straddling Fish Stocks and Highly Migratory Fish Stocks. gopher://gopher.un.org/00/LOS/CONF164/164_37.TXT.

Ward, T.J. (2000) Indicators for assessing the sustainability of Australia's marine ecosystems. *Mar.Freshwater Res.* **51**, 435-446.

Wilson D. C. (2002) Examining the Two Cultures Theory of Fisheries Knowledge: the Case of Bluefish Management *Society and Natural Resources* forthcoming.

Wilson, D. C. and Degnbol, P. (2002) The Effects of Legal Mandates on Fisheries Science Deliberations: The Case of Atlantic Bluefish in the United States *Fisheries Research* in press.

Wilson, J.A., Acheson, J.M., Metcalfe, M. and Kleban, P. (1994) Chaos, complexity and community management of fisheries. *Marine Policy* **18**, 291-305.

Chapter 3

THE ECONOMICS OF CO-MANAGEMENT

SUSAN HANNA
Department of Agricultural and Resource Economics, Oregon State University, Corvallis, Oregon, USA

1. INTRODUCTION

Economics is the study of how people allocate scarce resources among competing ends. This allocation addresses the production and distribution of goods and services over time and space. It takes place within the context of both human behaviour and organizational structure. In publicly owned fisheries, economics is further concerned with the allocation of access to the resource and with managing the interaction of people with the resource. Fisheries co-management, as an arrangement where responsibility for resource management is shared between the government and user groups, is one approach to allocating access and to managing the interaction.

From the perspective of economics people are, for the most part, rational in what they do. In fishery management as in all other endeavours they respond to the incentives they face. The incentives of fishery management have often encouraged both government managers and user groups to take short-term perspectives and act against long-term ecological and economic interests. In doing so, their actions create costs for management that affect its performance.

Co-management of fisheries is often proposed as a means to realign the incentives of fishery management and to contain its costs. It is sometimes implemented to remedy costs imposed by more centralized management approaches. Co-management is a process that can take many structural forms (Sen and Nielsen, 1996), within which any of the full range of management tools may be employed. Gear controls, effort controls, territorial use rights or individual quotas may all be applied within co-management.

The extent to which a co-management process will be effective in the application of these tools depends importantly on two economic concepts: transactions costs and incentives. These concepts have strong theoretical antecedents in the economics literature.

2. THE ECONOMIC BASIS OF FISHERY CO-MANAGEMENT

There are several functions that fishery management of any type must perform to be effective. It must coordinate multiple objectives and organizational tasks. It must generate information and develop decision processes that are legitimate and flexible. It must take a long-term perspective to accommodate social time horizons that include intergenerational claims to resource services (Hanna, 1998b). All these functions must be fulfilled within a set of specific legal economic, and social objectives. The economic question of management is whether it achieves its objectives and performs its functions in a cost-effective way. These same questions apply to fishery co-management.

Critical to the determination of fishery co-management's cost-effectiveness are the transactions costs it creates and the incentives it employs. These general concepts have been recognized in the economic literature for some time.

2.1. Transactions costs

Costs and their containment are central to the consideration of organizational structure and performance fishery co-management.

Arrow (1974) describes the purpose of organizations as being able to exploit the fact that decisions require the participation of many individuals. Organizations serve as coordinators of people and information in an environment of uncertainty. The costs associated with this coordination are transactions costs; the costs to gather information, design regulations, organize participants, monitor conditions, and enforce regulations (Mathews, 1986). Transactions costs are an inevitable part of organizations, including organizations of fishery management but their magnitude and distribution can be affected by the way the coordination is organized (Eggertsson, 1990).

Kuperan and Pomeroy (1998) identify three categories of transactions costs affecting fishery management: information costs; collective fisheries decision making costs, and collective operational costs. Hanna (1994) groups transactions costs on the basis of the sequence of fishery management decisions. Transactions costs are incurred in four stages of management: two *ex ante* stages (resource documentation and management programme design) and two *ex post* stages (implementation and enforcement).

The magnitude of some transactions costs of fishery management, such as the cost of generating scientific information, do not vary with the structure of fishery management. Other costs will vary with management structure and with the relative position of users and government in making decisions. The legitimacy of fishery management is an important determinant of the magnitude of transactions costs (Jentoft, 1989). The major avenue of this effect is through the costs of compliance. Fishery management requires the compliance of resource users, and when the management process and the regulations it produces are considered to be legitimate, compliance with regulations is likely to increase. As compliance increases, enforcement costs decrease.

How transactions costs are distributed among the *ex ante* and *ex post* categories can be very sensitive to management structure. By design, co-management distributes management responsibilities among government and user groups. Co-management is therefore more likely to be associated with higher *ex ante* costs, because data collection and regulatory design may be done at too small a scale to realize information economies. At the same time, co-management is likely to have lower *ex post* costs when active user-group participation enhances regulatory implementation and compliance (Hanna, 1994; 1995a).

The relative cost-effectiveness of co-management depends on the extent to which the gains from lower *ex post* transactions costs exceed the losses from higher *ex ante* transactions costs (Hanna, 1994; 1995a). Kuperan and Pomeroy (1998) note that transactions cost economic theory suggests that if given a choice, people will choose the set of institutions, contracts or transactions that will minimize costs.

2.2. Incentives

The need to have private incentives to achieve social objectives has been a continuing thread throughout the fishery management literature. As early as 1911, Jens Warming discussed the negative changes in incentives to protect the Danish coastal eel resources that would result if rights to use coastal eel weirs were converted to open access (Andersen, 1983).

Over forty years later, Gordon (1954) again raised the issue of the importance of incentives in fishery management, arguing that the poverty of fishermen and the inefficiency of their production stems from the common property (by which he meant open access) nature of fisheries. He noted that most fishery control measures designed by biologists pay no attention to the costs of fishing. The method of control has been to place limits on output, creating an incentive for fishermen to compete catch the fish before his competitors. This has led to the competitive race for fish, the rising average cost of fishing effort, the poverty of fishermen and the dissipation of economic rent. Economic rent is the payment over and above the opportunity cost of using the resource in its next best use.

The race for fish has common-sense origins and destructive results. When ownership of fish is possible only by capture, fishermen and seafood processors alike compete by investing in capacity that far exceeds levels sustainable by the fishery resource over time. Concomitant with over-investment in fishing and processing capacity tends to be an under-investment in management capital - the decision skills, knowledge of management tools, and understanding of monitoring and evaluation systems.

Gordon notes that fishermen are trapped in their profession because their immobility offers them few alternatives. Interestingly, the reasons he supplies for fishermen immobility are also those that can favour co-management: living in isolated communities with little knowledge of alternative opportunities, little financial capital, and educational and romantic ties to the sea.

The incentives of open access frustrate attempts to achieve optimal exploitation of fisheries. Gordon observes: 'There appears to be some truth in the dictum that everybody's property is nobody's property. Wealth that is free for all is valued by none because he who is foolish enough to wait for the proper time of use will only find that it has been taken by another. The fish in the sea are valueless to the fishermen because there is no assurance that they will be there for him tomorrow if they are left behind today'. He concludes that the rent from fishing cannot be appropriated by anyone when fishermen cannot control access to the fishing grounds or ensure their tenure in the fishery. The outcome is the same in all cases where resources are owned in common and exploited under conditions of individualistic competition. Some sort of control is necessary, whether collective or private (Gordon, 1954).

Anthony Scott expanded on Gordon's conclusions (Scott, 1955). Agreeing about the inefficiencies of open access, Scott argues that the existence of private property is not sufficient to ensure the efficient management of natural resources. Property must be allocated on a scale adequate to ensure complete control of the asset. A condition for efficiency in a fishery is sole ownership of the fishery, where 'sole ownership' describes not monopoly but rather the full appropriation of the fishery in a particular location.

One of the central dilemmas of resource management is the contradiction between individual incentive structures and collective needs. Autonomous behaviour promotes individual well-being without regard to the collective good. Resource sustainability requires the development of incentives that promotes a subjugation of individual well-being to the collective good. The fact that individuals are dependent on the outcomes of collective group actions strengthens the need for structuring group outcomes and is one of the motivations for co-management. The policy question becomes how to exploit this mutual dependence for collective gain.

The problem of constructing group positions out of individual behaviours is one that has been discussed and analysed in various contexts. Schelling (1960) has analysed conflicts that contain an element of mutual interdependence - non-zero-sum games. These games, instead of being 'I win, you lose' situations, represent conflicts in which some joint dependence requires cooperation 'if only in the avoidance of mutual disaster.' In addressing such interdependencies, Schelling identifies the important role played by consistent expectations and the quality of communication.

Olsen (1965) identifies the use of incentives (economic, social and psychological) as one of two critical factors in the effectiveness of collective decision making. Small groups that can maintain face-to-face participation tend to be more effective in the use of incentives to construct consensus.

Runge (1984) also addresses incentives, highlighting the importance of institutions like fishery management in diminishing the incentive to 'free ride'. If management performs effectively, it will coordinate expectations, help predict behaviour and resolve the assurance problem. Interdependence creates feedback loops that reduce uncertainty about others' behaviour.

Under an uncontrolled fishery, the race for fish, fishing overcapacity, and management under-capacity combined to shorten the time frames of fishery managers and user groups. Short-term actions crowd out long-term strategies. Reactions to crisis can overwhelm planned management. Assurance about the future declines, and conflict among competing interests increases. The result is unsustainable management built around incompatible incentives and high transactions costs.

The continuing scientific uncertainties in the knowledge base of fishery management also lead to incompatible incentives. A good example is uncertainty's dual incentives with regard to social and private time horizons. Uncertainties about the drivers, interactions and constraints of ecological and human systems shorten the private time horizons of managers and user groups when those same uncertainties provide, from the social perspective, an incentive to take a precautionary approach (Hanna, 1998b). Uncertainties about the tenure of rights to resource use also cause people to focus on the short-term. Both types of uncertainties create substantial difficulties in providing incentives for long-term sustainability.

A combination of high levels of uncertainty and lack of assurance about rights to resources creates an incentive to race for fish, which creates, in turn, an incentive to emphasize short-term gains. Long-term incentives for stewardship are de-emphasized and, as a result, levels of fishing capacity far exceed the productive capacities of fish populations; there are few feedback controls from ecosystems, and economic and ecological outcomes are suboptimal.

Incentive problems of centralized management processes have generated interest in finding alternative institutional forms, particularly those such as co-management that shift authority further away from the centre. But it is important to note that these approaches can also create incentives incompatible with long-term sustainability if care is not taken in their design. The design challenge for co-management is to contain transactions costs and to promote incentive structures that are less vulnerable to short-term interests.

3. ECONOMICS OF FISHERY CO-MANAGEMENT IN PRACTICE

Co-management carries particular appeal in small-scale fisheries for several reasons related to the conditions under which such fishing takes place. The context of small-scale fishing means that the pre-conditions for cooperative management are often in place. Because of their proximity to shore, small-scale fisheries are often most in need of effective management. The traditional tools and processes on which small-scale fishery management is based are in many cases proving inadequate to withstand contemporary pressures from increases in entry, capitalization and exploitation. Additionally, small-scale fisheries may have local or regional importance disproportional to their size. In many areas, for example, small-scale fisheries are the basis for protein food security of low -income people who depend on fishery participation (Hanna, 1998a).

For all of these reasons, the direct involvement of small-scale fishery stakeholders in the planning and control of their fisheries offers the potential for improving fishery management. The idea behind co-management is that people vested in planning and decision making are more likely to benefit from long-term thinking than those who are not, and will act accordingly. But as with any form of management, the costs and incentives of management will determine co-management's effectiveness.

Several case studies of co-management in practice illustrate the importance of transactions costs and incentives to the performance of fishery co-management.

Lake Kariba, Zambia (Sen and Nielsen, 1996, based on Malasha, 1996): Co-management was implemented for the Zambian artisanal gill-net fishery in 1994 as a solution for declining resources, poor enforcement and poverty. Management authority over a fishing zone was vested in fishery management committees, which were charged with monitoring compliance and recommending the allocation of development funds. Many start-up problems illustrate some of the structural challenges of organizing co-management, including the establishment of arbitrary boundaries and communities through the forced settlement of itinerant fishers into villages, the lack of complete stakeholder representation on management committees, and the need to transfer skills of participation and organization in the absence of democratic traditions.

> Transaction costs include start-up problems and the need to transfer skills of participation and organization in the absence of democratic traditions.
>
> Incentive problems include incomplete stakeholder representation on management committees and the establishment of arbitrary boundaries and communities through the forced settlement of itinerant fishers.

San Miguel Bay, Philippines (Pomeroy and Pido, 1995): Co-management of near-shore fisheries in San Miguel Bay was implemented in 1991 to address problems of over-exploitation and user-group conflict. The San Miguel Bay Management Council was established in 1994 to cope with the problem of jurisdictional fragmentation within the Bay. The Council's responsibility is to coordinate local governments, provide policy advice, and work toward needed fishery management decisions. The government retains final authority for decisions, but consults with users. Although experiencing some start-up difficulties, the co-management of San Miguel Bay is layered against a background of democratic participation and is expected to eventually be effective in addressing the problems of resource overuse, fragmented management authority, and conflict between government and users.

> Transaction costs derive from difficulties in initial organization. However,

containment of transactions costs at acceptable levels will likely be possible because of the background of democratic participation.

Incentive problems result from the position of user groups as consultants to government, rather than as co-authorities.

Lake Chapala, Mexico (Pomeroy, 1994): Lake Chapala Mexico supports a small-scale commercial fishery that has two fundamental management difficulties: weak property rights and poorly defined boundaries of control. Although shared authority between government and user groups exists to some extent, background conditions and the decision making structure prevent effective co-management. Property rights are variously defined, leaving it unclear whether fishermen's unions, cooperatives or government authorities can define the set of legitimate users. In addition, these property rights are inconsistently enforced, leaving the fishery de facto open access. Structural inconsistencies also exist in the definition of management-area boundaries. Boundaries between management areas are variously defined in laws and policies, then subject to further variability in interpretation at the group, local, state and federal levels. Policies at different scales are poorly coordinated.

Transaction costs are increased by poorly defined boundaries of control that are variously defined and interpreted by different authorities. In addition, policies at different scales are poorly coordinated.

Incentive problems derive from weak property rights and a lack of clarity about the definition of legitimate users. Inconsistent enforcement of property rights created a lack of assurance about these rights.

James Bay, Canada (Berkes, 1989): The 1975 James Bay and Northern Quebec Agreement between the government of Canada, Cree and Inuit peoples was designed to promote greater native participation in resource management, establish exclusive harvesting rights for Native groups, and enable co-management of fish and wildlife. The agreement included cooperative research and decision making. It has been successful in strengthening Cree control over resource use, crafting broader representation in the construction of environmental assessments, strengthening Cree authority, modifying provincial authority over fishing and hunting regulations, and protecting subsistence fisheries. The agreement has also had problems, including many details of implementation, funding and the transfer of authority. Especially noteworthy was the uncertainty introduced by the lack of consistent high-quality scientific information, which introduced a constraint on making long-term resource-use decisions.

Transaction costs derive from difficult implementation and from uncertainty over funding.

Incentives are both positive and negative. Greater Cree authority over resource use has resulted in broader representation in environmental assessments and stronger controls over fishing and hunting. Negative incentives have been created by difficulties in the transfer of authority compounded by the uncertainty introduced by the lack of consistent high-quality scientific information.

Lofoten Islands, Norway (Jentoft, 1993; Jentoft and Kristoffersen, 1989): The Lofoten cod fishery is an example of a long lasting co-management process, having functioned successfully for 100 years. Fishermen elect inspectors and regulatory committees who are charged with dividing ocean areas into gear group districts. The government sets total quotas but the committees do not explicitly limit catch. A challenge for this co-management process over time has been the resolution of conflict between gear groups over the size of fishing areas. The long-term success of the co-management system is due in large part to the existence of a common interest and a pressing need to resolve conflict so that fishing can take place.

> Transaction costs are lowered by a central government quota-setting process, but raised by the conflict between gear groups over the size of fishing areas.
>
> Positive incentives are provided by the existence of a common interest in resolving conflict.

Soft-shell clams, Maine (Hanna, 1998c): Management of the community-based Maine soft-shell clam fishery has co-evolved with the larger economy, made possible by the nesting of property rights in community and state levels of governance, and by the co-management activities of communities and the State of Maine. Although well adapted to its social, economic and ecological context, the co-management process is vulnerable to increasing costs of biological monitoring and difficulties of exclusion. The fishery is now in many places threatened with overexploitation. Changing conditions both internal to and external to the fishery change the distribution of management costs, requiring management to realign and adjust if it is to be sustainable.

> Transaction costs are lowered by a consistent nesting of property rights from the community to the state. However, increasing costs of biological monitoring and difficulties of exclusion increase transaction costs.
>
> Positive incentives are created by the adaptation of the management system to its social, economic and ecological context. Negative incentives are created by threats of overexploitation.

Coastal Fisheries, Japan (Asada *et al.*, 1983): Japanese coastal fisheries are managed under a system of local property rights that reflect long-standing traditions and customs of conflict resolution. Inshore fishing rights assigned to both individuals and groups are defined for territories and protected by law. A focus of the coastal fishery rights system is the resolution of conflict between individuals, fishermen associations and neighbouring villages. Sustainability is also a central concern, achieved with more success for sedentary species and localized stocks than for migratory species that cross boundaries into several management territories. The system of local property rights and co-management authority has also been blamed for retarding technological progress, being short-sighted, and being vulnerable to local corruption.

> Transaction costs are lowered by the successful resolution of conflict and by a focus on sustainable levels of extraction. These elements rest on the strength of tradition, which can also be the source of increased transaction costs in the case of preventing adoption of low-cost technological progress
>
> Incentives: Assurance is provided by the definition and enforcement of property rights. But disincentives are created by the short-sightedness of management and its vulnerability to local corruption.

Coastal Fisheries, Turkey (Berkes, 1986): The coastal fisheries of Turkey have had mixed success in local-level management. The two unsuccessful cases are managed as open access and suffer from overcapitalization. The three successful cases are closed access, with clearly defined rights to fish. The near shore co-managed fisheries suffer from external factors in offshore fisheries that are outside its boundaries of control. The successful cases demonstrate the positive effects of relatively homogeneous user groups, requirements for local residence, and smaller numbers of participants. Turkish coastal fishery co-management also illustrates the vulnerability of management to uncontrolled entry and to the effects of decisions that are made outside of its area of control.

> Transaction costs derive from external factors in offshore fisheries that are outside its boundaries of control. Transactions costs are lowered by the existence of small homogeneous user groups who reside locally.
>
> Incentives: Open access and the accompanying overcapitalization are disincentives to co-management. Positive incentives derive from closed access and clearly defined rights to fish.

Coastal Fisheries, Mexico (McGoodwin, 1994): The coastal Pacific fisheries of the South Sinaloa region of Mexico are examples of the root causes of depletion and ineffective co-management. The fisheries were relatively stable in early years of development, managed by a combination of government control and near shore territories. Boundaries were mutually recognized by user groups. A combination of more efficient fishing technology, increasing coastal populations, poverty, strong export markets, inadequate pollution controls and ineffective enforcement led to the breakdown of territories and an influx of fishing effort. The fisheries were left in a depleted state, with small remnant populations of highly mobile fishermen opportunistically fishing small aggregations of fish. The rapid rate of change in the region created uncertainty that shortened time horizons and intensified competitive use. Attempts at co-management were eroded by new entry. The long-term effect of open entry and depletion was the erosion of confidence among fishermen that participation in management could lead to sustainable fisheries.

> Transaction costs were elevated as co-managed fishery control was eroded by new entry. The long-term effect of open entry and depletion was the loss of legitimacy of management for sustainability.
>
> Incentives to race for fish were created by a combination of more efficient fishing technology, increasing coastal populations, poverty, strong export markets, inadequate pollution controls and ineffective enforcement. Background uncertainty shortened time horizons and intensified competitive use.

Pacific US (Hanna, 1995b): The Pacific Fishery Management Council holds management authority for fisheries out to 200 miles off the coasts of Washington, Oregon and California. In its structure, the Council represents shared decision making authority between federal agencies, state agencies and user groups. Since its inception in 1977 the Council has co-managed fisheries through a system of advisory committees that include user groups, environmental organizations and scientists. The benefits of this form of co-management have been both structural and procedural: the development of participation skills within user groups, contributions to learning among all participants, close attention to the fit between regulations and fishing operations, effective representation and a greater attention to equity. Costs of this co-management include overly complex management programmes and eroded boundaries caused by spillover effects between fisheries. The approach has demonstrated the importance of full representation, consistent processes, incremental change, and the learning accorded by ongoing participation.

> Transaction costs have been lowered by the development of participation skills within user groups, learning among participants, a consistency between design of regulations and fishing operations and a greater attention to equity. Transactions

costs have been increased by attention to equity to the point of handicapping programme flexibility.

Incentives have been positively affected by strong representation of user groups, consistent processes and incremental change.

Denmark (Nielsen *et al.*, 1997): Whereas in the past Danish fishery management has been oriented toward conservation, under the European Union two other management goals ascend in importance: the need to restore management legitimacy and to coordinate management with markets. Co-management is seen as the best route to achieving these goals. Basic requirements will apply to consideration of co-management in the Danish or other fishery settings. There must be a legal framework that enables the establishment of rules that are flexible to changing conditions in the resource, market or industry. User groups must have basic management capabilities that include financial resources, design of regulations, and ability to be resilient to changing conditions.

Transaction costs will be affected by the degree of human capital brought to management, and the financial resilience of user groups.

Incentives will be affected by the degree of legitimacy and flexibility to respond to change.

4. CONCLUSION

Implementing co-management in fisheries is often done to redress problems created under more centralized management approaches. Co-management is also seen as a way to achieve social goals of empowering resource users and building legitimacy of fishery regulations. Improving the economic condition of fisheries and their management is usually not at the fore of justifications offered for fisheries co-management. Yet economic problems are important reasons why more centralized management is deemed unsustainable, and economic concepts are fundamental to the performance of co-management and to its suitability to a given fishery context.

This paper has discussed two of the most important economic concepts influencing the success of fisheries co-management: transactions costs and incentives. These economic properties of a co-management process often go unrecognized, yet they are critical to management success. The degree to which fishery co-management can contain the costs of transactions and create incentives that elicit desired behaviour among co-managers determine whether it will be a sustainable process or just another failed experiment in fishery management.

ACKNOWLEDGEMENTS

This chapter was funded by the NOAA Office of Sea Grant and Extramural Programmes, US Department of Commerce, under grant number NA16RG1039 (project number RCF-10), and by appropriations made by the Oregon State legislature. The views expressed herein do not necessarily reflect the views of any of those organizations.

REFERENCES

Andersen, P. (1983) On rent of fishing grounds: Translation of Jen Warming's 1911 article, with an introduction. *History of Political Economy* **15**, 391-96

Arrow, K. (1974) *The Limits of Organization*. New York: W.W. Norton and Company

Asada, Y., Hirasawa, Y. and Nagasakai, F. (1983) Fishery management in Japan. *FAO Fisheries Technical Paper 238*. FAO, Rome.

Berkes, F. (1986) Local level management and the commons problem: a comparative study of Turkish coastal fisheries. *Marine Policy* **10**(3), 215-229.

Berkes, F.,ed. (1989) *Common Property Resources: Ecology and Community-Based Sustainable Development*. Belhaven Press, London.

Eggertsson, T. (1990) *Economic Behaviour and Institutions*. Cambridge: Cambridge University Press.

Gordon, H.S. (1954) The Economic Theory of a Common-Property Resource: The Fishery. *Journal of Political Economy* **62**, 124-142.

Hanna, S. (1994) Co-management. Pp. 233-242 in *Limiting Access to Marine Fisheries: Keeping the Focus on Conservation*. Washington, DC: Centre for Marine Conservation and World Wildlife Fund.

Hanna, S. (1995a) Efficiencies of User Participation in Natural Resource Management. In S. Hanna and Munasinghe, M., eds. *Property Rights and the Environment: Social and Ecological Issues*. Washington, D.C.: World Bank.

Hanna, S. (1995b) User participation and fishery management performance within the Pacific Fishery Management Council. *Ocean and Coastal Management* **28** (1-3), 23-44.

Hanna, S. (1998a) Co-Management In Small-Scale Fisheries: Creating Effective Links Among Stakeholders. In Proceedings of the International Workshop on Community-Based Resource Management, The World Bank, Washington, D.C., May 10-14, 1998.

Hanna, S. (1998b) Institutions for Marine Ecosystems: Economic Incentives and Fishery Management. *Ecological Applications* **8**(1) Supplement, 170-174.

Hanna, S. (1998c) Managing for Human and Ecological Context in the Maine Soft-Shell Clam Fishery. Pp.190-211 in F. Berkes and Folke, C., eds. *Social and Ecological Linkages for Sustaining Natural Resource Systems*. Cambridge: Cambridge University Press.

Jentoft, S. (1989) Fisheries Co-management: Delegating Government Responsibility to Fishermen's Organizations. *Marine Policy* **13**(2), 137-154.

Jentoft, S. (1993) *Dangling Lines: The Fisheries Crisis and the Future of Coastal Communities*. Social and Economic Studies No.50, Institute of Social and Economic Research, Memorial University of Newfoundland.

Jentoft, S. and Kristoffersen, T. (1989) Fisheries co-management: the case of the Lofoten fishery. *Human Organization* **48**(4), 355-365.

Kuperan, N.M.R.A.K. and Pomeroy, R.S. (1998) Transactions costs and fisheries co-management. *Marine Resource Economics* **13**, 103-114.

Malasha, I. (1996) In search of a new management regime on the northern shores of Lake Kariba. CASS Occasional Paper, University of Zimbabwe.

Matthews, R.C.O. (1986) The economics of institutions and the sources of growth. Economic Journal 96 (December): 903-910.

McGoodwin, J.R. (1994) 'Nowadays, nobody has any respect': the demise of folk management in a rural Mexican fishery. Pp.43-54 in Dyer, C. L. and McGoodwin, J.R., eds, *Folk Management in the World's Fisheries*. University Press of Colorado, Niwot, Colorado.

Nielsen, J.R., Vedsmand, T. and Friis, P. (1997) Danish fisheries co-management decision making and alternative management systems. *Ocean and Coastal Management*.

Olsen, M. (1965) *The Logic of Collective Action*. Cambridge, MA: Harvard University Press.

Pomeroy, C. (1994) Obstacles to institutional development in the fishery of Lake Capala, Mexico. Pp.17-42 in Dyer, C. L. and McGoodwin, J:R., eds, *Folk Management in the World's Fisheries*. University Press of Colorado, Niwot, Colorado.

Pomeroy, R.S. and Pido, M. (1995) Initiatives toward fisheries co-management in the Philippines: the case of San Miguel Bay. *Marine Policy* **19**(3).

Runge, C.F. (1984) Institutions and the free rider: the assurance problem in collective action. *Journal of Politics* **46**(1), 154-81.

Schelling, T.C. (1960) *The Strategy of Conflict*. Oxford: Oxford University Press. 1977 reprint.

Scott, A. D. (1955) The Fishery: The Objectives of Sole Ownership. *Journal of Political Economy* **63**, 116-24

Sen, S. and Nielsen, J.R. (1996) Fisheries co-management: a comparative analysis. *Marine Policy* **20**(5), 405-418.

Chapter 4

TOWARD SPECIFICITY IN COMPLEXITY: *Understanding co-management from a social science perspective*

EVELYN PINKERTON
School of Resource and Environmental Management, Simon Fraser University, Burnaby, Canada

1. REVITALIZING AN OVERUSED TERM

The term 'fisheries co-management' has now been so broadly used in applied settings and in social science that it risks losing important aspects of its original thrust. In addition, as social science thinking about management in general has evolved over the last two decades, we have all refined and enriched the way we see this concept. For the concept to remain useful, I argue that it should become more specific and complex instead of more general and generic. In the discussion below I attempt to reevaluate, and reorganize a few key dimensions of this term into a form that is more theoretically useful for dealing with complexity. I use the evolution of my own research and thinking on fisheries co-management over the last 15 years as a means of attempting to hone and revitalize the term. Also, in dialogue with colleagues, I suggest key alternative perspectives about what meaning we should assign the phrase.

2. DEFINING PARAMETERS WITH A FULLY-DEVELOPED CASE

I draw particularly on my field experience spanning two decades with a case which in many ways has defined co-management for me. I have been struck by how many social scientists have an explicit or implicit definition of co-management based more on their field experience than the literature. Over-reliance on one's own field experience can be limiting, unless balanced by careful examination of other cases, because our sociological imagination is so deeply formed by the models we ourselves study in depth and over time. Not only do we know and trust their functioning best, but we often believe they define the limits of the possible. Comparing the above-mentioned field experience to the literature, however, leads me to believe that this particular well developed example can provide a valuable template against which to analyse other cases. Emphasizing a fully-developed case is useful if it helps prevent us from defining co-management in too limited a fashion. A limited definition would condemn us, and especially disempowered fishing communities or organizations, to meager aspirations which may not enable their survival. At the same time, *we need to rigorously define the conditions under which 'complete' or even adequate co-management can develop and survive.* This discussion is limited to only a few contentious aspects of this effort, since a review of propositions that I and others have been generating since 1988 about these conditions would be too lengthy here.

The case I have been fortunate to encounter first hand offers a highly-developed and continually evolving model (the Washington State treaty tribes) which I invite colleagues to consider for what it suggests about the scope, scale, nestedness, embeddedness, dynamism, and limits of various forms of co-management. The model is very advanced in that most of the vertical linkages between the fishing communities and local and senior levels of government have been institutionalized, so that the system is fully 'nested' at all levels of governance (Ostrom, 1990). That is, decisions made at one level interact with other levels so that there is both policy stability at higher levels of governance and also capacity to innovate at lower levels. The system is furthermore integrated horizontally into multiple multi-party processes at local, regional, state, and national levels. The system's development demonstrates how *changes in governance in a central key relationship can radiate out to effect the democratic functioning of civil society itself,* as discussed below. And it is a system whose key learnings have already been transferred to at least one other setting. I first studied it in collaboration with Canadian communities (Pinkerton, 1989b) who have now adopted, adapted, and further expanded the model into ecosystem management with the help of my PhD. student (see Chapter 9, Loucks *et al.*).

3. BACK TO THE ORIGIN OF THE TERM: A HIGHER-LEVEL COLLECTIVE CHOICE RIGHT

The term 'co-management' was first used in the late 1970s by US treaty tribes in western Washington State to describe the relationship they aspired to have with state managers. The tribes had won court recognition of their access and withdrawal rights to an allocation of fish over the decades preceding the landmark 1974 *US v. Washington* case, popularly known as 'the Boldt decision' (based on the court's interpretation of earlier treaties). However, tribes had been barely able to *exercise* these rights, because the harvest was managed by the state in such a way that little fish (in this case, mostly salmon) remained by the time this migratory species reached the marine and riverine territories in which the tribes could legally fish. The state managed the non-tribal commercial and sport fisheries so that all but some five per cent of the fish was harvested elsewhere. Therefore, Judge Boldt reasoned that only by recognizing the tribes' right to *participate in planning and regulating the entire harvest* (which he called 'concurrent management') would their allocation right ever be exercised. In other words, the treaty's promise that the tribes would share in access to the fish alongside the citizens of the territory was unrealizable without a second and higher-level right being granted: the right to participate in management decisions about how the harvest would be conducted. Institutionalists would say that it was necessary to grant a *collective choice* right (decision-making about harvest planning) to a group in order for it to exercise an *operational right* (taking an allocation of fish) in this particular context (Schlager and Ostrom, 1993).

This distinction is key, as there has been a tendency to apply the term co-management to mere operational rights, an inappropriate watering down of the term to a narrower, less powerful, right. (See Cohen, 1986, 1989 for an excellent treatment of the legal, ethnographic, and early historical context of this case). Co-management is misnamed unless it involves at least the right to participate in making key *decisions about how, when, where, how much, and by whom fishing will occur.* We need to distinguish this level of power from the right to simply access a pre-defined catch. Furthermore, certain collective choice rights may be too small in scope and/or in scale to confer meaningful power in the long run. The right to plan the manner of harvest in a local bay would probably not address the issue of a sustainable harvest, for example. As is proposed elsewhere (Pinkerton and Weinstein, 1995; Pinkerton, 2002), it is useful to array small-scope operational or collective choice rights within a matrix of other potential co-management rights/activities of varying levels of importance, such as policy-making, defining membership and boundaries, habitat protection, enforcement, etc. In addition, the degree and type of power held by the non-governmental co-manager in a particular small-scope activity could be expressed as a rung (or even assigned a number) on the ladder of participation (Berkes, 1994). The level of power might vary in different management activities, so ideally we would 'score' any arrangement specifically, noting both the scope of the arrangement, and how power is distributed in each arena of decision making. This type of matrix array opens the way to an analysis of what critical bundle of rights is sufficient to allow a co-management system to be effective in achieving the long term objectives of one or both partners, as discussed below. So let us start with the assumption that the power to participate as a partner in planning the harvest is a necessary but often not a sufficient right for a co-manager.

4. WHAT OTHER RIGHTS AND ACTIVITIES HAVE TO BE INVOLVED?

How can communities participate in planning a sustainable harvest without access to or the ability to produce data, the capacity to analyse that data, and access to dispute resolution over varying interpretations of data and tolerance of risk? By the time I began studying this example in the early 1980s (Pinkerton, 1988), the struggle was in full swing between the tribes (who now had hired biologists and built capacity) and the Washington Department of Fisheries (WDF), which resisted the exercise of the tribes' collective choice right. The conflict revolved around tribal access to WDF's stock abundance data, everyone's harvest data, and the very definition of conservation. This last meant how much fish should not be harvested, but rather allowed to spawn in each stream. The WDF did not want to reveal the paucity of its stock abundance data and the level of uncertainty surrounding its analysis and decision-making about the harvest. Furthermore, it did not trust the tribes more than any other fishermen to report their catch accurately, especially since some tribes had asserted their treaty rights through illegal fishing for decades. The final negotiation of a co-management agreement which resolved these problems, among others, took 10 years and resulted in a complex power-sharing relationship in which the state and tribes agreed to work jointly on every aspect of data gathering, data analysis, and harvest planning and eventually played complementary and mutually supportive roles. The tribal-state relationship was less a delegation of powers than a complex division of powers and a collegial collaboration in problem-solving as putative co-equals. The two parties eventually developed a high level of trust and learned to make the best use of limited funding by sharing (and sometime agreeing to trade off specializations) in virtually every aspect of management and every stage of planning, from international negotiations to collecting data on indicator streams. The sharing and improvement of data gathering and data analysis through mutual accountability provided the foundation on which trust was built in harvest planning. (This sharing then provided such a mutual benefit that it drove the building of cooperation in other areas of management). What has been learned is that *the exercise of the collective choice right depends on a series of other rights*, some of which may be narrow-scope or purely operational. Nonetheless these narrow-scope rights may be critical to the effective exercise of other small-scope and larger-scope rights. The court had not envisaged all the complexities of co-management, so many of the other rights were recognized only through negotiations after intense political struggles, or continued use of the courts and its extensions. Through the repeated exercise of their harvest management right, the tribes gradually learned what other rights were necessary for making this core co-management function operable. As discussed below, these learnings continued to evolve over time. They were learning what set of rights enabled sustainable co-management in their particular case.

5. VERTICAL AND HORIZONTAL GOVERNANCE BROADENS THE CO-MANAGERS' ROLES

As soon as harvest co-management protocols were agreed to in 1984, the next set of issues around co-management emerged. These included habitat protection, regional planning, setting broader policies at a higher level, and international allocation (interception) agreements. The co-management system set the stage for a complex multi-stakeholder exercise in watershed analysis, and eventually for the most challenging exercise in complex collaboration ever attempted, involving federal agencies regulating endangered species protection and water quality under federal statutes. Thus *harvest planning and regulation emerged as only a small part of what would eventually be involved in co-management*. The success of the tribes in asserting a full complement of collective choice rights revealed the range of political, legal, and social factors that affected management and how power would be shared with the state in each aspect. In-depth discussions of the case are published elsewhere (eg Pinkerton and Keitlah, 1990; Pinkerton, 1992; Singleton, 1998; Ebbin, 1998).

6. KEY ASPECTS OF COMPLETE CO-MANAGEMENT

My purpose here is to use this bare sketch of what I call 'complete co-management' (Pinkerton, 1989a) as a context for discussing seven aspects that are key to such collective choice arrangements. They provide a useful template against which to measure various less complete forms of co-management, and the barriers which limit them. I select these seven as *neglected and controversial aspects of co-management*, not as a complete list of defining or permitting conditions. These key aspects of complete co-management are as follows:

(1) Government as a co-manager plays a key and desirable role, and is ideally an engaged partner rather than a delegator.

(2) Co-management, like management itself, involves far more than the control of fishing effort.

(3) Sustainable co-management arrangements involve some control by community partners over the terms and conditions of sale to fish buyers.

(4) The successful exercise of rights on one level depends on the exercise of rights at higher and lower levels, including the right to participate in data collection/analysis and in setting policy agendas at the highest level.

(5) Co-management, as it matures, will ideally involve multiple horizontal negotiations leading to cooperative activities with other players and potentially greater democratization of civil society.

(6) The power to exclude from some defined territory is optimal.
(7) Complete co-management is based more on the collective rights of a group than on individual rights.

I now contrast these seven key aspects of 'complete co-management' with various ways the term has been applied in social science. The purpose here is to clearly identify the barriers to the full development of less developed forms of co-management.

6.1. Government is a key player in complete co-management

A number of social scientists have seen co-management as a small step from self-management of various types, a step which does not alter the fundamentally self-managing character of the arrangement, but adds more legitimacy or logistical support to it. For example, institutionalists tend to view the state as parasitic on self-managing communities, and thus fail to see the key role it plays. In his superb analysis of self-governing groundwater management boards in southern California, Blomquist (1992) shows the key role played by the 'water master', who is first a government employee and later works for the board. Yet his analysis neglects the legitimizing, human capital, and social capital aspects of the role of this key state employee who turns to working for communities. Similarly, many maritime anthropologists and sociologists, extending a tradition of analysing the norms and self-regulating capacities of isolated fishing communities and the extent to which the state serves the interests of corporate capital, tend to see ideal co-management as a situation in which the state is involved in management as little as possible. While there are is no lack of evidence the state has largely supported capital at the expense of communities (Marchak *et al.*, 1987; Durrenberger, 1992), the ultimate goal of sound co-management is ideally to balance power so that the state plays a mediating and levelling role among interest groups, as demonstrated in cases discussed below. In the case of industrialized states, fish are not the most significant commodity, and fisheries management agencies are not the most significant players in state governments. The larger political battles are likely to be over major industrial developments which affect fish habitat in watersheds and in the coastal zone (eg oil drilling, intensive net-cage aquaculture, industrial and major agricultural effluent, intensive logging, dams, etc.). In the battles over fish habitat, the state fisheries management agency has the potential to be an ally of co-managing communities or groups (as it is in the Washington case, Pinkerton, 1992).

But there are far more important reasons why government is a key player in complete co-management, which apply to developing countries as well. While it is true that some forms of co-management have developed by incorporating existing self-management regimes into a new regime, this very incorporation alters the self-managing aspect in fundamental ways. I agree with Holm *et al.* (2000) who point to the 'modernity' of the era in which the co-management contract is made and the fact that career paths and social controls have changed in at least some important ways. However, I would go further than these authors in emphasizing the importance of the role played by government. What is most helpful to co-managing communities is that government can be the provider of technical support, credit, marketing assistance, or protective legislation, such as has occurred in the Philippines (Chevalier and Buckles, 1999; Berkes and Pomeroy, 1997). It is worth noting, however, that external or non-fishing NGOs such as academics, international bodies, or advocacy groups may be the chief sources of legitimation for the co-management relationship, as in the Dominican Republic (Stouffle *et al.*, 1994; Chevalier and Buckles, 1999), not necessarily government, even though government must eventually provide protection for co-management to work. Legitimation may be the least of the roles played by government. This has important implications for the power communities may be able to exercise, as discussed below.

Even more important in how co-management fundamentally alters self-management is the nature of the contract itself. The co-management relationship creates a series of dilemmas for self-managers in that, post co-management, they are constrained by the timetables and decision-making modus operandi of national and international governments and planning bodies. For example, they may no longer have the luxury of reaching consensus or the level of certainty they desire in their own time. They risk becoming bureaucratized and oligarchized in ways that run counter to the values and goals of the community they serve and must constantly trade off internal accountability and externally imposed timetables and efficiency (Pinkerton and Keitlah, 1990). They may have staff or board members who do not necessarily communicate with community members in a regular and democratic way (Kofinas, 1998). The co-management relationship thus *transforms the traditional community*, even as it attempts to express its values and concerns. The trade-off is positive for the community, but co-management does come at a cost.

Furthermore, government is unlikely to be a neutral disinterested party in its dealings with the community, and may even be thought of as a 'stakeholder', given that it has a relationship with many affected actors and is itself affected by the outcome (Jentoft and McCay, 1995; Mikalsen and Jentoft, 2001). In some multi-party arrangements, government may even formally play a dual role as stakeholder and sponsor of the arrangement, who sets the initial conditions or rules which give the parties incentives to come to the table together (Pinkerton, 1996). Whatever the hazards of confusing its sponsor role and its stakeholder role, governmental participation is key to the well-being of both simple and multi-party co-management contracts, as discussed below, because government is the only body with the authority to protect the interests of the co-managers against other parties (Pomeroy and Berkes, 1997; Tyler, 1999).

Even communities which struggle to assert more 'local control' will recognize at some point that they need government protection. Recognizing that there may be many areas/activities within an agreement which need not involve government, government is still a key player. In mature co-management such as the Washington tribes' case, government became a key ally with whom the tribes agreed to trade off certain management functions, depending on who is best placed to do the job. In other words, in complete and mature co-management, *the relationship with government is seen as a partnership delivering a net benefit.*

This concept of partnership is rather different from the way many analysts have seen co-management, as primarily a matter of delegating powers to users (Jentoft, 1989), 'in which communal property and [community based resource management] are always embedded in state property systems and derive their strength from them' (Pomeroy and Berkes, 1997), as further discussed below. However, what may have been finally blessed as 'delegation' may not have so originated nor be understood as such at the community level. In Japan, for example, where local fishing cooperatives hold a form of sea tenure to local fishing space, a large number of decisions are made locally. The system developed *first* at the local level and was later integrated into regional and national governance, and protected in national legislation (Yamamoto and Short, 1992; Pinkerton and Weinstein, 1995). However, periodic efforts by the state to establish formal ownership of resources or to abolish local marine tenure boundaries were always abandoned (Matsuda and Kaneda,

1984). It might be more accurate to characterize many co-management situations as a standoff in which parties agree to disagree, and partnership is forged out of the need to work together. Enabling legislation can lay the groundwork for such a partnership, but it is in the implementation of the legislation that one finds the 'proof of the pudding'. In Alaska, for example, the regional aquaculture associations created under enabling legislation in 1976 took on increasing power as their managers and staff gained seniority and stature and they were able to take increasing leadership and initiative in addressing broader management questions. Their de facto power came to outstrip their de jure power.

6.2. Complete co-management is about more than effort control for conservation

A large number of analysts of self-regulation and co-management have defined co-management narrowly as being based on forms of self-regulation such as allocation of fishing space, size limits, seasonal limits, time limits, or even access to mooring space which have the effect, intended or not, of controlling fishing effort. However, as many have also noted, these regulations often also affect the distribution of access and the efficiency of operations. I differ from Holm *et al.* (2000) in that I do not see this as a reason to dismiss these regulations as forms of management, albeit narrow-scope. Even conventional fisheries managers usually recognize that management involves far more than direct effort control and state that 'we have to manage fish by managing people' (Larkin, 1988). If we see management (and 'complete co-management') as including allocation (which may be either a side-effect of another regulation or a formal plan) as well as harvest planning, and if we assume harvest planning is often influenced by considerations of efficiency as well as conservation, then management in practice has seldom been uniquely focussed on effort control. Government managers often consciously use efficiency and allocative regulations to achieve conservation indirectly, but recognize that because they are managing people, they may more effectively achieve a conservation goal through a strategic use of indirect tools.

I argue therefore that it is more useful to conceptualize management itself broadly, because in fact the problems encountered by managers and communities inevitably require solutions beyond the straightforward restricting of fishing effort. This is because, as Schlager and Ostrom (1993) conclude in a literature review of 30 cases of self-regulating rules devised by fishermen's groups/communities, even small-scale and pre-industrial fisheries typically need to deal with three common pool resource dilemmas which they may think of in terms of efficiency as much as conservation: appropriation externalities (too much fishing effort, increasing everyone's costs per unit harvested), technological externalities (crowding and gear conflicts, decreasing efficiency of operations), and assignment problems (allocation conflicts over access to the most productive locations). Although Schlager and Ostrom did not exhaust the available literature on self-regulation (and I believe we do have cases of self-regulation for conservation), it is notable that none of the examples they encountered *required* an explanation for self-regulation related directly to conservation. So pre-industrial and small-scale fisheries also use multiple and indirect tools which have the effect of restricting fishing effort perhaps as much as modern managers do. My point is that managers and co-managers are stuck with this situation. As a result, modern managers wishing to bring in new conservation measures often try to find a constituency which supports its conservation measures for allocative reasons. So let us consider the utility of *conceptualizing management and co-management systems broadly, even if we consider the sole legitimate objective of management to be conservation*. This would be consistent with the finding that resource users accept conservation regulations more readily if they believe the allocative effects of them are equitable, and that equity has a direct impact on efficiency (Oakerson, 1992). I continue to argue (Pinkerton, 1989a; Pinkerton and Weinstein, 1995) that it is useful to consider all the management activities in which we know communities to be involved as 'real' aspects of management and co-management which have attached rights or duties. Our challenge should be to *analyse how these management activities, rights and duties interact* and contribute (or not) to sustainable management.

6.3. Sustainable complete co-management arrangements involve some control by community partners over the terms and conditions of sale to fish buyers

Importantly, I believe that Holm *et al.* (2000) implicitly acknowledge that management should be defined broadly when they argue later on that returning optimum value to small-scale coastal fishermen through the Mandated Sales Organization (MSO) system is fundamental to the fisheries management system in Norway. Norway has long been an outstanding example of a modern jurisdiction which did not permit vertical integration of buyers and fishermen-sellers, and thus prevented the domination of the industry by corporate capital...even if this was partially in support of the national goal of promoting coastal settlement (eg Jentoft, 1993; Brox, 1993; Wadel, 1972). As has been argued elsewhere (Pinkerton and Weinstein, 1995), it is useful to conceptualize returning optimum value to fishermen as a fundamental aspect of management, because its impact on the way a management system operates is as fundamental as allocation. In Japan, harvest planning is affected by this consideration (advantageously for the local fishermen's cooperatives), and in Washington State, this is the one area in which co-management is incomplete. The Washington co-management situation survives because of federal support to implement treaties, which pays for management costs. The Washington treaty tribes have no control over terms of sale, no supportive legislation or other overarching policies supporting them in this regard. During recent years of low fish abundance and/or prices, many tribal fishermen cannot sustain themselves. Sustainable co-management, then, requires a situation in which fishing communities can capture a large enough share of the benefits to support at least some management costs. Captured benefits should be conceptualized as an essential aspect of complete co-management, alongside adequate control of fishing effort.

6.4. The successful exercise of rights on one level depends on the exercise of rights at higher and lower levels

As discussed above, an operational right can be a weak and narrow right unless buttressed by a higher-level right to decide under what conditions the operational right can be exercised. Similarly, co-managers are not in a position to make good decisions if they don't have access to data and the capacity to analyse it. Access to data might be considered a narrow-scope operational right, even narrower than access to an allocation of fish. Very weak forms of what is sometimes mis-named co-management often involve fishing communities collecting data for government, but having no power to make decisions or even have access rights to fish based on the data, even if they can analyse it. Nevertheless, *possession of these narrow-scope operational rights (data collection and analysis) is crucial to the exercise of the higher-level collective choice right*, since it is a necessary precursor to harvest planning. This is a crucial point that has been missed by institutionalists focussing solely on the hierarchy of rights of access/withdrawal, management, and exclusion (Schlager and Ostrom, 1993).

Similarly, the narrow-scope collective choice right to plan the harvest will become meaningless if large-scope policy decisions are made at a higher level by the state that another nation will be allowed to intercept the fish, or that the state has sole rights to define conservation, or that the private property rights of landowners whose activities impact fish habitat will take precedence over the rights of fish harvesters to good fish habitat. Judge Boldt in the second phase of *US v. Washington* in 1980 reasoned that both the access/withdrawal right and the management right were meaningless unless the tribes also had the right to protect fish habitat. In practice, this meant that the tribes became part of the multi-agency council of the governor of the state of Washington, negotiating with line agencies which regulated land use practices affecting fish abundance. Exercising rights at this level involves participating in setting the agenda for how issues are defined and acted upon, as well as the timing of when certain issues will be given priority attention. Agenda setting also occurs at the Regional Council level, ie the Pacific Fisheries Management Council (see McCay and Creed, 1999; Loucks *et al.* Chapter 9), as discussed below. More than having the right to protect fish habitat specifically, having the right to participate in setting policy agendas means that the tribes have an opportunity to influence the relative importance of fishing and fish habitat issues, including what level of resources are proposed in specific legislative packages. They also eventually got a tribal seat on the Forest Practices Board, since forest practices (logging and silvaculture) have a key effect on fish habitat. The exercise of rights at this higher level means that co-management activities at a lower level have a greater chance of having their intended outcome. Therefore we can generate the proposition that to be effective, *power-sharing in co-management needs to be scaled up and down to the level capable of affecting its operation.*

6.5. Complete co-management involves multiple horizontal negotiations leading to cooperative activities with other players

Not only does complete co-management involve the scaling up and down of activities of various scopes and their attached rights, it also ideally involves co-operative planning, research, education, and monitoring with other fishing and water-using parties and jurisdictions. Such parties might be thought of as being in 'horizontal' relationships with fishing communities, since they do not normally have jurisdiction or rights over harvest management, but rather have potentially competing rights to fish or water.

In some cases the state plays a direct role in bringing these parties into joint harvest planning with co-managing communities. For example, on a statewide basis the Washington State treaty tribes participate in joint pre-season cooperative data analysis and harvest planning with the WDF, sport fishing groups, commercial fishing groups, and the federal National Marine Fisheries Service, which has jurisdiction outside three miles from shore, in the Pacific Fisheries Management Council (PFMC). These parties first negotiate what level of conservation is appropriate for various salmon runs, then what overall fishing strategies should be used to allocate various stocks to various parties without compromising conservation. The PFMC is not a co-management body in itself, but for its role in setting the parameters for the more detailed planning, it is an essential piece of vertical (federal and state) and horizontal (commercial and sport fishermen) negotiation for the tribes.

On a regional level, three types of multi-party watershed planning activities are particularly noteworthy horizontal negotiations: those affecting water quality, water quantity, and forest practices, respectively. Because of their co-management rights to protect fish habitat, the tribes are always represented in these processes, and sometimes are the lead agency. Each type of process affords opportunities to jointly create rules for more flexible and creative fish habitat protection. Each also constructs good will, trust, and social capital among parties who would not normally associate with one another. In the best circumstances, this social capital continues to contribute to evolving improvements in joint problem solving in the watershed, which is likely to be the place where 'social learning' occurs and is maintained because the processes are embedded in communities of place. Social learning refers to processes which transform social relations and generate less conflictual ways of addressing difficult joint problems. I briefly outline the essentials of each process to illustrate the role and importance of different types of horizontal negotiated agreements for the co-management process. (More in-depth discussions are in Pinkerton, 1991, 1992, 1994; Pinkerton and Baril, 2001).

Water quality watershed planning, especially to protect shellfish-rich estuaries or valuable salmon runs, began in the late 1980s under a state superagency which empowered key counties to convene water quality planning processes. The most successful processes used a trained facilitator/mediator who chose qualified local representatives (including tribal and non-tribal fishermen, environmental groups, farmers, marina operators, real estate developers, and county commissioners) to meet regularly over a year and produce a plan for achieving the state water quality standard. The plan set target community standards and methods and timetables for achieving them. The process involved in-depth education about water quality problems and their causes, volunteer clean-up projects, and volunteered strategies by the polluters and others to contribute to solutions. The planning and implementation process was effective not only because consensus was achieved, but also because the county had a mandate to require the other agencies to comply with the plan, once it had been approved by the Department of Ecology, and funding existed to produce the plan through the superagency. What we need to draw from this example is not that such processes would necessarily be absent without legal backing: in this case a social movement in response to pollution in Puget Sound was in fact the driving force. But this movement was aided by the pre-existence of fisheries co-management. More importantly for this discussion, its existence buttressed the tribes' co-management rights, because it created or extended the capacity of communities or regions to implement habitat protection. Like the PFMC example discussed above, it is not an example of complete co-management itself. It involves the rights to data and data analysis, but not the right to extract benefits from the result of work done, other than the benefit of a healthier environment. Perhaps the overriding importance of this type of process is that it focuses ordinary citizens on their *duty to protect public resources* rather than their right to extract benefits. This concept of duty is key because it highlights what is often not understood as a key aspect of complete co-management: that co-managing communities and their horizontal partners do the work not just because of the benefits they extract in this generation, but because of their *duty to future generations, and the survival of their communities*. It is one of the reasons communities of place are so crucial, as discussed below. It is in such places that citizens are able to link environmental benefits to the health of their families and communities and act out of a sense of duty. Otherwise put, the community is able to capture the benefits of its own stewardship when it occurs on a watershed scale. So *the pre-existence of co-management aids in the building of or reinforces horizontal institutions of stewardship at the community/watershed level*.

A second type of watershed planning exercise occurred in the early 1990s under the statewide Chelan agreement to pilot water quantity planning, in particular how irrigators, hydroelectric companies, and tribes would share scarce water. In this agreement the tribes had further asserted their rights to be considered *full partners in governance*. Not only were they the convenor of this process, as the county had been of the water quality process, but they had more power. The Chelan agreement produced a decision rule in which all three local 'governments' - the county, the tribe, and the state agencies led by the Department of Ecology - had to agree to any rule, in addition to the majority of the other non-governmental stakeholders in the planning process. In the Dungeness-Quilcene watershed pilot, the stakeholders decided they wanted to operate on consensus, but the decision rules would clearly have been a fallback position had they been unable to achieve consensus. This arrangement highlights the fact that a full co-manager has more authority and legitimacy than other 'user groups' and acts more like a government than a user. So this process exemplifies both a horizontal supportive agreement and an extension of tribal authority to veto power in water quantity planning. This agreement also created the incentives necessary for parties to craft new strategies to reward water conservation efforts. For example, rather than blame irrigators for dewatering streams, a major collaborative effort has created a new legal construct - the State's first 'trust water right' - which rewards and encourages water conservation by all sectors in the watershed during critical salmon spawning months. Irrigation districts signed an agreement with the State and tribes to cut back on water diversions by 40% during August and September when chinook salmon return to the Dungeness River (Seiter *et al.*, 2000; Pinkerton and Baril, 2001).

The third type of watershed planning, called watershed analysis, began in the mid-1990s and was to regulate the cumulative impacts of forest practices (chiefly logging) on mid-size basins so that salmon habitat would receive adequate protection. Evolving from the statewide Timber, Fish, Wildlife agreement of 1986 between tribes, timber companies, state agencies, and environmental groups, watershed analysis involved applying a set of analytical 'modules' to predict the cumulative effects of logging in a particular watershed. Analysis focussed particularly on the effects of mass wasting (landslides from slope instability), road building, and riparian forest buffer width on in-stream habitat conditions. The producers of the modules (and of the watershed analysis manual indicating how to integrate them) were scientists working for tribes, the timber industry, and state agencies. Representatives from these three groups also participated in generating prescriptions for how logging could be safely conducted in a particular watershed. In this process, the tribes held rights of access to data and data analysis and the key right to jointly set the parameters within which logging would occur in the watershed. This was the most ambitious horizontal process of all, since it constitutes the *right to co-regulate major industrial activity affecting fish habitat*. It is also perhaps the process which produced the most significant social learning. Social learning has occurred in some tribal areas where the process went smoothly and increased not only the understanding of watershed processes and the quality of forest practices, but in general increased the cooperation between tribes, state regulators, and timber companies. In other areas, conflicts were not resolved at the watershed level, and scientific discourse was shuttled to the Cooperative Monitoring, Evaluation, and Research arm of the Timber, Fish, Wildlife Agreement in which all the parties participate. This process, recently reformed, is intended to jointly monitor the results of new forest practices and conduct research which would answer the difficult unresolved questions. The parties agree to implement research findings. This *right to monitor, research and implement research findings* is, as in the fisheries example of the importance of lower-level rights, likewise key to effective co-regulation of major industrial activities. The research element makes this right even broader, however.

These three examples of watershed planning/analysis taken together illustrate not only mixtures of horizontal and vertical processes, but also the *potential of co-management to stimulate broader reforms toward more participatory democracy in civil society.* Although early analysts, including the author, have often lumped co-management with larger reforms of democratic process leading to more direct democracy, the Washington case demonstrates the utility of distinguishing co-management processes which are limited to specific stakeholders holding collective choice rights at multiple levels from processes which engage a broad spectrum of citizens with weaker rights and interests in protection of common pool resources as public goods. In other words, complete co-management involves a complex set of rights which may not directly reform civil society. However, the less direct horizontal negotiations which emerge from the core of complete co-management pull in a broad spectrum of citizens and involve a broad democratizing of civil society, at least in the area of common pool resource management. Mikalsen and Jentoft (2001) have noted the (largely horizontal) spectrum of 'stakeholders' potentially involved in fisheries planning and management. The foregoing discussion was consistent with their findings and also noted the piggy-backing of many forms of stakeholder involvement on the core co-management relationship.

6.6. The power to exclude from some defined territory is optimal for creating complete co-management

Much of the literature on self-management as well as co-management concerns place-based groups with clearly defined membership which exclude outsiders either from membership and/or from access to and decisions about some clearly-defined local territory or local stocks. They have incentives to improve the resource because their investments do not have to be widely shared, as has been broadly discussed by institutionalists and by maritime anthropologists and sociologists.

It is worth developing this point more, because of recent suggestions that organizations of holders of ITQs can act as co-managers. Bonnie McCay (NRC, 1999) has made a valuable distinction between communities of place and communities of interest, the latter being a term which could characterize at least some ITQ holders. Can communities of interest (groups which are not place-based) be co-managers? Can a group co-manage a fish stock that is not territorially based?

The answer to these questions depends on the incentives driving the community of interest. It is certainly theoretically possible that a community of interest could have incentives to steward fish and fish habitat in a local territory, as occurs, for example, with the Alaska regional fishermen's associations practising salmon enhancement and harvest planning, which include a number of non-local fishermen who have rights to fish local areas only (Pinkerton and Weinstein, 1995). But this incentive is likely to be less than the incentives of local residents, simply because non-locals have more options, and locals are more highly dependent on local resources. A co-management board which included non-locals, but in which locals had the major voice might (such as occurs in the Alaska case) still has a good chance of making decisions in response to the strong stewardship incentives discussed above. The watershed-based examples discussed above suggest that the local specificity of problem-solving provides a key incentive for parties to work together, and that the frequency of contact in addition to place identification increases opportunities to build trust which in turn enhance social learning and problem-solving ability. Rules and norms have a good chance of becoming embedded in local and regional social life (Apostle *et al.*, 1998). A community of interest managing a territorially-based stock will always have far fewer incentives and less capacity to steward than does a community of place or a mixed community of place and interest. In the case of a transferable access right (such as in an ITQ fishery), there are even more incentives to free-ride, since the real impacts of overfishing (especially on complex ecosystem relationships) may not be evident for some time.

There are also fewer incentives for a community of interest to sustainably co-manage a non-territorially-based stock. Such a group has neither territorial exclusion nor community of place membership in its favour, which creates more opportunity for free-riding and non-compliance. This may be part of the explanation for the resource stress in many ITQ fisheries, which have been forced to adopt strict state dockside monitoring and marketing regulations to diminish these problems. The individual (vs collective) action which is possible in such communities of interest, when/if it coincides with a fishery that is not place-based, may be lethal to sustainability. Whereas it is theoretically possible for communities of interest to exercise collective choice rights as collectives of ITQ fishermen, the incentives to do this in practice are weak relative to the incentives to maximize individual short-term interests.

6.7. Complete co-management is based more on the collective rights of a group than on individual rights

As discussed earlier, there is a tendency to talk about co-management as being an arrangement between the state and users or user groups. Users are sometimes conceptualized as individuals who may or may not be organized into fishing associations, and are sometimes spoken of as synonymous with civil society. Co-management is thus often seen in its broadest sense as a reform promoting greater participatory democracy, vs indirect electoral democracy. It is simply making things work more as they are ideally intended to work.

As we saw in section 6.4 above, the horizontal processes catalysed by complete co-management which engage other actors do indeed reform civil society, and may even promote stewardship, social capital, and the identification of shared values. However, when it comes to collective choice rights to make decisions about harvesting and higher order policy issues, I question whether it is useful to conceptualize these rights as individual rights. Rights of this order are essentially collective because by their very nature they imply the ability to decide as a group on issues involving value judgments about risk, priorities about research, and the distribution of benefits. They are based fundamentally on the ability of the group with the rights to act in its collective interest. This means that the group exercising collective choice rights must have at least one institution (eg a tribal council or a board) which is empowered to act in the collective interest. Individuals are empowered *because* they can act in the collective interest, and the implication is that this institution has some life of its own beyond the life of the individuals who populate it or the individuals affected by its decisions. My point is that if we think of co-management as being about collective rights and collective action, we are in a very different world than the world of individual users, or even 'user groups'. Co-managers, whether they are tribes with constitutionally-protected rights or not, take on some aspects of government when they make collective choice decisions. As such they express group values and act for the good of the group. They go beyond the rights and activities of highly participatory citizens.

What are some of the implications of this for co-management research? A fruitful area might be the collective ownership of licences as a solution to various collective action problems. For example, the Canadian Department of Fisheries and Oceans has been experimenting for at least 20 years on the Canadian west coast with the collective ownership of 'N' and 'F' licences. In the first case, salmon gillnet 'A' licences owned by the major processing company in the province of British Columbia, B.C. Packers, were sold to Northern Native Fishing Corporation, made up of three Tribal Councils. 'A' licences, once freely transferable on the market, were converted to 'N' licences, non-transferable out of the corporation and leased to individual tribal members who met certain criteria such as responsible boat maintenance and fishing behaviour. Although individual tribal members leasing 'N' licences have access rights to fish, they do not have the right to transfer these rights, a collective choice right held only by the corporation, run by a board of the three tribal councils. The board allocates licences and provides training in boat maintenance and access to capital for its members to acquire vessels. What is key in this arrangement is that the goal of the board is to maximize the access rights of local members who can perform in a reasonably consistent manner and to spread access as equitably as possible among the tribal councils and within each tribal council. The board thus exercises collective choice rights of allocation and exclusion. Individual members who lease licences from the corporation exercise access and withdrawal right not as individuals, but only by dint of their membership in the corporation and their performance according to standards set by the corporation. A similar configuration is found in 'F' licences held by an individual tribe or band, and leased to individuals or groups within the tribe. An Experimental licence is held by a group practising an innovative fishing method such as a live-capture selective fishery. Importantly, the innovation allows for collective fishing and collective decisions about sharing work and benefits. These innovative forms of access rights demonstrate some of the ways in which co-managers could hold both access and harvest management rights collectively. They can and are being used to experiment with more conservation-oriented technologies, and also with group ownership of ITQs (Schwimmer *et al.*, 2000). Future co-management research could explore questions about whether the exercise of collective rights in this manner improves or impedes the achievement of the goals of the collective body. Such arrangements may offer considerable potential for co-managing communities to play a role in innovation, especially if they are linked to preferential access. Communities are well-placed to play this role because of their potential to realize equitable allocation of opportunity among members, and to link opportunity to performance.

7. CONCLUSION

It has been argued that complete co-management offers opportunities to respond in an appropriately flexible, adaptable, and precautionary manner to aquatic ecosystem variations (deYoung *et al.*, 1999) and as such offers fisheries managers tools not available under other institutional arrangements. I have argued here that it is analytically useful to distinguish the core aspects of co-management arrangements which create these opportunities, and to array rights and duties within an analytical framework which permits us to distinguish different levels of power and necessary bundles of rights permitting a co-management system to be effective. A case of 'complete co-management' was used to illustrate levels of power potentially held by fishing communities as expressed in specific rights, as a template for comparison with other co-management situations. The capacity of different arrangements to generate agreement and stewardship was discussed. The discussion of rights and power builds on an earlier framework for classifying the scope and scale of co-management activities in any particular system (Pinkerton and Weinstein, 1995).

A case of complete co-management was also used to deepen discussion of some contentious issues in the definition of co-management, and generated seven propositions, some with corollaries.

(1) In complete and mature co-management, the relationship with government is seen by fishing communities and groups more as a partnership delivering a net benefit than as a delegation of powers. If fish are less important than competing uses of aquatic habitat, governmental fish regulators can potentially ally themselves with co-managers once initial power struggles are settled.

(2) Complete co-management will involve rights and activities that go beyond sustainable harvest management, and are likely to include activities such as allocation, habitat protection, and policy making.

(3) An analysis of co-management must consider a broad array of harvest and non-harvest co-management activities, rights and duties, and how they interact, eg how the absence of some rights affects the exercise of other rights.

- (3a) Some degree of collective choice (vs operational) decision-making is essential to complete co-management.
- (3b) The successful exercise of rights on one level depends on the exercise of rights at higher and lower levels.
- (3c) To be effective, power-sharing in co-management needs to be scaled up to the level capable of affecting its operation.
- (3d) The right to protect fish habitat ideally includes the right to co-regulate major industrial activity affecting fish habitat.
- (3e) The right to monitor, research and implement research findings is key to effective co-regulation of major industrial activities.
- (3f) The power to exclude from some defined territory is optimal.

(4) Complete co-management arrangements have the capacity to stimulate broader reforms toward more participatory democracy in civil society around fish management issues.

- (4a) The pre-existence of co-management aids in the building of or reinforces horizontal institutions of stewardship at the community/watershed level.
- (4b) Horizontal agreements between co-managers and linked community-based processes are ideally based on the duty of both these parties to protect public resources rather than their right to extract benefits, ie their duty to future generations, and the survival of their communities.

(5) Complete co-management is based more on the collective rights of a group than on individual rights.

(6) Sustainable co-management arrangements involve some control by community partners over the terms and conditions of sale to fish buyers.

(7) The fishing community is able to capture the benefits of its own stewardship when it occurs on a watershed scale.

These propositions do not pretend to be a complete or linked set of findings, but are rather what has emerged in a discussion of a particular case of complete co-management and what it illustrates about contentious issues in the definition of co-management.

It may well be the case that complete co-management is 'difficult to find and sustain' (McCay, 2000), and that social scientists will be tempted to dismiss the case presented here as too unusual to be useful. I have presented it, however, not to diminish the importance or potential of less developed cases, but to draw attention to the question of what outcomes can rationally be predicted from what degrees of power-sharing in which arenas of decision making. When a paradigm such as co-management becomes better known, it risks being 'captured', co-opted, and misapplied to situations where there is in fact little power-sharing.

Co-management as a strategic approach to problem-solving in fisheries management will be judged to perform very poorly if social scientists predict good results from small-scope or small-scale arrangements in which fishing communities enjoy only minor degrees of power or an inadequate linking of power over different arenas of decision making. If we are to develop a more powerful predictive model which deals with the complexity of different arrangements, we must develop a more comprehensive framework for comparing very different situations and for distinguishing what different arrangements can actually deliver.

REFERENCES

Apostle, Richard, Gene Barrett, Petter Holm, Svein Jentoft, Leigh Mazany, Bonnie McCay, and Knut Mikalsen (1998) *Community, State, and Market on the North Atlantic Rim: Challenges to Modernity in the Fisheries.* Toronto. Un. of Toronto Press. 363p.

Berkes, Fikret (1994) Co-management: Bridging the Two Solitudes. *Northern Perspectives.* **22**(2-3), 18-20.

Blomquist, William (1992) *Dividing the Waters.* ICS Press. Berkeley.

Brox, Ottar (1993) Let us now praise dragging feet! In Nordal Åkerman (ed.): *The necessity of friction.* Springer Verlag, New York:123-134.

Chevalier, Jacques and Buckles, Daniel (1999) Conflict Management: A Heterocultural Perspective. p.13-41. In Buckles, Daniel, (ed.), *Cultivating Peace: Conflict and Collaboration in Natural Resource Management.* International Development Research Centre. Ottawa. 285p.

Cohen, Fay (1986) *Treaties on Trial.* American Friends Service Committee. University of Washington Press.

Cohen, Fay (1989) Treaty Indian Tribes and Washington State: The Evolution of Tribal Involvement in Fisheries Management in the US Pacific Northwest. In E. Pinkerton (ed.) *Co-operative Management of Local Fisheries: New Directions in Improved Management and Community Development.* Vancouver: University of British Columbia Press: 37-48.

de Young, B., Peterman, R. M., Dobell, A. R., Pinkerton, E., Breton, Y., Charles, A. T., Fogarty, M. J., Munro, G. R., Taggart, C. (1999) Canadian Marine Fisheries in a Changing and Uncertain World. *Canadian Special Publication of Fisheries and Aquatic Sciences* No. 129., 199p.

Durrenberger, E. Paul (1992) *It's All Politics. South Alabama's Seafood Industry.* Chicago. Un. of Illinois Press. 216p.

Ebbin, Syma (1998) Emerging Cooperative Institutions for Fisheries Management: Equity and Empowerment of Indigenous Peoples of Washington and Alaska. PhD. dissertation. Yale University. 662p.

Kofinas, Gary P. (1998) The Cost of Power Sharing: Community Involvement in Canadian Porcupine Caribou Co-Management. PhD. dissertation. Faculty of Graduate Studies. University of British Columbia. 471p.

Holm, Petter, Hersoug, Bjorn, and Ranes, Stein Arne (2000) Revisiting Lofoten: Co-Managing Fish Stocks or Fishing Space? *Human Organization* **59**(3), 353-364.

Jentoft, Svein (1993) *Dangling Lines. The Fisheries Crisis and the Future of Coastal Communities. The Norwegian Experience.* Institute of Social and Economic Research. St. John's, Newfoundland. 161p.

Jentoft, Svein and McCay, Bonnie (1995) User Participation in Fisheries Management: Lessons Drawn from International Experiences. *Marine Policy* **19**, 227-246.

Larkin, P. (1988) The Future of Fisheries Management: Managing the Fisherman. *Fisheries* **13**(1), 3-9.

McCay, Bonnie J. and Carolyn F. Creed (1999) Fish or Cut Bait: A Guide to the Federal Management System. 2nd, revised edition. Fort Hancock, NJ: New Jersey Marine Sciences Consortium. 20 pp.

McCay, Bonnie (2000) Sea Changes in Fisheries Policy: Contributions from Anthropology. pp. 201-217 In Durrenberger, E. Paul and King, Thomas D. (eds) *State and Community in Fisheries Management: Power, Policy, and Practice.* Westport, CT.: Greenwood Press.

Marchak, P., Guppy, N., McMullan, J. (eds) *Uncommon Property: the Fishing and Fish Processing Industries of British Columbia.* Vancouver: University of British Columbia Press.402p.

Matsuda, Y. and Kaneda, Y. (1984) The Seven Greatest Fisheries Incidents inJapan. In Kenneth Ruddell and Tomoya Akimichi, eds. *Maritime Institutions in the Western Pacific.* Osaka: National Museum of Ethnology, 159-182.

Mikalsen, Knut and Jentoft, Svein (2001) From User-Groups to Stakeholders? The Public Interest in Fisheries Management. *Marine Policy* **25**, 291-292.

NRC (1999a) Sustaining Marine Fisheries. Washington, D.C.: National Academy Press. (National Research Council [Committee on Ecosystem Management and Sustaining Marine Fisheries, Ocean Studies Board, Commission on Geosciences, Environment, and Resources]).

NRC (1999b) Sharing the Fish: Toward a National Policy on Individual Fishing Quotas. Washington, D.C.: National Academy Press. (National Research Council [Committee to Review Individual Fishing Quotas, Ocean Studies Board, Commission on Geosciences, Environment, and Resources]).

Oakerson, R. (1992) Analysing the Commons: A Framework. pp.41-59. In Bromley, Daniel, (ed.) *Making the Commons Work: Theory, Practice, and Policy*. San Francisco: ICS Press.

Pinkerton, E.W. and Weinstein, Martin (1995) *Fisheries That Work. Sustainability Through Community-Based Management*. David Suzuki Foundation. #219, 2211 West 4th Ave., Vancouver, B.C. V6K 4S2. 199pp.

Pinkerton, E.W. (1996) The Contribution of Watershed-Based Multi-Party Co-Management Agreements to Dispute Resolution: the Skeena Watershed Committee. *Environments* **23**(2), 51-68.

Pinkerton, E.W. (1994) Local Fisheries Co-Management: A Review of International Experiences and Their Implications for British Columbia Salmon Management. *Canadian Journal of Fisheries and Aquatic Sciences* **51**(10), 2363-2378.

Pinkerton, E.W. (1992) Translating Legal Rights into Management Practice: Overcoming Barriers to the Exercise of Co-Management. *Human Organization* **51**(4), 330-341.

Pinkerton, E.W. (1991) Locally Based Water Quality Planning: Contributions to Fish Habitat Protection. *Canadian Journal of Fisheries and Aquatic Sciences*. **48**(7), 1326-1333.

Pinkerton, E.W. and Nelson Keitlah (1990) The Point No Point Treaty Council: Innovations by an Inter-Tribal Fisheries Management Co-operative. U.B.C. Planning Paper DP#26. Vancouver: UBC School of Community and Regional Planning and Nuu-chah-nulth Tribal Council. 54pp.

Pinkerton, E. and Katherine Baril (2001) A Model for Community-State Co-Management of Fish Habitat Protection in the Coastal Zone. Presentation to the Conference on Integrated Coastal Zone Management, Universidad de Oriente, Santiago de Cuba, Cuba. May 14th.

Pinkerton, E.W. (1989b) Rural Resource Planning in Coastal British Columbia: Can Fishing Communities Plan the Future of Their Fisheries? *Plan Canada* **29**(2), 80-86.

Pinkerton, E.W. (1989a) Attaining Better Fisheries Management Through Co-Management: Prospects, Problems, and Propositions, in E. Pinkerton, ed. Co-operative Management of Local Fisheries: New Directions in Improved Management and Community Development. Vancouver: University of British Columbia Press:3-33.

Pinkerton, E.W. (1988) Co-operative Management of Local Fisheries, a Route to Development, in John W. Bennett and John Bowen, eds, Production and Autonomy: Anthropological Studies and Critiques of Development. Lanham, MD.: Society for Economic Anthropology, and University Press of America:257-273.

Pomeroy, Robert and Berkes, Fikret (1997) Two to Tango: the Role of Government in Fisheries Co-Management. *Marine Policy* **21**(5), 465-480.

Schlager, E. and Ostrom, E. (1993) Property Rights Regimes and Coastal Fisheries: An Empirical Analysis. pp. 13-41 In Anderson, Terry and Simmons, Randy (eds) *The Political Economy of Customs and Culture: Informal Solutions to the Commons Problem*. Rowman & Littlefield. Lanham, MD.

Schwimmer, Eric, Houle, Caroline, and Breton, Yvan (2000) La Coexistence Precaire de la Peche Mondialisee et de la Peche Coutumiere: le Cas des Maori de la Nouvelle-Zelande. Departement d'anthropologie, Universite Laval. Quebec.

Seiter, A, Newberry, L., and Edens, P. 2000. Cooperative Management of the Dungeness Watershed to Protect Salmon in Washington State. *Journal of the American Water Resources Association* **36**(6), 1211-1217.

Singleton, Sara (1998) *Constructing Cooperation: the Evolution of Institutions of Co-Management*. Ann Arbor: Un.of Michigan Press. 165p.

Stoffle, Brent W., David Halmo, Richard Stoffle, and Gaye Burpee (1994) Folk Management and Conservation Ethics Among Small-Scale Fishers of Buen Hombre, Dominican Republic.pp.115-138. In Christopher L. Dyer and James R. McGoodwin, (eds) *Folk Management in the World's Fisheries. Lessons for Modern Fisheries Management*. Niwot: University Press of Colorado.

Tyler, Stephen R. (1999) Policy Implications of Natural Resource Conflict Management.pp.263-280. In Buckles, Daniel, (ed.) *Cultivating Peace: Conflict and Collaboration in Natural Resource Management*. International Development Research Centre. Ottawa.

Wadel, Cato (1972) Capitalization and Ownership: the Persistence of Fishermen-Ownership in the Norwegian Herring Fishery. pp. 104-119 In Andersen, Raoul and Wadel, Cato (eds) *North Atlantic Fishermen: Anthropological Essays on Modern Fishing*. Institute of Social and Economic Research. St. John's, Newfoundland.

Yamamoto, Tadashi and Short, Kevin (1992) International Perspective on Fisheries Management: with special emphasis on Community-Based Management Systems Developed in Japan. National Federation of Fisheries Cooperative Associations (ZENGYOREN) Tokyo. 527p.

EXPERIENCES WITH FISHERIES CO-MANAGEMENT

Chapter 5

EXPERIENCES WITH FISHERIES CO-MANAGEMENT IN AFRICA

MAFANISO HARA
Programme for Land and Agrarian Studies, University of Western Cape, South Africa

JESPER RAAKJÆR NIELSEN
Institute for Fisheries Management and Coastal Community Development Hirtshals, Denmark

1. INTRODUCTION

Since the 1950s when the wave of de-colonization began to sweep through Africa, the continent has gone through several changing perspectives of 'development' approaches. Hyden (1993) characterises the discourse on development as having passed through four cycles; modernization (1955-65), dependency (1965-75), popular participation (1975-85) and enabling environment (1985-onwards). This moved the paradigm from development being measured purely in terms of economic stages of development to the wholesome term of human development (Hyden, 1993) specifically, development should imply improving the readiness and ability of societies to 'problematise' issues. In other words development becomes meaningful to people when they have a chance to wrestle with end/means relations in ways that are relevant to their own predicament. It was symptomatic of this shift that by the end of the 1980s terms such as *people centred development* (The World Commission on Environment and Development - WCED, 1987), *sustainable development* (Hyden, 1993) and *sustainable livelihoods* (Chambers and Conway, 1992) became increasingly common in development language[1].

Towards the end of the 1980s, several development co-operation agencies carried out evaluations of fisheries projects in order to draw lessons from the previous four decades of failures in fisheries projects (Spliethoff *et al.*, 1990). The main lesson was that the failures were attributable to interventions that did not take into account the special characteristics of capture fisheries, given the complexity of the fisheries sector, the special bio-ecological characteristics of fisheries and the prevalent low social and political status of fishing communities in terms of government priorities. The conclusion from this evaluation was

---

[1] In the early 1990s, the World Bank also adopted participatory approaches to project identification, formulation and implementation (World Bank, 1994). This followed the dismal failure of the Integrated Rural Development Programme, that it had launched in 1970, which was based on a top-down interventionist approach (Keare, 2001).

that 'there was need for more careful and comprehensive preparation involving wider consultation and active participation of beneficiaries, flexible and phased approaches with emphasis on the development of human resources' (Chambers and Conway, 1992: 2).

The end of the cold war era brought with it increasing demands for Africa to democratise and implement Structural Adjustment Programmes (SAP) from its traditional western donors, led by the World Bank and the International Monetary Fund (IMF). This also came at a time when there was a marked shift in resource management paradigm towards participatory approaches, devolution of authority and decentralization of powers. Thus by the early 1990s, *user participation* (and Participatory Rural Appraisal - PRA) had become almost a *given* requirement for donor funded development projects. Given this context, the international donor agencies pressured African countries to introduce co-management or at least establish more democratic processes in the formulation of fisheries management objectives and the decentralization of fisheries governance.

This chapter provides an overview of co-management in Africa and the historical, political and paradigmatic reasons for the shift. The historical context is important when analysing the performance of the regime. The main reasons why co-management is being increasingly adopted in Africa are explained by analysing the objectives hereof. The chapter evaluates what is meant by co-management in the African context using the variety of types of user involvement in practice and the standard continuum of possible arrangements under the co-management regime. Next, it will look at how co-management is being implemented, including whether it is achieving the objectives it is supposed to achieve. The final section will discuss and draw some lessons from the co-management experience on the continent.

The chapter draws on experiences from Southern Africa (Malawi, Mozambique, South Africa, Zambia and Zimbabwe) East Africa (Lake Victoria grouping Kenya, Tanzania and Uganda) and West Africa (Benin, Congo, Cote de Ivoire, Senegal, etc.) where co-management arrangements in fisheries have been or are in the process of being implemented[2]. In most of these countries, fisheries co-management is a relatively new approach that has only been formerly introduced in the last six to nine years. The comparative analysis of the cases at this early stage could give indications of what seems to be the critical issues in the planning and implementation of fisheries co-management arrangements in Africa.

2. REASONS FOR ADOPTING CO-MANAGEMENT

While the arguments and decisions leading to adoption of the co-management approach might vary in specific cases, the most common and powerful reason has been the failure of governments to effectively manage capture fisheries, resulting in over-exploitation (Hara *et al.*, 2002; Donda, 1998; Jackson *et al.*, 1998; Sowman *et al.*, 1998; Horemans and Jallow, 1998; Kponhassia and Konan, 1998; Kebe, 1998; Geheb and Crean, 2000). In other cases, governments view co-management as a way of controlling fishing effort by establishing

---

[2] The case studies from Southern Africa and West Africa are mainly drawn from volumes based on workshop proceedings of the Institute for Fisheries Management and Coastal Community Development (IFM)/International Centre for Living Aquatic Resources Management (ICLARM)/National Aquatic Resource Systems (NARS) Co-management Research Project by Viswanathan (ed.) (Forthcoming) and Normann *et al.* (eds), 1998. Those for East Africa are based on the experiences of the Lake Victoria Fisheries Research Project Phase II, Geheb and Crean (eds), 2000. In addition, specific analysis of emerging initiatives in South Africa, Hauck and Sowman, 2001 are included.

property rights for some groups in order to forestall future problems of over capacity. This is one of the major reasons on Lake Kariba (Hachongela *et al.*, 1998; Machena and Kwaramba, 1995), in Lake Nokoue in Benin (Atti-mama, 1998) and Lake Chiuta in Malawi (Donda, 2001). Co-management is also seen as a tool for conflict mediation among various stakeholders. This was the particular objective in Senegal (Kebe, 1998), on Lake Kariba (Hachongela *et al.*, 1998; Jackson *et al.*, 1998), the Oliphants River (Sowman *et al.*, 1998) and in Mozambique (Lopes *et al.*, 1998).

The implementation of Structural Adjustment Programmes (SAP) in most African countries[3] and changing economic development funding priorities in the donor community has made decentralization and delegation of management responsibilities a very attractive proposition. Thus, user-participation or co-management has become mainstream in management of fisheries in Africa.

It should be noted that decentralization and delegation of authority for the management of natural resources is not new in Africa. The British had tried it under an approach that was called 'indirect rule' (Mamdani, 1996) in their colonies in Africa[4]. A similar system of decentralization was also practised by the French in their African colonies based on the 'cercle' (Mamdani, 1996). These approaches were based on authority being vested in local chiefs who were empowered to make by-laws and collect local taxes as basis for control in their local areas of jurisdiction. In Mozambique, the Portuguese tried to use the traditional leaders, the *mwenes*, for their own political and economic interests. Their functions were primarily to collect licence fees, taxes and dues from fishing communities (Lopes *et al.*, 1998). Such approaches were contrary to what the colonialists practised in their own countries where centralised control was seen as the solution to the management of'common property' resources. On gaining independence, most Africa governments passed legislation putting the management of fisheries solely in the hands of government, mainly for reasons of wanting to have total political control (Hviding and Jul-Larsen, 1995). The new states felt that they needed all the sources of power they could muster. To leave the control of important natural resources in the hands of partly competing political institutions at the local level was considered unacceptable (Hviding and Jul-Larsen, 1995). The centralized approach was also well suited to the different political regimes that had succeeded colonial rule in Africa, whether they were socialist planned economies as in Mozambique, Tanzania, Zambia and Zimbabwe or the autocratic rule which for many years was the situation in Malawi and Kenya or under apartheid as practised in South Africa until 1994. In addition, the management problems and requirements for intervention at the time did not necessarily call for a consideration of the need for change in the management approach. Generally, the centralized approach was adopted in African countries after their independence because this approach was being applied by industrialized countries, whose influence continued after independence through financial and technical aid and the Western training of African fisheries managers. In reality though, the ability of governments for practical intervention never really matched the laws in the statutes due to budgetary constraints and the prevailing characteristics of African artisanal fisheries, which tend to be small-scale operations spread out through the coastal areas, whereby fishers usually land

---

[3] South Africa is the only country of the countries from which these case studies are drawn that is not officially implementing SAP programmes.

[4] Hara (2001) and Malasha (2001) elaborate attempts at applying 'indirect rule' in the specific cases of fisheries administration in Malawi (then called Nyasaland) and Zambia (then called Northern Rhodesia) respectively by the British colonial governments.

their catch in their home villages or constantly migrate following better catches.

The arguments for 'indirect rule' under British or French colonial era were similar to those currently being used for decentralization and devolvement of authority; Since the central government is far removed from the day-to-day aspects concerned with the exploitation of fish resources, its management interventions have usually been ineffectual. Given this context, it would be best to decentralise and devolve authority to the lowest possible level.

3. OBJECTIVES FOR CO-MANAGEMENT

In most cases the primary stakeholders (the users and the state) in a co-management arrangement share a broad common goal, ie, the recovery and/or sustainable exploitation of the fishery in order to enhance the social and economic benefits of the user communities. This commonality of objective may or may not be reflected in the way co-management is used as a strategy for achieving the goal.

The government objectives are usually ensuring biological sustainability of the resource and maintaining biodiversity (Hauck and Sowman, 2001; Hara, 2001; Geheb and Crean, 2000; Machena and Kwaramba, 1995; Hachongela *et al.*, 1998). Governments see co-management as an alternative strategy for controlling fishers in the wake of evidence of mounting problems under the centrally directed regimes. Thus the same existing regulations (technical regulations such as mesh size, net length and closed seasons) are reformulated under the *central* direction of the government, only this time with the supposed participation and support of fishers under more democratic and transparent arrangements. The content and structure of the regulations remains largely unchanged. Inputs from fishers, especially those that might seem contrary to government's conservation objectives, are usually ignored. Governments hope that the communities will take responsibility for the enforcement of regulations in their areas, a task that government itself has been unable to fulfil adequately. Thus while the decision making powers largely remain with governments, communities are being expected to implement the decisions with greater zeal.

Whereas most rural fishing communities formerly fished for subsistence, the integration of rural communities into the market economies has shifted fisheries exploitation into the realm of the profit motive. The organization of fishing operations, ie, who is fishing and what their objectives are for fishing, has become more diffuse and complex. Usually, a fishing unit owner employs crew members or assistants without active participation in the actual fishing himself. As a result, the unit owner has little influence on the decisions at the operational level out on the fishing grounds. For most of the crew members or assistants, fishing is usually an economic safety net, enabling them to make a daily living. For fishing communities in Africa in general, the short-term economic objectives tend to be higher than long-term (government) conservation objectives because of the dire economic conditions in rural areas.

From the early 1990s, donors have demanded political democracy and transparency as essential conditions for development aid. In addition, donor projects in the resource management sector have drawn inspiration from some precepts of the World Bank Structural Adjustment Programmes, including the notion that the central state should play a reduced role in directing and managing economic activity (Lawry, 1994). The assumption is that local institutions are more accountable and function at a level where self-interest and responsibility for sustainable resource management are greater. Co-management is expected to improve the efficacy of fisheries management because it is assumed that acceptance of management measures will be higher if fishing communities have been involved in the decision making process. The contents of management measures will be better reflected if users' knowledge has been included in their formulation (Jentoft, 1994). Thus user-participation in resource management has increasingly been one of the conditions for donor aid as it is believed that this will result in greater accountability for management decisions. It is also seen as part of the general drive towards empowering the formerly disenfranchised populace[5]. The donors want the best possible use of their funds that would result in positive social and economic outcomes for the target communities. Donors seem to believe that the subsidiarity principle being commonly applied in the west should also be applied in developing countries and that political empowerment of user communities in the resource management process would result in improved resource management and thus positive economic effects on user communities.

While these enforced moves towards democracy and greater participation for the local communities are laudable, they have, in most instances, caught government officials and user communities unprepared. The former have had to accept user participation to supplement their cash strapped budgets while the latter are suspicious as to why governments are turning around and telling them that they have to participate in the management process, at times to the extent of being forcibly co-opted.

4. WHAT PASSES FOR CO-MANAGEMENT IN AFRICA?

There are quite a number of forms of user involvement in the management of fisheries in Africa. Whether all these should be called co-management, is a matter for debate. In South Africa, it is enshrined in the new Constitution of 1996 and in other policies and legislation concerned with natural resource management that people must be involved in decisions concerning their lives and natural resources (Hauck and Sowman, 2001). The revision of apartheid era laws and regulations thus goes through formal political processes aimed at making provision for inputs from stakeholders and the public. During apartheid, the commercial industry in South Africa had established arrangements with government that involved the exchange of information (Hutton and Pitcher, 1998). Formal structures were created by the Sea Fishery Act of 1988 that stated that the minister could recognise any industry body or interest group to advise and make recommendations to government (Hauck and Sowman, 2001). A formal structure that has been put in place following the new Act of 1998 is the Consultative Advisory Forum, which is supposed to provide management advice directly to the minister. This forum is comprised of invited members from industry, the public and outside research institutions.

In some countries that adopted socialism at independence such as Mozambique and Tanzania, co-operatives and villagization were adopted as official policy respectively. Engagement with communities for the management of natural resources seems to have started with these policies. On the Zambian side of Lake Kariba, the government's approach to limiting entry into the fishery and engaging with fishers started with the moving of fishers into centralised villages. In Zimbabwe, co-management is based on the Communal Areas Management Programme for Indigenous Resources (CAMPFIRE) principles that

---

[5] According to Lawry (1994), Western donors argue that better resource management will result from policies that extend clearer property rights to users and give greater authority to local institutions, believed to be more accountable to the public.

give 'appropriate authority' to Rural District Councils. In most of the countries in Africa, though, modern era user participation has its basis in a new philosophy of sustainable rural development through the extraction and exploitation of 'their' natural renewable resources (Derman, 1995; WECD, 1987). The problem is for governments to ensure that this is achieved without endangering the biological sustainability of the resources. Most countries have revised their Fisheries Acts in order to provide for enabling legislation that would make provision for decentralization and devolution of powers.

In almost all co-management arrangements in Africa[6], governments have the legal authority, and have voluntarily decided to devolve some aspects of management to fishing communities according to very discrete agreements concerning specific functions. This type of approach is the top-down, vertical 'contracting out' model as described by Berkes (1997), Sen and Raakjær Nielsen (1996); Pomeroy (1995); and Berkes *et al.* (1991)[7]. This model views power as moving from the state to communities. With the exception of a few cases such as Lake Chiuta in Malawi (Hara *et al.*, 2002) and some inland lakes in West Africa (Horemans *et al.*, 1998), most African co-management arrangements are heavily top-down (Sen and Raakjær Nielsen, 1996).

5. HOW HAS CO-MANAGEMENT BEEN IMPLEMENTED?

Introduction of co-management arrangements in Africa have to a large extent been initiated as donor funded projects[8], with the assumption that national departments could take over the activities after the end of the projects. This has not been without its problems, because in many cases the assistance has been short-term and the process has lacked flexibility because of specific donor requirements.

In most cases, organization of user communities has been the first step of the implementation process. Government or NGOs have initiated or facilitated the mobilization of the communities for user involvement in the new management regime. New institutions are created, with the facilitation of these external agencies, usually using the western democratic principles of electing committees as vehicles for participation. These community organizations in co-management arrangements have different names in different countries. In Malawi, they are called Beach Village Committees, in Zambia Fishing Village Management Committees, in Tanzania Beach Management Units or Management Committees while in Zimbabwe these are referred to as Management Committees for Exclusive Fishing Zones (based on the specially demarcated areas of the inshore fishery by

[6] These arrangements primarily involve small-scale fisheries. Except for the formal co-management in the South African hake fishery (Hutton and Pitcher (1998)), we have not come across any well documented examples of co-management in the industrial fisheries, although the authors are aware that consultative arrangements are in place in some countries, eg the shrimp fishery in Mozambique.

[7] Berkes (1997) points out that all these are variations of Arnstein's (1969) 'ladder of public participation'.

[8] The Lake Malombe project in Malawi had been funded from several sources namely the German Technical Foundation (GTZ), the United Nations Development Programme (UNDP), British Overseas Development Administration and a World Bank loan. On Lake Kariba the co-management project came under the Zambia/Zimbabwe SADC Fisheries Project funded by the Norwegian Agency for Development Cooperation (NORAD) and the Danish International Development Agency (DANIDA). The Lake Victoria Fisheries Research Project was funded by the European Development Fund. In Mozambique the implementation of co-management in Moma/Angoche districts was part of Nampula Artisanal Fisheries Project funded by the International Fund for Agricultural Development (IFAD). In West Africa, the initiatives came under the Integrated Development of Artisanal Fisheries in West Africa (IDAF) funded by DANIDA. Hauck and Sowman (2001) point out that in South Africa most of the initiatives had been funded by external agencies.

government). Although new institutional structures have been created, the local institutional structures based on traditional authority and customary law are in most cases still the basis for organization of the fishing communities. A few exceptions can be found. In Zimbabwe, the government sought to dilute the influence of traditional authority based on the principles of scientific socialism. In this context, the traditional authority (chiefs) have been ignored or declared improper for the management task. In the case of the Oliphants River project in South Africa, the 'coloured' community in question do not have traditional authority structures.

Hauck and Sowman (2001) point out that in South Africa most of the initiatives have been driven by external agents outside the responsible government department and that thus far there has been a lack of buy-in into the idea of co-management by government. This lack of total acceptance of communities as equal partners (or slow warming to the idea of co-management) by government fisheries management departments also applies to most of the other countries.

In general, the fact that community structures had been, in most cases, organized by or through the influence of agents from outside the communities has had a large bearing on the ownership of the process and the acceptability of the new committees meant as representative bodies for communities in a co-management arrangement. In turn, the stability of the elected co-management structures has had great influence on the sustainability of the new regime.

6. LESSONS FROM THE AFRICAN EXPERIENCE

According to records, most formal co-management arrangements on the continent have been going on for less than ten years. Most of them have been introduced and driven by governments and remain largely top-down. All the same, it is possible to start drawing some lessons concerning the African experiences with co-management arrangements meant to govern fisheries exploitation. In the discussion of lessons, focus will be given to objectives for co-management; issues related to participation and representation; the role of traditional authority and the institutional capacity for co-management.

6.1. Conflicting objectives: conservation vs economic subsistence

In most of the African cases we have investigated, co-management seems to be a government driven initiative with underlying conservation objectives. Governments generally perceive co-management as an alternative strategy to pursue the same old conservation objectives. The government approach is instrumental based on co-opting users into the process, without really relinquishing management authority and control of decision-making power. Governments hope to achieve better outcomes in terms of sustainable patterns of fisheries exploitation by changing the time preference for exploitation of fish resources within communities from short to long term through user participation in the management process. The expectations within communities are likely to be contradictory those of the government.

Most people in fishing communities focus on economic objectives. In most cases, the effects of poverty and lack of alternative economic opportunities for the communities due to the overall macro-economic situation in Africa drives up their rate of time preference to the point where only consumption today matters. The question then becomes whether co-management alone can change their economic strategies. This is one of the main problems on Lake Malombe (Hara *et al.*, 2002) and Lake Victoria (Geheb and Crean, 2000).

This highlights the fact that achieving sustainable exploitation of the fisheries in most water bodies is likely to be dependant on the broadening of economic opportunities and general economic development in concerned rural communities. The point is that as populations grow and economic needs increase, pressure on fisheries is bound to increase from the horizontal growth of effort unless alternative outlets in terms of employment outside the fishery can be created. In this sense, one cannot separate fisheries management issues from economic development issues.

Jul-Larsen *et al.* (2002) have emphasized the need for caution about the general assumptions that there is need to control effort in some small and medium fresh water lakes in the SADC region, because in the case of Lakes Chirwa, Mweru and Kariba, there are no clear links between the level of fishing effort and the biological resilience of the resource and ecological systems. There is very little biological evidence to support the need for control of fishing effort, so the use of co-management to meet conservation concerns might be based on a misconception on the part of government.

6.2. Property rights, exclusion and limited access

From a government perspective co-management has commonly been seen as a vehicle for establishing or ensuring property rights for local fishing communities. Granting property rights has been one of the major reasons for adopting co-management for many fishing communities. The underlying perception is that better resource management can be promoted through policies that give the communities stronger incentives for sound resource use through reforms that extend clear property rights to them. Having acquired exclusive rights, it is expected that fishing communities would limit access and fishing effort. Whereas in the past, community members had been free to enter and leave the fishery, limited entry systems identify specific groups of individuals that should have exclusive rights to the resource with the ability to keep others out. In the face of shrinking job markets, dependency on natural resources to meet basic needs is increasing. This increasing dependency on natural resources has lead to a decline in the value and strength of local informal controls that might have existed, leading to increasing tendencies towards open access to natural resources. This also probably explains why in some instances fishing communities have been reluctant to accept proposals to introduce limited entry and limited access due to implications of these as forms of privatization of a common pool resource in which everyone has been free to fish. Secondly, fishing communities have felt that such proposals risk them being excluded from other areas to which they seasonally migrate during closed seasons as other communities give them tit-for-tat (Hara *et al.*, 2002; Geheb and Crean, 2000 and Horemans, 1998). In this context, propositions aimed at limiting entry or access have been difficult to implement. In the words of Brox (1992:232), 'the Common Property Theory exposes the tragic potential of natural resources being free and accessible to all, but it easily prevents one from seeing that commons involve opportunities which are far from being tragic for the people involved, but rather necessary for the maintenance of local communities. Of special importance are the opportunities that common resources like fish offer to anybody looking for subsistence or resources that can be turned into cash'. Brox's observation reflects the situation in Africa and is supported by Jul-Larsen *et al.* (2002) in their study of small and medium freshwater lakes in southern Africa.

A link between fisheries and the general economy and the fisheries ability to act as a sink for excess labour from the fishing community is becoming clear. As long as employment opportunities in the other sectors of the economy remain low[9] and fisheries continue to act as the major economic sector in rural areas, limiting entry to fisheries is going to be difficult if not impossible. This is also a warning that any new management regimes, whether co-management or any other type, that bring any unpalatable changes to the existing social and economic order of the fishing communities, might be impossible to implement even if they reflect the governments' conservation objectives. In any case, official government worries about limiting effort as a management objective might be misconstrued in some of the small water bodies in the SADC region as Jul-Larsen *et al.* (2002) have demonstrated. The authors show that in some of these water bodies there might be no reason to control effort and that the importance of such resources lay in their ability and resilience to act as a buffer for absorbing the social and economic needs of constantly changing local macro-economic situations of the dependent communities without endangering the productivity of the resource in question.

6.3. Level of participation

The practical adaptation of co-management in Africa has been to involve fishing communities in the implementation process - a rather instrumental approach to co-management. In a few cases, fishing communities/user groups have been consulted in the decision-making process concerning determination of technical regulations such as gear type restrictions (minimum mesh size and maximum length of seine or gill nets), closed seasons and protected areas. However, the knowledge base for these management decisions has been 'scientific knowledge' provided by government departments. The justification or reasoning behind such knowledge is seldom clearly explained to fishers. In addition the scientific rationale for the applied technical regulations remain questionable (Jul-Larsen *et al.*, 2002). Thus it is clear that government officials from fisheries agencies generally do not perceive co-management as a means of introducing more democratic principles into fisheries management, but as an instrument for achieving their own management objectives more efficiently through involving communities in the implementation and monitoring activities. Governments have generally not been prepared to include the setting of management objectives as part of the co-management process and the determination of what knowledge from fisher communities should be included in the management decisions. In the fishers' experience, the government always sets the rules and regulations and has the responsibility for enforcement. Thus most co-management arrangements in Africa exclude fishing communities from the collective choice rules (Ostrom, 1990) concerning who should participate in making the operational rules, which the co-management arrangements were supposed to be about (Sverdrup-Jensen and Raakjær Nielsen, 1998). Unless users are genuinely allowed and empowered to participate in the setting of management objectives on equal terms with government, co-management cannot really be considered as a serious institutional innovation. If co-management is put forward as a process of empowerment and self-determination, communities might believe that it will lead to powers of decision making, including decisions that might be contrary to those of government.

Control and law enforcement continues to be mainly undertaken by government departments. In some cases such as in Malawi, one of the arguments for co-management

---

[9] In most rural areas of Africa, the levels of formal employment remain low. Usually, people rely on seasonal Agriculture for food and income. The rate of unemployment in such areas is extremely high, usually in excess of 50%.

was to make the BVCs responsible for enforcing operational rules in conjunction with the Department of Fisheries. Apart from governments' willingness to devolve responsibility for this type of task, another problem has been the capacity and willingness on the part of local communities to undertake such tasks.

Irrespective of the type of political system that has been in place in the past, fishers' trust in government authorities has always been moderate at best. Fishers have hardly ever found themselves at the winning end of relationships with government. Wherever initiatives to establish co-management have been taken by government authorities, these have been met with profound scepticism from fishers who, with good reason are suspicious of the motives and sincerity of government authorities when they propose collaboration and the sharing of management responsibilities. Unless governments show willingness to build trust and bring about the democratization of the processes of fisheries governance, the incentive to give collaborative management arrangements with government a try is likely to decline.

Although some governments are genuinely trying to introduce real institutional changes, one still finds that in most instances fishing communities are not legally empowered and their negotiating position versus that of governments is still comparatively weak. As Chirwa (1998:69) points out *'The local user communities are the recipients rather than the initiators of decisions. They, themselves, are managed, together with their resources'*. This statement seems to be applicable to most of the examples of co-management from Africa and serves to emphasize the need for enabling legislation regarding co-management or operationalizing such legislation where it exists, in order to empower user groups. Governments generally seem reluctant to devolve power and bestow legal rights and authority for fisheries management to user groups. Devolution of management authority is obviously a sensitive issue for most governments and one that does not seem to be easily resolved, as it requires changes in laws, policies and administrative procedures, a process that can be both cumbersome and time-consuming.

As explained in chapter 11, the scale issue is crucial, and not all problems and issues can be solved at the local level. The African experiences support this argument. The scale issue is somewhat easier to deal with in small freshwater lake fisheries such as in lakes Nokoue, Mweru, and Chiuta, whereas it is more complicated in the large freshwater lakes like Malawi and Victoria or marine fisheries, eg the shrimp fishery in Mozambique or the pelagic fishery in South Africa. The latter fisheries are good examples of resource systems being too large to be controlled entirely by a few communities. Fisheries (co)-management institutions must therefore be able to address problems of resource access and distribution above local level. One of the solutions to this problem of scale is representation (Mikalsen, 1998 and Chapter 16). Representation, however, raises a new set of problems related to the mechanisms that could ensure its genuineness in order to avoid alienation between communities and management. The fact that the co-management institutions (local committees) exist within nested systems compounds the problem. The African examples clearly indicate, that local communities have difficulties in getting their views articulated through the different layers of the decision-making hierarchy in larger scale environments. In addition the local communities are as in the case of the Mozambican shrimp fishery and the pelagic fishery in South Africa, up against very well organized associations representing larger industrial companies. Thus local communities are often at the losing end, when decisions are taken.

6.4. The role of Traditional Authority

As explained above, African co-management institutions have generally been established at the local level and most often been closely linked to existing traditional power structures. Traditional power systems in Africa play a very prominent role in relation to resource exploitation. In many African fisheries co-management arrangements, traditional authorities serve as the link between the government and communities/user groups. As a consequence, co-management arrangements are in many cases rooted in traditional customs, practices and beliefs, religious institutions and myths that influence the decision-making process.

It can be argued that the heavy involvement of traditional authorities in co-management arrangements adds a twist to an arrangement that, in the strict Western understanding, should be between direct users and government. On the other hand, such tripartite arrangements build upon and involve institutions that are considered legitimate by fishing communities. To use Weberian terminology (Selznick, 1992), the legitimacy of this type of co-management arrangement is likely to be based on a combination of traditional and charismatic authority. The sustainability of such an arrangement will, to a large extent, rely on the personality of the chief and the regard of his or her subjects. In order to make co-management arrangements able to apply sanctions on their own at local level, they often will need the civil and customary power structures that reside in traditional authorities. In any case, traditional leaders and elders have historically and are still highly respected in rural areas.

The strong link between traditional authorities and co-management in Africa is not without its problems. Empowering user groups has created tensions between traditional authorities and the co-management institutions in their areas (Hara *et al.*, 2002; Jackson *et al.,* 1998). In addition, traditional authorities view the process of democratization with suspicion because it may undermine their authority. This will have direct impact on the resiliency of the co-management institutions. Co-management initiatives may change perceptions within the fishing community and the government about what are considered to be legitimate management institutions. A major challenge for co-management in Africa is how to include the traditional institutions. It is obvious that they cannot remain in the driver's seat if co-management is concerned with empowerment of communities. At the same time, they are also too important to exclude.

6.5. Capacity of local communities and governments for co-management

Not all tasks inherent in a co-management arrangement, can be undertaken by community organizations in all situations. In many cases fishing communities may be neither willing nor capable of taking on particular fisheries management responsibilities. Furthermore, not all elements of fisheries management can, or should, be allocated to the communities due to the fact that the incentive(s) - economic, social and/or political - to undertake major fisheries management responsibilities may not be present within the community. In cases such as Lake Malombe, the incentives were initially monetary. Since the financial incentives could not be sustained because they were being drawn from short-term donor project funds, willingness to accept responsibilities on the committees declined as the donor projects came to an end. Another problem is related to the low levels of education and the poor organization among fishing communities.

Capacity building in the communities and self-sustenance has been found to be important when it comes to successful introduction of co-management. One of the major problems with the implementation of co-management in Africa is that in most cases, this is being done as donor funded projects of limited duration. Where government cannot take over the funding of the activities, it has been common to observe that the activities are collapsing, leading to serious problems in the long-term success of the regime. The solution to this is long-term commitment of human and financial resources to change the management regime. It is crucial that key-persons or resources are not withdrawn before institutions have matured and the regime has taken root. Where this has happened such as in the Lake Malombe and Oliphants River, it has been observed that the arrangement was becoming increasingly under strain and appears to be moving towards collapse. The lack of capabilities and/or aspirations among fishing communities to participate in the fisheries management process might also explain the lack of participation of true fishers in the decision-making process in some of the cases studied. As stated by Pinkerton (1989), strong local institutions with human and financial capacity are a pre-condition for co-management. In most small-scale fisheries in Africa, such self-reliant institutions for management purposes are difficult to find or create.

Support for co-management initiatives from governments is inadequate. One reason for this might be that there has not been real adaptation in the organizational structures of government departments to cope with the change in concept and philosophy from centralised management towards co-management. The shift to co-management from centralised regimes creates new demands on government with respect to the type of support that they need to provide to fishing communities. The situation is complicated due to the fact that most departments were built and geared around attempting to ensure biological sustainability of fisheries resources. Their duties have mainly been seen as biological research to come up with regulations, top-down delivery of extension messages and enforcement of regulations. As emphasized by Donda (2001), it is a problem that government departments and their fisheries research and extension have not been reorganized as part of the process. Fisheries departments in Africa have usually recruited people from natural sciences, with no or very few economists or social scientists. Although the need for new type of skills is widely recognised, very little has been done to broaden the skills base of fisheries departments.

One of the reasons for this is that government departments are being asked to downsize under SAPs. This means that it has not been easy to create new positions that would accommodate the requisite skills and qualifications. It has thus not been easy to re-orientate departments especially when such dramatic change means re-organization, change in management philosophy, new ways of interacting with fishing communities and the possible threat to the jobs for the existing staff. It therefore means that there is usually no clear link between the organizational structure of the responsible departments and resources (human and financial) at the their disposal to match the requirements for the facilitation and implementation of the new regimes.

7. THE CHALLENGE FOR THE FUTURE OF FISHERIES CO-MANAGEMENT IN AFRICA

The implementation of co-management in Africa does not seem to differ from what is going on in other parts of the world. Institutions regulating access to fisheries have in many - if not most - cases been in place long before modern concepts of fisheries management were developed (Hviding and Jul-Larsen, 1995). The problems that African fishing communities are facing are therefore not necessarily a result of an absence of management institutions, but rather the result of the inadequacy of these institutions to deal with the evolving social, economic and political systems and situations. Revitalization of such existing institutions might not therefore necessarily lead to solutions to the problems. The institutions may have lost their significance exactly because they were set up to solve other problems and are inadequate to deal with the present situation. The basic challenge to governance of African fisheries seems to be how to establish and maintain co-management institutions (norms and rules guiding decisions including a formal framework for decision making) that will enable both government and communities to deal with the new challenges and complexities that come with responsibilities for sustainable exploitation of their fisheries resources nested in modern economic systems increasingly influenced by the effects of SAP and globalization.

Modern approaches to fisheries management, including some varieties of co-management, have developed within the modern rationality of industrialised societies. This approach is inherently unable to address the present problems of fishing communities in general and Africa in particular due to the ways objectives are defined and limitations in the knowledge on which they are based. The overriding problem is the differences in objectives for fisheries management between government and fishing communities, and the fact that this aspect is seldom addressed. What we observe in Africa is the instrumental type of co-management approach, which does not differ significantly from the centralised management approach. If co-management in an African context continues to be instrumental as is likely, co-management might damage relations between government and fishing communities because the frustrations among the latter are likely to increase due to the lack of genuine empowerment and government support as expected at the local level. We have observed the first signs of collapse of what was considered as a successful co-management arrangement such as that on Lake Malombe in Malawi according to Donda (2001) and Hara (2001) because when co-management was launched, it created huge expectations of genuine participation and empowerment. In practice though, the adoption of co-management seems to be turning out to be business as usual for governments as no real institutional reforms are adopted and carried through.

If co-management in Africa moves towards genuine participation and empowerment it is likely to challenge the objectives of mainstream international conventions for fisheries management such as the 'Code of Conduct for Responsible Fishing' or the 'precautionary approach for fisheries management'. These conventions attempt to reduce uncertainties with respect to stock estimation and fisheries management so as to distribute and spread the risk of fish stocks collapsing among fisheries managers. The problem, though, is how to take real cognisance of the needs or wishes of fishing communities. In this sense, the approach taken at these international conventions is very similar to the conventional fisheries management approach that is increasingly being discredited. In this context, the international decision-making arenas might have under-estimated the limitations of these conventions when it comes to how they are to be related to the intended empowerment of fishing communities through co-management. In an African context and many places elsewhere, a balance between conservation, social and economic concerns has to be struck and if they are not struck, these conventions will be more harmful than helpful. Co-management that results in empowerment can actually facilitate this process because governments are likely to be confronted with and forced to address the impact of international conventions on the livelihoods of fishing communities when the communities participate in the definition of management objectives for specific countries, locations or fisheries.

This experience could assist national governments to present and balance various objectives when they participate in forums where decisions on such international conventions and agreements are made. The empowerment of fishing communities through co-management should be seen as a way of giving fishing communities a chance to influence their own future and to cope with the impacts from phenomena such as globalization, competing use of coastal environments and other fisheries related issues. Empowerment through the co-management approach is likely - to a higher degree than previous or present management approaches - to influence positively the overall management objectives, even in cases where these objectives might differ from those set by their governments.

An important lesson to be drawn from implementation of co-management in Africa so far is that co-management should not be seen as a question of 'constructing' local institutions that can co-operate with government authorities in managing fish resources. That kind of approach was what had been attempted by the colonial powers under 'indirect rule'. Co-management ideally implies a process of mutual adaptation for both government and fishing communities for the co-ordination of the existing access regulating mechanisms (and the underlying mutual interests they represent) through some form of coherent mediating system. Any management regime is political in the sense that it is bound to include some and exclude others from access to valuable resources. This points to the practical dilemma, discussed above, that commodification of the fisheries intensifies uncertainties over who is and who is not a fisher, and hence over the related question of who should and should not be included in co-management arrangements. The lack of security of tenure for the crew members means that there is great fluidity in movement in and out of crews and fishing units. Some people participate in fishing only during particular times of the year such as when farming is out of season. An added twist is the relationship between gear owners and crew members and traders who might provide them with capital or loan the fishers money during times of stress. In such complex organizational webs, who is fisher and their level of influence on decisions and fishing activities they have, is never straight forward.

What is clear is that in the African context, most people within fishing communities or in the vicinity of fish resources, are potential fishers. If they are not fishing at a specific given time, it might simply be that they have better opportunities elsewhere. The cases we have reviewed show that there are good reasons to maintain things this way. Co-management systems have had a tendency to define fishers on the basis of who is present at a specific moment and a specific place in question. This is not likely to be a very helpful definition or starting point for analysis. Malasha (2001) shows that when various groups of the local population on the Zambian part of Lake Kariba were able to establish an alliance of interests to effectively stop newcomers entering the fishery, it happened as an outcome of a government initiative for co-management. Since then the number of fishers has fallen dramatically, which might mean that probably a large amount of people have lost a valuable opportunity to make a living. Chapter 11 contains a detailed discussion of the problem of this kind of boundary creation in co-management arrangements.

8. LESSONS LEARNED

Co-management has been applied and implemented instrumentally by governments. This has created a situation whereby the process, in contrast to the intention, has not lead to the empowerment of the local fishing population. The incentives for co-operation are primarily on the side of government, where fishers/fishing communities have realised that they continue to be recipients of instructions rather than equal partners in the decision-making process. A good example is that government departments and their fisheries research extension services have not been reorganized and the inputs to be accommodated by government have not been changed as a result of the changes in emphasis towards a more social and economic fisheries management (Donda, 2001).

In an African context, the control of people seems to have been a more important concern than considerations for the resources. Co-management has in many cases been going back to the roots and at most, differs only slightly from what the colonial powers attempted under the label of 'indirect rule'.

Based on the past experiences and the present challenges for African fisheries management, it is generally accepted by fisheries managers that institutional reforms in the governance structures for fisheries management are required and the co-management is considered the way forward. The major challenge for fisheries co-management in Africa is to improve the knowledge base for management and we argue that participation of the fishing communities is a precondition for this. Communities are also important in the quest to implement reliable monitoring systems. Based on the experiences from Africa it is clear that the way forward on this continent is away from the classical management paradigm towards a more pragmatic and adaptive approach to management. If co-management remains advanced 'social engineering', as many donor agencies and NGOs tend to favour, without involving and excluding the direct users in the process, it is then an illusion. Co-management needs to be a mutual adaptation that tries to establish a convergence between government policies and the local institutional structures. We are aware given the institutional landscape in Africa that this not going to be an easy task, but rather a long-term process with a lot of 'muddling through'. Based on African experiences, we argue that co-management in the region needs to be a learning process for each specific fishery, with specific tailor-made organizational structures according to each case in question.

Finally, it is important to emphasize that it will be a troublesome exercise, without any guarantee for success. It is, however, the only realistic alternative route to proceed in order to improve the efficacy of fisheries management in Africa.

Empowering user groups/fishing communities raises the question of whether these groups can act as independent participators, their organization having been facilitated by government. In addition, to what extent are government agencies shirking or abrogating responsibilities in the delegation of some powers - but not others - to user groups. In Africa co-management and user participation are also seen as part of the introduction of political pluralism. The question then becomes whether the real objective is to enforce participatory democracy or to achieve better fisheries management.

What should be the objectives of co-management? Do they include Institutional capacity building? Can co-management achieve the objectives of all players, given that these might not be always exactly the same and may often contradict one another? How can the balance between government conservation objectives and the social and economic needs of fishing communities be achieved? What is understood by *sustainable exploitation* by the various role players? These are some of the questions that implementation of co-management in Africa raises and governments, donors and researchers must wrestle with. Until we start finding bold answers to them, co-management appears more as an illusion than as the empowerment of fishing communities.

REFERENCES

Arnstein, S. (1969) A Ladder of citizen participation. *Journal of American Institute of Planners* 4, 216-224.

Atti-Mama, C. (1998) Trends in the management of continental fisheries in Benin: The case of Lake Nokoue. in Normann, A.K. *et al.*, 1998 pp255-272.

Berkes, F. (1997) New and not-so-new directions in the use of the commons: co-management. *The Common Property Resource Digest* 42. Indiana: International Association for the Study of Common Property pp5-7.

Berkes, F., George, P. and Preston, R.J. (1991) Co-management. Alternatives **18**(2), 12-18.

Brox, O. (1990) The Common Property Theory: Epistemological Status and Analytical Utility. Human Organization, Vol. **49**, No. 3 pp227-235.

Chambers, R. and Conway, G. (1992) Sustainable rural livelihoods: practical concepts for the 21st century, IDS Discussion Paper 296. Brighton:IDS.

Chirwa, W. C. (1998) The lake Malombe and Upper Shire River Fisheries co-management programme: An assessment. In Normann *et al.*, 1998, pp 61-77.

Derman, (1995) Environmental NGOs, Dispossession, and the State: The Ideology and Praxis of African Nature and Development. Human Ecology. Vol. **23**. No.2 pp199-215.

Donda, S.J. (1998) Fisheries Co-management in Malawi: Case Study of Lake Chiuta fisheries. In Normann *et al.*, 1998, pp 21-39.

Donda, S. J. (2001) Theoretical Advancement and Institutional Analysis of Fisheries Co-management in Malawi: experiences from Lakes Malombe and Chiuta. Unpublished PhD Thesis. Aalborg, Aalborg University.

Geheb, K. and Crean, K. (2000) The Co-management survey report: an executive summary. In Geheb, K. and K. Crean (eds). The Co-management Survey: Co-managerial perspectives for Lake Victoria's fisheries. Lake Victoria Fisheries Research project Technical Document No.11. LVFRP/TECH/00/11. The Socio-economic Data Working Group of Lake Victoria Fisheries Research Project, Jinja, pp 1-10.

Hachongela, P. J. Jackson, Malasha, I. and Sen, S. (1998) Analysis of emerging co-management arrangements in Zambian inshore fisheries of Lake Kariba in Normann, A.K. *et al.*, 1998 pp79-98.

Hara, M.M. (2001) Could co-management provide a solution to the problems of artisanal fisheries management on the Southeast Arm of Lake Malawi? (PhD thesis). University of the Western Cape. Cape Town. South Africa.

Hara, M. (1998) Problems of introducing community participation in fisheries management: Lessons from the Lake Malombe and Upper Shire River (Malawi) Participatory Fisheries Management Programme in Normann *et al.*, 1998 pp41-60.

Hara, M.M., S.J. Donda and F.J. Njaya (2002) Lessons from Malawi's Experience with fisheries co-management initiatives In Geheb, K. and Sarch, M.-T. 2002. Africa's Inland Fisheries: The Management Challenge. Kampala, Foutain Publishers.

Hauck and Sowman (2001) Coastal and Fisheries co-management in South Africa: An Overview and Analysis. *Marine Policy* **25**, 171-185.

Horemans, B. and A. Jallow (1998) Current state and perspectives of marine fisheries resources co-management in West Africa in Normann *et al.*, 1998 pp233-254.

Hutton and Pitcher (1998) Current directions in fisheries management policy: A perspective on co-management and its application to South African Fisheries. Benguela Dynamics. In Pillar, S.C., C.L. Moloney, A.I.L. Payne and F.A. Shillington (eds.). *South African Journal of Marine Science*. No.19. Pp471-486.

Hviding, E. and Jul-Larsen, E. (1995) Community Resource Management in Tropical Fisheries. (1st edition). University of Namibia. Windhoek.

Hyden, G. (1993) Shifting perspectives on development: Implications for research. In Masst, M, T. Hylland Eriksen and J.Helle-Valle (eds): State and locality . Proceedings of the NFU Annual Conference 1993. Centre for Development studies, Univeristy of Oslo, Oslo pp21-40.

Jackson, J., Muriritirwa, W., Nyakahadzoi, K. and Sen, S. (1998) Analysis of emerging co-management arrangements in Zimbabwean inshore fisheries of Lake Kariba in Normann *et al.*, 1998 pp99-124.

Jentoft, S. (1994) 'Decentralization of fisheries management: a quest for subsidiarity' paper presented to the VIth Annual Conference of the European Association of Fisheries Economists. Crete.

Jul-Larsen, E., Kolding, J., Overå, R., Raakjær Nielsen, J. and Zwieten, P. A. M. (2002) Management, co-management or no-management? Major dilemmas in the sustainable utilization of SADC freshwater fisheries. Synthesis Report. FAO Fisheries Technical Paper 426/1, Rome.

Keare, D.H. (2001) Learning to Clap: Reflections on Top-Down versus Bottom-Up Development. *Human Organization* **60**, No. 2. 59-165.

Kebe, M. (1998) Artisanal fisherfolk's involvement in fisheries rehabilitation in Senagal: Co-management perspectives in Normann *et al.*, 1998 pp291-302.

Kponhassia, G and Konan, A. (1998) The traditional management of artisanal fisheries in Cote d'Ivoire: The case of Aby Lagoon in Normann *et al.*, 1998 pp273-290.

Lawry, S.W. (1994) Structural Adjustment and Natural Resources in the Sub-Saharan Africa: The Role of Tenure Reform. *Society and Natural Resources* **7**, pp383-387.

Lopes, S., Poisse, E., Wilson, J., Kromer, J.L., Manuel, L., Cululo, C., and Pinto, M.A.R. (1998) From no management towards co-management? A case study on artisanal fisheries in Angoche district, Nampula Province, Mozambique in Normann *et al.*, 1998 pp125-153

Machena, C. and Kwaramba, R. (1995) Creation of Property Rights to promote sustainable development in the Lake Kariba Inshore fishery in Zimbabwe. Paper presented at the Fisheries Co-management workshop (May

1995), Hirtshals, Denmark.

Malasha, I. (2001) Fisheries Co-management: A Comparative Analysis of the Zambian and Zimbabwean Inshore Fisheries of Lake Kariba, Draft D.Phil thesis, University of Zimbabwe.

Mamdani, M. (1996) Citizen and Subject: Contemporary Africa and the Legacy of Late Colonialism. Princeton University Press. New Jersey.

Mikalsen, K. H. (1998) Regulation and Representation: Institutional challenges in fisheries management. In Gray, T. (ed.). *'The politics of Fishing'*. MacMillan. London. pp.100-115.

Normann, A.K., Nielsen, J.R. and Sverdrup-Jensen, S. (eds). Fisheries Co-management in Africa: Proceedings from a regional workshop on fisheries co-management research. Held at Boadzulu Lakeshore Resort, Mangochi, Malawi from 18-20 March 1997. Institute for Fisheries Management. Hirtshals, Denmark. Fisheries Co-management Reseacrh Report No.12.

Ostrom, E. (1990) Governing the commons: evolution of institutions for collective active. Cambridge University Press. Cambridge. England.

Pinkerton, E. (ed.) (1989) Co-operative Management of Local Fisheries: New Directions for improved Management and community Development. Vancouver: University of British Columbia Press.

Pomeroy, Robert S. and Pido, Michael D. (1995) Initiatives towards fisheries co-management in the Philippines. The Case of San Miguel Bay. *Marine Policy* **19** No. 3, 213-226.

Selznick, P., The Moral Commonwealth - A Centennial Book, University of California Press 1992.

Sen, S. and Raakjær Nielsen, J. (1996) Fisheries Co-management: a comparative analysis. *Marine Policy* **20**, No. 5, 405-418.

Sowman, M., Beaumont, J., Bergh, M., Maharaj, G. and Salo, K. (1998) An Analysis of emerging co-management arrangements for the Olifants River harder fishery, South Africa in Normann *et al.*, 1998 pp177-204.

Spliethoff, P., MacPherson, N. and Pena, M. (1990) A Review of Experiences in and Guidelines for Fisheries Development Assistance. CEC Evaluation Series No.1. Commission for European Communities Directorate General for Development, Brussels.

Sverdrup-Jensen, S. and Raakjaer Nielsen, J. (1998) Co-management in small scale fisheries. A synthesis of Southern and West African experiences. In Normann, *et al.*, 1998. pp 303-322.

Viswanathan, K. (ed.) (forthcoming). Proceedings from International Workshop on Fisheries Co-Management, Penang, Malaysia. 23-28 August 1999.

World Commission on Environment and Development (1987) Our common future. Oxford and New York: Oxford university Press.

World Bank (1994) World and participation. Washington, D.C.: World Bank, Operations Policy Department.

World Commission on Environment and Development (1987) Our common future. Oxford and New York: Oxford university Press.

Chapter 6

EXPERIENCES WITH FISHERIES CO-MANAGEMENT IN SOUTHEAST ASIA AND BANGLADESH

ROBERT S. POMEROY
University of Connecticut, Avery Point, Groton, USA

K. KUPERAN VISWANATHAN
WorldFish Center, Penang, Malaysia

1. INTRODUCTION

The countries of Southeast Asia, ie Brunei, Cambodia, Indonesia, Laos, Malaysia, Philippines, Taiwan, Thailand and Vietnam, have a current population of over 510 million, of whom approximately 35 percent live below the poverty line. The population of these nine countries is expected to reach 650 million by the year 2020. The average fish consumption for the region is relatively high at 22 kg per capita per year and is higher in coastal communities. In some countries and coastal communities, such as in the Philippines and Indonesia, fish provides the main source of animal protein. Fishing and the extraction of coastal resources provide the main livelihood for millions of families. It is estimated that the demand for food fish in the year 2010, calculated at a constant per capita consumption rate of 22 kg/year, would be 18 to 19 million metric tons. Production from marine capture fisheries is not expected to keep up with the demand, causing concerns for food security in the region. The increasing demand for fish from the expanding population will create more stress on the already depleted coastal and inshore fishery resources in the region.

It is now almost universally accepted that most of the coastal fisheries in Southeast Asia are overfished. Coastal resources and ecosystems are degraded and in decline from a variety of factors. While the governments of these countries are working to attain sustainable development of coastal and marine resources and to improve the social and economic conditions of coastal residents, funds and other resources for these purposes are limited. This is not new information. However, new actions must be taken to deal with these issues. With limited government resources, the resource users will need to take more responsibility for finding solutions to their problems and needs. The resource users must be involved in making management and development decisions. They will need to be educated, informed and empowered to take action. New governance arrangements for fisheries and coastal resources must be examined and put into place. Resource management policies must shift from a use orientation to a conservation and resource management orientation. Attention must be given to policies that address issues of food security and people's well-being and livelihood, not just fisheries management.

To prevent further overexploitation and degradation of coastal and marine resources for those who depend on them, there is an imperative for better management. Many present resource management arrangements in Southeast Asia have not succeed in the effort to coordinate and restrain the many users. They have not kept pace with the technological ability to exploit the resource or with the driving incentives to exploit and degrade - economic returns, population growth, food, and employment. Management systems have focussed on development but have not succeeded in addressing the issues of economic efficiency, equity and user conflict. Increasing competition for and conflict over scarce resources will further stress coastal and marine management systems.

Mixed with concerns about improved resource management and conservation is the need to directly address problems of poverty, unemployment and decreasing quality of life in fishing communities in Southeast Asia. The main brunt of such economic and social distress is borne by women, children and unskilled fishers, as well as by those unskilled people who are directly and indirectly dependent on the fishing industry. Integral elements of this prevailing scenario are a high level of unemployment or underemployment, lack of ready alternative and supplemental employment and livelihood opportunities within the fishing community, a growing population and pressure to find additional fisheries resources, lack of credit and markets, and the paucity of institutional mechanisms to undertake system-wide development.

There is a growing consensus among many fisheries researchers and managers working in Southeast Asia that solutions to the current problems in the sector rest outside its traditional realm. This dilemma calls for a broader vision of the fisheries system, going beyond fisheries sector-specific policies to the vast array of seemingly unrelated policies that may have beneficial side effects for the fisheries sector. The broader policy context is justified by the interlinkages among fisheries resource management issues, on the one hand, and issues of economic and community development, on the other hand.

Co-management and community-based natural resource management (CBNRM) strategies are increasingly seen as an approach for such linked development and management initiatives. Community-centred co-management can serve as a mechanism for not only resource management, but for social, community and economic development by promoting participation and empowerment of people to actively solve problems and address needs in their community. Throughout Southeast Asia, co-management and CBNRM has reemerged, through the initiatives of the people, NGOs, government and international agencies, as a way to involve resource users, provide greater localized control over resources, and utilize indigenous institutional arrangements and knowledge in fisheries management.

This chapter discusses current approaches to co-management for the sustainable governance of coastal fisheries in Southeast Asia. Specific examples of co-management taken from a number of Southeast Asian countries are discussed. While not a Southeast Asian country, a brief review of co-management activities in Bangladesh are included in this chapter due to the significant number of experiences in co-management, especially in inland open-water fisheries.

2. WHAT IS CO-MANAGEMENT IN THE CONTEXT OF SOUTHEAST ASIA?

The history of co-management in the region shows a shift from community based fisheries management (CBFM) to co-management and finally to issues of decentralization. Community based initiatives to rehabilitate, conserve and protect natural resources based on the use and enhancement of local knowledge and skills of the people actually began in irrigation programmes in a number of Southeast Asia (SEA) countries in the 1960s. People oriented forestry programmes started in the early 1970s. CBFM started in SEA in the early 1980s primarily through the work of NGOs and some donor supported projects. CBFM, as defined by the NGOs, was people centred and community focussed and often had very little government involvement. CBFM practioners often viewed government in an external role only to be brought into the activities at a later stage or as needed. This often lead to misunderstandings and lack of full support from government for these initiatives. In early 1990s, co-management emerged in the region focussing not only on people and the community but also on a partnership arrangement between government and the local community and resource users. CBFM was then considered an integral part of co-management. Community based co-management emerged in the region as a new approach which included the characteristics of CBFM and co-management, ie, people centred, community oriented, resource-based, and partnership-based. Thus community-based co-management has the community as its focus, but recognizes that to sustain such action, a horizontal and vertical link is necessary. Successful co-management and meaningful partnerships can only occur when the community is empowered and organized. Community-based co-management focuses not only on resource management, but community and economic development and social empowerment issues as well. This approach will be more complex, costly and time-consuming to implement than just CBFM due to the need to develop partnerships early in the process and to maintain them over time.

In early 1990s, at about the same time co-management was emerging, there was also a movement in Asia towards decentralization. This refers to the systematic and rational dispersal of power, authority and responsibility from the central government to lower or local level institutions to states or provinces in the case of federal countries, for example, and then further down to regional and local governments, or even to community associations. While decentralization was addressing general government administrative restructuring, it was also undertaken in support of government polices and programmes which stressed the need for greater resource user participation and the development of local organizations to handle some aspects of fisheries management.. For example, the Local Government Code of 1991 in the Philippines sought to decentralize government functions and operations to local government units. However it also included specific provisions which addressed fisheries such as defining municipal waters and supporting resource user rights.

More recently, at a meeting of senior fisheries officials of ASEAN countries held in Bangkok in late 2001, it was concluded that co-management mechanisms and granting of exclusive fishing rights to community-based institutions should be promoted for small-scale fisheries and coastal fisheries under a decentralized fisheries management system.

3. CURRENT APPROACHES TO COMMUNITY-BASED RESOURCE MANAGEMENT AND CO-MANAGEMENT OF COASTAL FISHERIES IN SOUTHEAST ASIA

Several countries in Southeast Asia and Bangladesh are now recognizing the important potential role that community-based resource management and co-management systems can play in contemporary fisheries management. Each country is taking a different approach to co-management. In this section, the approaches being taken in the Philippines, Thailand, Malaysia, Indonesia, Laos, Cambodia, Bangladesh and Vietnam will be discussed.

3.1. Philippines

The Philippines has a long history of traditional fisheries rights and allocation (Kalagayan, 1991; Lopez, 1983). However, under both the Spanish and American colonization of the country, community authority and rights were superseded by municipal government control of local fishing grounds. This administrative structure of municipal authority remains in place today.

Although for centuries natural resource management in the Philippines has been centrally determined, top-down and non-participatory, in reality local level decisions and willingness to recognize the laws and regulations has played a strong role in the use of natural resources such as forestry, fisheries, mining and water. The increasing deterioration of natural resources in the country threatened its ability to pursue sustainable development. Starting in the 1960s, alternative methods of resource use and management were explored in an attempt to reverse these negative trends. Consequently, there has been a shift to forward-looking policies and strategies that advocate resource management and conservation over a use orientation through community-based initiatives to rehabilitate, conserve and protect natural resources based on use and enhancement of local knowledge, skills, responsibility and accountability. The irrigation sector was the first to evolve an institutional development scheme for mobilizing the active participation of water users in 1968. People-oriented programmes in the forestry sector started in the early 1970s (Sajise, 1995). Community-based coastal resource management (CBCRM) started in the early 1980s. To date, well over 180 CBCRM projects have been implemented by government, NGOs, fishing communities, and academic and research institutions for near shore areas but not including the commercial sector. No country in the world has the range of experience with CBCRM and co-management as exists in the Philippines (Pomeroy and Carlos, 1997).

Between 1975 and 1998, fisheries management in the Philippines has been guided by Presidential Decree (PD) 704, otherwise known as the Fisheries Act of 1975. Under PD 704, fisheries management is the responsibility of the government, both national and municipal. The management measures (mainly through regulatory instruments) undertaken by the government during this time, however, have been ineffective in promoting the sustainable development and management of the country's fisheries. The first small marine protected area implemented through a local government unit, but with limited participation of the resource users, in the country was established in 1974 at Sumilon Island in the Central Philippines. This model provided feedback on the benefits of having closed reef areas to fishing. In the early 1980s various projects replicated this island-based model but with more effective community involvement.

The Aquino administration provided some impetus for community-based resource management when in 1989 a Presidential Commission on Anti-Illegal Fishing and Marine

Conservation or the Bantay Dagat committee, was created. The Commission called for increased coordination among government agencies in enforcement of fisheries laws and increased participation of fishers in management (Kalagayan, 1991). This commission however did not succeed in getting much going in the area of increased fisher participation. In 1991, the government recognized the need to increase participation in management and to devolve control over resource access to local levels through policy and institutional reforms. The Local Government Code (LGC) of 1991 devolved most authority to local government units (LGUs). The basic tenet of the LGC is decentralization. A general operative principle is a provision that the LGUs may group themselves, consolidate or coordinate their efforts, services and resources for purposes commonly beneficial to them. Section 35 specifically states that LGUs may enter into joint ventures and such other cooperative arrangements with people's organizations (PO) and NGOs to engage in the delivery of certain basic services, capability building and livelihood projects, and to develop local enterprises designed to diversify fisheries, among other things. The LGUs and local communities are also given certain privileges and/or preferential rights. Municipalities shall have the exclusive authority to grant fishery privileges in municipal waters, up to 15km from shore, and impose rentals, fees or charges. In terms of fishery rights, the organizations or cooperatives of marginal fishers shall have preferential rights to fishery privileges within the municipal waters such as the erection of fish corrals and gathering of fish fry free of any rental, fee or charge (de Sagan, 1992; Tabunda and Galang, 1992).

Through several initiatives, the government now actively promotes community-based resource management to conserve the coastal resources and diversify the income sources of the low-income small-scale fishers. The 1993-1998 Medium-Term Philippine Development Plan has among its strategies: to implement a community-based fishery management strategy; to regulate fishing effort within maximum sustainable yields; to promote territorial use rights for small fishers; to intensify aquaculture and the optimal utilization of offshore, deep sea resources, and to provide diversified occupational opportunities among marginal fishers (NEDA, 1993). The core programme for fisheries development implemented under the Plan was the Fisheries Sector Programme from 1990 to 1995. Among the policy and institutional reforms instituted through the Fisheries Sector Programme were: (1) decentralization of authority and simplification of procedures for clearance of local fisheries management ordinances subject to national laws and/or policies; (2) strengthening the enforcement of fisheries laws through municipal-based inter-agency law enforcement teams; (3) promoting community-based initiatives to rehabilitate, conserve and protect the coastal resources and to diversify the sources of income of small-scale fishers; (4) NGOs were to be engaged to assist and undertake community organizing; and (5) shifting to limited access in concerned fishing areas. At the core of the resource and rehabilitation thrust of the Fisheries Sector Programme was coastal resource management. Fishers, local government units and other concerned agencies in the area were given the opportunity to determine the specific problems in their areas and to identify the management strategies to counteract these problems. The Fisheries Sector Programme ended in 1997 and was replaced by the Fisheries Resource Management Project which has similar objectives..

In 1998, Republic Act No. 8550 or the Philippine Fisheries Code was signed into law. Under the Fisheries Code, several sections of the LGC were supported such as the devolution of the function of fisheries management to local government; the designation of municipal waters up to 15 km from shore, and the granting of preferential rights to fishing privileges in municipal waters to registered fisher organizations and cooperatives.

In addition, the Fisheries Code endorsed the establishment of Fisheries and Aquatic Resources Management Councils (FARMC) at the national and municipal levels. The FARMCs are mandated to carry out a number of management advisory functions in close collaboration with the LGU. These functions include assisting in the preparation of Municipal Fishery Development Plans, recommending the enactments of fishing ordinances, assisting in enforcement, and advising the LGU on fishery matters. The FARMCs are formed by fisher organizations and cooperatives and NGOs with assistance from the LGU. Now the country is reviewing a new national policy that would integrate coastal management. It will help classify the values of national government agencies in supporting local governments to deliver coastal resource management as a basic service in all coastal areas of the country.

3.2. Thailand

Hinton (1985) speculates that traditional community-based aquatic resource management systems once existed in Thailand but no firm evidence exists. Since Thais historically exploited freshwater and not marine waters, it is possible that CBCRM systems may have existed for Thai freshwater fisheries. On the other hand, there were traditional village justice systems through which people settled conflicts. These justice systems were, however, not considered legitimate by the ruling class (Bangkok Post, 27 August 1998).

A centralized fisheries management system currently exists in Thailand and this system has, in general, not been effective in addressing problems of fisheries overexploitation, low incomes of small-scale fishers, and conflicts between small-scale and commercial fishers (Juntarashote, 1994). A DOF review of past experiences in the fisheries sector revealed that government development programmes alone cannot achieve the long term objectives as long as the fisheries is left as an open access resource and the enforcement of fisheries regulations is ineffective. Recognizing the benefits of 'bottom-up' rather than 'top-down' fisheries planning and management, the Thai government has initiated a new programme which advocates the involvement of the fisher in the planning, management and implementation process.

During the Sixth (1987-1991) and Seventh (1992-1996) National Economic and Social Development Plans, the strategies covered the reinforcement of various marine fisheries management measures, including the construction of artificial reefs in many coastal areas; fishery resource conservation; and the setting up of pilot projects on coastal area management with more active participation of small-scale fishing communities in the management of the fishery resources in waters adjacent to their villages to ensure sustainable yields. The Eighth National Economic and Social Development Plan (1997-2001) and the recent promulgation (1997) of a new constitution have brought about more changes in Thailand as to the fundamental basis for participation and community-based resource management. The Eighth plan has whole sections related to 'popular participation in natural resource management' (Part VI), 'popular governance' (Part VII), and incorporation of local knowledge in management. The new Thai Constitution also contains several provisions with regards to rights and participation of people. Chapter III, Section 46, states '...traditional community shall have the right to conserve and restore their custom. Local intellect, art or good culture of their community and the nation and participate in the management, maintenance, preservation and exploitation of natural resources and the environment in a balanced fashion and persistently as provided by law'. From the perspective of resource management, the new constitution is revolutionary as it requires government agencies to restructure their policies and practices to work directly as a local level and have people participation in planning and management.

Priority fisheries management strategies of the DOF include community-based resource management, fishing rights system, encouraging the formation of fisher associations, and encouraging better cooperation between resource managers and fishers. In 1993, the DOF, with the collaboration of the Department of Fisheries Management, Faculty of Fisheries, Kasetsart University, established a community-based fishery management programme to evaluate development of a fishing rights system. In early 1994, the DOF set up several committees to establish the fishing rights system, pilot project preparation and draft a new fishery law. The fishing rights over specific areas are granted by DOF at the community level. Local fishers organizations or cooperatives are awarded exclusive fishing rights over a designated sea area. The fishers who are members of the organization have a right to fish in the area. The members of the organization establish rights and rules for operating in a designated fishing ground. The fishing rights programme has had a slow start but implementation is continuing at selected sites in the country (Juntarashote, 1994; Department of Fisheries, 1994). Through the enactment of the new Constitution, there is now a legal basis for the implementation of the fishing rights system.

The present fishery law was enacted in 1947 with some minor amendments in the 1980s. The law, Fisheries Act, B.E. 2490, had been enacted on the basis of freshwater fisheries which was the leading fishery at that time. Several sections of the fisheries law can be employed as a legal basis for CBRM. For example, the exclusive user rights currently in use in Thailand for stationary gear and coastal aquaculture will provide a basis for the establishment of the fishing rights system. In principle, the fishing rights system must not conflict with the constitutional law. According to the Fisheries Act of B.E. 2490 of 1947, the country's basic fishery law, several sections are valid for the enactment of a law or a ministerial decree with regards to fishing rights (Lhaopadchan, 1993; Piumsombun, 1993; Tokrisna and Duangsawasdi, 1993).

A number of CBRM projects have been undertaken in recent years, particularly in Southern Thailand, many with the support of NGOs. These projects have emphasized awareness creation among members of the community about sustainable management; building of local organizations and capacity for conservation and rehabilitation of coastal resources; and encouraging coordination among resource users, government and NGOs (Tokrisna, Boonchuwongse and Janekarnkij, 1997). In line with the new Constitution and the Eighth Development Plan, it has been recommended that the Sub-district Administration Authority (*Or-Bor-Tor*) has been utilized as a core unit of CBRM. The Or-Bor-Tor, consisting of the sub-district head, village head and sub-district council, could serve to help organize fishers and manage conflicts. Institutional support from national government will be needed to support this activity.

3.3. Malaysia

The political and legal framework in Malaysia strongly favours central control of fisheries management. Since independence from Britain in 1957, the Malaysian government has pursued policies for managing the fisheries that are developed at the national level and implemented at the community level with minimal participation of fishers or industry. Participation of fishers occurs through local level consults where fisheries department staff meet with local fisher organizations (cooperatives, associations) to inform them about new policies and to solicit feedback on the policy. This structure has removed almost all forms of traditional community-based management that has any serious support from fishing communities. Bailey (1991) concluded that in fishing communities, moral economy values have long been relegated to the 'cultural dust bin'. He found that fishing communities did not have pre-existing organizational capacity around which to build cooperatives, which are often seen as an important institution for CBRM. Yahaya and Yamamoto (1988) conclude that Malaysian fishers in general, are not familiar with the self-management concept. Although generally supporting the fishing right principle, the fishers were not willing to accept the role of guardian of the fishery resources nor as the enforcer of laws and regulations.

Education of fishers is needed for fishers to accept self-management. Under the Malaysian fishery management policy, four fishing zones were established through a limited licensing scheme whereby rights and rules in each zone were designated for specific fishing method, class of vessel, species caught and ownership pattern (Majid, 1992). Although the policy has produced progress towards meeting its stated objectives of achieving optimum yield, eliminating conflict between small-scale and commercial fishers, and a more equitable distribution of catches, enforcement problems still exist. In 1997, the government indicated a policy shift in fisheries management towards a more decentralized community-based management approach, with greater fisher participation in fisheries policy and management. Under one proposal, an alternative monitoring, control and surveillance (MCS) system would allow for sharing of responsibility with the community. Under the proposed MCS system, the community (fisher organizations and NGOs) would have increased responsibility for monitoring and surveillance and the federal government would maintain responsibility for control and enforcement (Sulaiman, 1994). The most recent national agricultural policy, however, stresses the need for more community involvement in fisheries management and the setting up of pilot sites for the implementation of community based management initiatives.

3.4. Indonesia

Community-based management systems have a long history in Indonesia and are the most long-enduring in the region. These traditional systems are by-and-large localized practices found in geographical pockets throughout the country. The traditional fishing rights and community-based management systems are based on 'restriction' which closely parallel the modern management concepts of closed areas and seasons. Like other such traditional systems around the world, they have adapted to change over time (Wahyono *et al.*, 1992; Bailey and Zerner, 1992; Nikijuluw and Naamin, 1994).

Current national laws in Indonesia do not recognize local community-based resource management systems in coastal fisheries. Fisheries administration and governance is centralized through the Directorate General of Fisheries and provincial fisheries services. There is no policy that gives legal mandates to provincial and district governments to manage coastal and marine resources, with the exception of small-scale fisheries. Provincial governments are authorized to license vessels of less than 30 gross tons. The provincial governments are also expected to protect against any inappropriate practices that may threaten the interests of small-scale fishing and local communities. Traditional rules and norms potentially support the written laws in order to promote community participation in managing fisheries. By law, the existence of traditional laws (unwritten laws) are recognized as complementary to the existing written laws and both national and local interests. Explicit recognition of local authority and the concepts of customary law and local territorial rights would require amendments to both the National Fisheries Law No. 9/1985 and the National Administrative Law No. 5/1979, the law which authorizes the structure of village government (Bailey and Zerner, 1992). One reason for not adopting traditional laws into written laws is that such an action could cause disintegration of legal systems, particularly with regard to traditional village boundaries with new boundaries established by national law. It should be noted that in 1999 the Ministry of Home Affairs has proposed draft regulations (local government administrative law) as part of amendments to Law No. 5/1979. Those related to fishers management include: Chapter II, paragraph 3, mentioning that the Local Government Level I (province) be given authority for inland and sea territory up to 12 miles from shore; and Chapter IV, paragraph 9, states that the Provincial, Regency and Municipal governments in coastal regions be given authority to manage: exploration, exploitation, conservation and management of coastal fisheries.

There are positive signs for the effective development of community management institutions. The establishment of fisher organizations have allowed fishers to participate in fisheries management conducted by the Directorate General of Fisheries such as the Collective Management System (Pengelolaan Bersama) and the Forum of Coordination for Fisheries Resources Utilization (Forum Koordinasi Pengelolaan Pemanfaatan Sumberdaya) (FKPPS). This FKPPS consists of representatives of national and local governments, private sector and local fisher representatives. The main activity of the FKPPS is to organize fisheries exploitation for the interest of all stakeholders and monitor the implementation of action plan established by the FKPPS. Indonesia's long-term (25 year) development plan was completed in early 1994. Policy and strategy in the form of the State's Main Guidelines (Garis-Garis Besar Haluan Negara) for the first five-year development plan (1993-1998) was laid down by the People's Consultative Council. The State's development plan (Repelita) stresses the need to alleviate poverty and consider environmental protection in order to maintain sustainable development of coastal and marine resources. During the current Repelita, the government will stress regional development in order that the livelihood of fishers and their families will be improved through integrated coastal community development that will involve other economic sectors in the community. In 1994, a new programme of poverty alleviation, called Inpres Desa Tertinggal (IDT) (Presidential Instruction on the Less Developed Village) was initiated. This programme aims to promote economic growth in fishing and farming villages through decentralization and active participation of the local community. Existing social and economic institutions, such as cooperatives, at the village level will be utilized as vehicles for greater participation of target groups and the community. The programme will focus on generation of income and employment opportunities and improvement in social structure. The programme objectives stress a reformulation of the basic approaches to fisheries and agricultural development, from a production approach to one of enhancing fisher and farmer income and welfare. This will lead to more sustainable resource management (Cholik and Ilyas, 1994). In addition to the government, a number of NGOs and communities have implemented CBRM activities. These initiatives have involved community organizing and conservation and rehabilitation activities. The process of decentralization is continuing in Indonesia. The ratification of Law No. 22/1999 by the regional authorities in early 2001, provides the mandate for local governments to exercise responsibility over their natural resources. The local authorities can now work closely with their stakeholders in formulating the policy for the management of natural resources. The law now gives authority at the Kabupaten/Kota or municipal level for the exploration, exploitation, conservation and management of marine resources within four nautical miles of the province's jurisdiction (Khan and Fauzi, 2001).

3.5. Laos

The fisheries sector of Laos is dominated by capture fisheries from the Mekong River, its tributaries, flood plains, wetlands and reservoirs. Aquaculture in ponds and integrated aquaculture-agriculture farming systems are now on the increase to supplement harvests from capture fisheries. Fisheries management and development is the responsibility of the Ministry of Agriculture and Forestry, Department of Livestock and Fisheries (DLF) and its provincial offices. Recent policies of the DLF support the sustainable use and conservation of aquatic resources by the communities (Sinkham Phonvisay, 1997).

Laos has a history of informal community-based capture fisheries management. In some areas, closed and semi-closed wetland fisheries, fish ponds and other water bodies are managed through rules and regulations based on community-based management. In some communities, water bodies are managed by community members for the purpose of community development including temple and school construction and other infrastructure projects. These communities have developed management plans and rules for catching fish and use of the catch (AIT Aqua Outreach-Lao PDR, 1997). Recently, the Lao government has shown support for community-based management and co-management strategies for both capture and culture fisheries as a way to utilize traditional knowledge of the resource by fishers and to allow the resource users to establish conservation and management strategies. The DLF has a policy to use 'bottom-up' approaches to policy formulation and implementation.

Torell (1998) has reported that, in principle, the Lao constitution supports community-based management initiatives. Article 8 states that 'All ethnic groups have the right to protect, preserve and promote the fine customs and cultures of their own tribes and of the nation'. Article 14 states that 'collective and individual ownership' is recognized. Torell further states that there are, however, a number of articles which, if backed up by laws to that effect, could imply reduced rights for local communities and 'top down' resource management. Again, if the laws and policies of the government are supportive than the constitution could also be supportive of community-based management and customary law.

Two recent fisheries projects in Southern Laos have utilized co-management. Since 1993, the Lao Community Fisheries and Dolphin Protection Project (LCFDPP) has been working to establish village-level aquatic resources conservation and management strategies in Khong District, Champasak Province in southern Laos (Cunningham, 1998). These strategies include the establishment of fish conservation areas and village-level rules governing the harvest of certain aquatic resources in the Mekong River. Originally started to protect the Irrawaddy dolphin, the project has also worked on community development, fisheries research, and environmental awareness activities (Baird, 1996). As of 1998, 63 villages had developed aquatic resources management strategies. This includes established sets of rules designed to conserve and sustainably manage aquatic resources in the mainstream Mekong River, wetland areas, streams and seasonally irrigated and rain fed paddy rice fields adjacent to rural villages.

A recent evaluation of the LCFDPP found that 'villagers regard the initiative to establish the conservation zones as their own' (AIT Aqua Outreach-Lao PDR, 1997). The evaluation reported that the LCFDPP represented a good example of successful co-management. 'In this case the local people are taking an active role in defining management practices and implementing them. The local government is supporting these activities by giving them official status. The national government has contributed by establishing policies that prioritize conservation and management of natural resources, with special emphasis given to fisheries' (AIT Aqua Outreach-Lao PDR, 1997).

Over a two-year period beginning in February 1997, an Indigenous Fisheries Development and Management Project funded by outside donors was working in Champasak province. In addition to identifying indigenous fisheries management strategies for different fisheries and aquatic resource ecosystems, the project developed an interactive, participatory, study-based process of management and fishery enhancement involving the community, district and provincial officials, and the project team (Hirsh, 1998).

It is, however, not clear how the lessons from these specific cases can be extended to other parts of Laos. In parts of Laos where the history of community change have been unstable and where social religious, kinship, ethnic and linguistic conditions are less homogenous, these experiences from the Khong in the Champasak province may not be easily extended to the less homogenous areas (Baird, 2002). However, the flexibility possible in the arrangements for co-management in which communities, NGOs and government participate in various levels of power sharing and responsibilities for managing natural resources, does represent an important option for improving the management and equitable distribution of natural resources in Laos.

3.6. Cambodia

Community-based resource management systems do not exist in Cambodia. Fisheries, or rather 'commercial' fisheries, are managed by the state through a 'top-down' approach in which the Department of Fisheries and the provincial fisheries offices are the primary decision-making bodies (Ahmed and Tana, 1996).

The sectoral policy for fisheries in Cambodia states that, '... the Government will strengthen the control, monitoring and surveillance on fisheries resource utilization to sustain the production by reducing overfishing in inland and marine waters. Strictly prohibiting the illegal fishing operation and the destruction of inundated forest that may affect to inland and marine fishery environment. Encouraging aquaculture and livestock investment.'

Fisheries in Cambodia are divided into 'commercial', 'small-scale' and 'family' fisheries. Family (subsistence) fishing is allowed in inland water bodies at all times of the year, although restrictions are placed on effort (ie, gear and mesh size). Middle scale (small-scale) and large scale (commercial) fishing are allowed only in the open season of October to June. Large-scale fishing takes place in leased fishing lots that are controlled by the rich and their agents, while subsistence and small-scale fishers have limited access to good fishing grounds. In contrast to the rules being applied to commercial and small-scale fisheries, family fishing is treated as open access, in spite of the fact that production figures from subsistence fishing are quite substantial (van Zalinge *et al.*, 1998). Increasing commercialization of subsistence fishing and lack of an effective mechanism to distinguish it from small-scale fishing have resulted in a virtually uncontrolled situation in the small-scale freshwater capture fisheries of Cambodia.

The current fisheries legislation (the Fiat-Law on Fisheries Management and Administration of 1987 and its sub-laws) based on fisheries law and regulations designed during the French colonial period, emphasizes inland capture fisheries and revenue orientation. The manner in which the present fishery law is applied, conservation issues and fisher participation are not considered. However, applied in a different way, the fishery law (or revision thereof) could be one of the best existing instruments for sustainable fisheries management. In principle, the fisheries law gives a mandate to the Department of Fisheries to manage all wet and flooded areas (including mangroves, flooded forests, etc.) as fishing areas or 'fisheries domains' for the purpose of sustaining the resource base for fisheries. This mandate is at present not being followed (Cambodia Working Group 1998). Fisheries management relies on an extensive regulatory regime that uses control and enforcement by the fishery department. There is no fisher participation in management nor are there any fisher organizations. The political and economic turmoil that has existed in Cambodia since the mid-1970s has been a deterrent for fishers to organize and engage in community-based resource management. This lack of participation and 'top-down' management approach has brought about low compliance with regulations and has been a major factor in resource overexploitation and lack of recognition of the family fisheries (Ahmed and Tana, 1996). More recently, the Government of Cambodia is moving more aggressively into community based management of the fishery with the release of the fishing lots based system to organized groups and communities to fish and manage the resource. There are, however, many challenges ahead as resources for the development of community organizations and training required to develop effective community organizations for managing the fisheries resources are not available within the government budgets to the Department of Fisheries.

Torell (1998) points out that, in principle, the necessary elements are present in the Constitution of Cambodia to support community-based resource management. The Constitution states that 'All persons, individually and collectively, shall have the right of ownership' (Article 44). Several new laws and regulations, such as the Royal Decree on the 'Creation and Designation of Protected Areas' of 1993 and the Law on Environmental Protection and Natural Resource Management of 1996 are concerned with conservation of aquatic resources and the coastal environment. A number of international development and resource management projects and papers by Cambodian government officials have emphasized the need for resource user participation in management and co-management (Cambodia Working Group, 1998).

More recently, the Cambodian government has put in motion reforms for the management of fisheries. The government has decided to reduce the areas reserved for the fishing lot system and allow for broader community management of the fisheries. Fishing community organizations are being established in the areas previously reserved for auctioned fishing lots systems. In 2001 it is reported by the Department of Fisheries in Cambodia that some 165 fishing community organizations were established in the fishing lot areas. The rules and regulations for the enabling of co-management is being drafted and deliberated by the various arms of the government in Cambodia. The real challenge is in the limited capacity of the Department of Fisheries to provide the necessary inputs in terms of training and community development to ensure proper development of community organizations capable of managing the aquatic resources. A community development unit has been set up within the Department of Fisheries to establish fishing community organizations to enable effective co-management of the fisheries. How successful these developments will be in addressing the resource degradation and user conflict issues needs to be seen.

3.7. Viet Nam

Vietnam has centuries old traditions and customary practices for fisheries management. Many of these are intertwined with religious and other cultural activities and many were very active before 1945. Once Vietnam began establishing fishing cooperatives (from 1960 in the North and after 1975 in the South), many traditions and customs were lost. The government labelled some of the ceremonies as superstitious and banned them. As the role of the central government was increased, traditional communal activities were restricted or eliminated. With the beginning of *doi moi* (reform) policies in the late 1980s and as the control of the central government was loosened, some traditional practices began to reappear (Thong and Thieu, 1998). One example is an association established by fishers called van, which is similar to those in agriculture, to preserve village or community social structure and to provide mutual assistance related to the fishing activities. These associations, headed by an elected fisher, performed several functions including conflict resolution, establishment of the method of sharing catch, and, to lesser extent, establishment of management rules. Like other organizations in Vietnam, the van was banned in the 1970s. It is now slowly reemerging in fishing communities throughout the country.

The Ministry of Fisheries (MOF) has overall responsibility for formulation of policies, planning and regulation for the sector, registration of large vessels, and foreign fishing activities. Provincial fisheries departments have responsibility related to capture and culture fisheries within their own jurisdiction including planning, collection of statistics, guidance of fishers on government policies, registration of smaller vessels, control and inspection of fisheries activities, protection of the resources, and control of local fisheries enterprises. Thus, the province is given a great deal of autonomy for management and development of fisheries. Provincial officials consult with district officials about planning, management and development (Ministry of Fisheries, 1995a).

In light of recent political and economic reforms in Vietnam, the government is currently developing new policies for the fisheries sector. The recently completed Master Plan for Fisheries to the Year 2010 (Ministry of Fisheries, 1997) sets out several programmes and projects for the fisheries sector. There is recognition that coastal or near shore fisheries are overexploited due to the high levels of fishing activity, destructive fishing practices, and lack of enforcement which occurs in these waters. Protecting the aquatic environment and preserving the fisheries resources of Vietnam will be undertaken through a programme called 'Using the Environment and Fisheries Resources'. One of the four projects under this programme is 'Allocation of Resource User Rights and Obligations'. The specific objectives of this project are to gradually, but firmly, move away from the 'open access' nature of fisheries and aquatic resources to a managed approach to the allocation of resource user rights and user-obligations. A sub-project of this project is 'operationalizing Participatory Approaches to Resource Management'. The specific objectives of this sub-project are to operationalize the new systems for resource use and management at the community level. This is expected to include the introduction of participatory approaches to both the management and use of habitats and aquatic resources. It is also expected to include systems for the valuation and the permanent distribution of resource user rights and user-obligations (Ministry of Fisheries, 1997). The new policies will constrain fishing activities in near shore waters and emphasize aquaculture as an alternative to near shore fishing and encourage offshore fishing. The state enterprises and collectives through which fisheries were developed under the centrally planned economic system will be replaced with an emphasis on private business enterprises and the household and fishers organization. Coastal areas may be privatized, in essence a fishing right, and managed by the household or fisher organization for fishing and aquaculture (Ministry of Fisheries, 1995a).

The Ministry of Fisheries has endorsed fisheries co-management as a strategy for managing near shore and estuarine areas and the introduction of a regulatory system of resource user rights and obligations. Recognizing the diversity of the coastal zone and regions in the country and the difficulty of effective monitoring and enforcement, the Ministry will undertake a programme to delegate resource management functions to local institutions, including fishers' organizations. This programme will be implemented in several steps beginning with pilot sites to develop models and gain practical experience in co-management. The lessons learned from these pilot sites will be integrated into national policies and laws to support co-management. Pilot site activities are currently underway (Thong, 1995; Ministry of Fisheries, 1995b).

A recent review of the legal framework for fisheries co-management in Vietnam found that the existing legal system does not prevent the introduction of co-management (Chircop and Torell, 1997). It is possible to find a basis for co-management and customary law in the constitution of Vietnam, although these provisions are not stated specifically. Fisheries law reform is needed to fully support co-management.

3.8. Bangladesh

While not a Southeast Asian country, a brief review of co-management activities in Bangladesh is being included because of the significant number of experiences in co-management in the inland, open water fisheries.

Administrative arrangements for inland fisheries management in Bangladesh from 1950 to 1986 involved only the leasing of water bodies. The Ministry of Lands (MOL) has, since 1950, managed state-owned inland open water fisheries with the objective of raising revenues by dividing up the fisheries into jalmohals (fishing estates) and leasing fishing rights to these fisheries to the highest bidder for short-term (one to three years) periods. Despite measures in 1973 to restrict leasing to registered fisher cooperatives, this revenue-oriented strategy led to overfishing and exploitation of the poor by leaseholders and their intermediaries.

A combination of recommendations from the Land Reforms Committee, pressure from the Department of Fisheries (DOF), and lobbying from the National Fishers Association, resulted in the introduction of the New Fisheries Management Policy (NFMP) on an experimental basis in some 270 water bodies (out of a total of over 10,000) in 1986. The main objectives of NFMP were: 1) to free the fishing people from exploitation by intermediaries, leaseholders and financiers; 2) to redirect the major benefits of fisheries to genuine fishers; and 3) to ensure the conservation and propagation of fisheries resources (Aquero and Ahmed, 1989; Ahmed, Capistrano and Hossain, 1992). The strategy of NFMP was to abolish gradually the system of leasing rights in public water bodies to middlemen and replace it with individual licenses for 'genuine fishers'. Management authority over NFMP-designated water bodies was transferred from the MOL via the Ministry of Fisheries and Livestock to DOF for the duration of the experiment. It was expected that direct cooperation between government departments and fishing communities would be established - a co-management arrangement (Ahmed, Capistrano and Hossain, 1997).

Under the NFMP, the DOF and a number of NGOs became involved in fisheries management and some gains in a more equitable distribution of fishing access were achieved (Aquero and Ahmed, 1989). However, where there was no project or NGO support for the individual fishers, the ex-leaseholders often were able to retain control by advancing funds to pay license fees. No further water bodies were handed over under NFMP, and government policy changed in 1995 when a MOL circular abolished revenue collection from open waters (meaning rivers) and was interpreted as ending NFMP (Farooque, 1997). This has resulted in open access fishing in rivers (where before either leaseholders or licensing had limited effort to some extent), and reinforced the revenue oriented competitive leasing strategy in other 'closed' water bodies. Exceptions are the few water bodies currently being managed by the DOF under various existing fisheries development projects, most of which follow licensing. Upon completion of the projects, these too will be leased out (Capistrano, Hossain and Ahmed, 1997), except for about 20 oxbow lakes which have been handed over to DOF for group management by fishers (members receive licenses) for 50 years (Apu *et al.*, 1999).

Although this indicates a contradiction and dilemma in government policies, the DOF, NGOs and development agencies continue to utilize leasing and licensing mechanisms to develop co-management models for fisheries. While many fisheries have lacked any management arrangements since 1995, and many are leased to individuals, there are still many which are currently managed through a wide range of co-management arrangements between the government (DOF), NGOs and fishing/landless people. Ahmed, Capistrano and Hossain (1997) have identified three broad categories of co-management arrangements: 1) NGO-led strategy (fishery leased to NGO or its group); 2) Government-led strategy (government licenses fishers or leases to a fisher cooperative); and 3) Government and NGO partnership (support from government and NGOs either through licensing or a 'community-based' or group approach). A number of international development and research projects have supported the co-management in inland water bodies, and the most recent trend is for multi-stakeholder co-management bodies involving more than just target groups of fishers, and which involve other wetland users and government and are usually facilitated by NGOs (Thompson *et al.*, 1998).

4. AN EXAMPLE OF CO-MANAGEMENT IN THE PHILIPPINES

The fishery of San Salvador was showing signs of overexploitation in the late 1970s. The fish catch was in decline; illegal fishing using cyanide and explosives was rampant. The fishery was *de facto* open-access, with virtually no law enforcement. From an average reported catch per fishing trip of 20 kilograms in the 1960s, the catch had declined to barely three kilograms in 1988. Many reef fishes, such as groupers, snappers, and damselfish, had become scarce. In 1988, living coral cover had declined to an average 23 percent for the entire island. Though the San Salvador fishers knew that action was needed to protect their livelihood and the resource, the central government of the Philippines was too distant to control the situation and the fishers themselves were too fragmented to embark on any collective action to avert resource degradation.

San Salvador Island, with an area of 380 ha, forms part of Masinloc municipality in the province of Zambales, on the western coast of Luzon, about 250 km north of Metro Manila. In its population of 1 620 persons, the majority of the 284 households depend on fishing for their livelihood. The island has been home to three generations of residents, the first of whom came from the mainland and were largely farmers who fished part-time using hand lines and nets. During World War II, Japanese troops occupying the island sometimes used explosives to catch fish. After the war, villagers continued with their non-destructive subsistence fishing. Until the late 1960s, resource-use conflicts were rare and the resource remained in good condition. But the early 1970s saw an influx of fishers from the central Philippines who brought illegal fishing methods such as cyanide, fine mesh nets, and explosives. The new fishers also integrated the village economy into the international market for aquarium fish.

In the late 1980s, when resource overexploitation, degradation and use conflicts reached a crisis point in San Salvador, residents went in search of solutions to their problems. External change agents were instrumental in initiating new resource management measures. A Peace Corps volunteer who arrived in San Salvador in 1987 conceived of the Marine

Conservation Project for San Salvador (MCPSS), a community-based coastal resource management (CBCRM) project for coral reef rehabilitation. In 1989, a local non-governmental organization led a project to establish a marine sanctuary. The project featured biological (sanctuary and reserve) and governance interventions (management plan, community organizing, income-generation, rules and regulations, education, and training). That same year, the core group members made an exchange visit to a successful marine sanctuary in the central Philippines. The visit increased motivation and support for the idea of a sanctuary and reserve, and resulted in the drafting of a local ordinance to ban fishing within the sanctuary and allow only non-destructive fishing methods in the reserve. In July 1989, the Masinloc Municipal Council passed an ordinance for the marine sanctuary and reserve. Core group members became increasingly active in monitoring illegal fishing activities and guarding the sanctuary. Other resource users participated in village consultations, endorsed local ordinances, adhered to the rules, and adopted non-destructive fishing methods.

While the MCPSS was not conceived of as a co-management project, the increased role and participation of the government brought about a resource management partnership between government and fishers. The Masinloc municipal government, which has political jurisdiction over San Salvador, was drawn into the picture in a number of ways. The municipal government: 1) passed enabling legislation that provided a legal basis for the sanctuary and for apprehending rule violators; 2) mediated conflicts between local and outside resource users; 3) provided boat and equipment for patrolling coastal waters; 4) created a government patrol team to enforce laws; and 5) provided a political environment that allowed for the pursuit of community-based initiatives. Thus, co-management can be considered to date back to mid-1989, prompted by the political dynamics in San Salvador and the village fishers' lack of resources to run enforcement activities. The main government partner in this co-management arrangement was the municipal government; the support of the national government through its fisheries agency was not as visible.

In 1991, policy and legal support for co-management was strengthened in the Philippines through passage of the Local Government Code, which gave the municipal government jurisdiction over near shore waters. Following the turnover of the San Salvador Island project from the NGO to the village-based fisher organization in 1993, co-management became increasingly visible. Fishers and government shared responsibility for law enforcement, and the government provided funds for local enforcement operations.

San Salvador Island co-management was a win-win solution. For the resource users, co-management was prompted by their dependence on fishery resources for livelihood, recognition of resource management problems, and legitimacy and enforceability of rules. The government was motivated by its concern for improved living conditions for the fishers and their families and for sustainable resource management. The path to co-management was not trouble-free, however. Fishers using destructive methods, and those displaced from the sanctuary, became alienated and resentful. Over time, however, tangible benefits in the form of higher fish catch from San Salvador's fishing grounds helped to encourage rule compliance and non-destructive fishing practices.

A large number of factors contributed to the success of the project. These included the resource stakeholders' participation and sense of ownership in project planning and implementation, clarity of objectives, supportive leadership, the partnership between fishers and government, specification and legitimacy of user rights and enforcement, capability building, and tangible benefits such as redefined resource access, a shift to non-destructive fishing methods, improved enforcement, and observable biological, economic and social changes. Tangible benefits were clearly important. Fish catches went from barely three k in 1988 to six to 10 k in 1998, accompanied by an improvement in the diversity of fish species. The extent of living coral reef cover increased from 23 percent to 57 percent for the whole island. Fishers perceived gains in equity, knowledge, household income, empowerment, and conflict reduction. The San Salvador co-management project gave the village residents a reason for optimism, a motivation for collective action, and pride in their resource management achievements.

5. CONCLUSIONS

The idea of active participation of local resource users and communities in development and management is not a new one; it has been part of the development process in Southeast Asia since the 1960s. What is different is the increasing commitment of governments to policies and programmes of decentralization and community-based resource management. This is seen in a variety of policies and programmes in the Philippines, Thailand, Malaysia, Indonesia, Laos, Vietnam and Bangladesh.

Community-based resource management systems cannot succeed in isolation. The planning and implementation of these systems will require the development of new legal, administrative and institutional arrangements to complement contemporary political, economic, social and cultural structures. Community-based fisheries management will move forward in Southeast Asia in a form of co-management with the willingness of governments to accept the greater roles of communities and the need to provide the enabling legislation that foster the partnerships needed between communities and governments to address the serious issues of resource degradation and user conflicts. This new management philosophy once again makes the fisher a part of the resource management team, balancing rights and responsibilities, and working in a cooperative, rather than antagonistic, mode with government managers. Such co-management is a rational extension of evolutionary trends in fisheries management over the past decades.

How co-management will proceed in the future will depend on the success achieved in the different projects and approaches used by the different actors involved in implementing co-management programmes in the countries of Southeast Asia. What is clear, however, is that many of the programmes are still new and are learning along the way the skills needed for effective co-management. Increased support from government institutions to enable co-management to become part of the national and local fisheries management programmes will be required for co-management to succeed. In particular the changes that existing government institutions will have to undergo to share power, authority and information for managing fisheries will be crucial for success. Co-management will have to shift from being donor driven and dependent to local and national resource and policy dependent for long term sustainability.

REFERENCES

Ahmed, M. and Tana, T.S. (1996) Management of Freshwater Capture Fisheries of Cambodia - Issues and Approaches. *NAGA, the ICLARM Quarterly* 9(1), 16-19.

Ahmed, M., Capistrano, D. and Hossain, M. (1992) Redirecting benefits to genuine fishers: Bangladesh's open water fisheries management policy. *NAGA, the ICLARM Quarterly* 15(4), 31-34.

Ahmed, M., Capistrano, D. and Hossain, M. (1997) Experiences of partnership models for the co-management of Bangladesh Fisheries. *Fisheries Management and Ecology* 4(3), 233-248.

Apu, N.M., Sattar, M.A., Nathan, D., Balarin, J.D. and Middendorp, H.A.J. (1999) Fisheries co-management and sustainable common property regimes based on long-term security of tenure in oxbow lakes in Bangladesh. In. H.A.J.Middendorp, P.M. Thompson and R.S. Pomeroy (eds) Sustainable inland fisheries management in Bangladesh. ICLARM Conf. Proc. 58.

Aquero, M. and Ahmed, M. (1989) Economic Rationalization of Fisheries Exploitation through management: experiences from the open-water fisheries management in Bangladesh. P.747-750. In R. Hirano and I. Hanyu (eds) The Second Asian Fisheries Forum. Asian Fisheries Society, Manila, Philippines.

Asian Institute of Technology Aqua Outreach-Lao PDR (1997) Participatory evaluation workshop of the LCFDPP. Unpublished report. Savannakhet, Lao PDR.

Bailey, C. (1991) Social relations of production in rural Malay society: comparative case studies in rice farming, rubber tapping and fishing communities. In J.J. Poggie and R.B. Pollnac (eds) Small scale fishery development: sociocultural perspectives. ICMRD, University of Rhode Island, Kingston.

Bailey, C. and Zerner, C. (1992) Community-based fisheries management institutions in Indonesia. *MAST* **5**(1), 1-17.

Baird, I. (1996) Inland Community Fisheries Management in Southern Laos. *NAGA, the ICLARM Quarterly*. January. P.13-15.

Cambodia Working Group (1998) Draft Synthesis Report on Management of Fisheries, Coastal Resources and the Coastal Environment in Cambodia: Institutional, Legal and Policy Perspectives. Wetlands International, Phnom Penh.

Capistrano, D., Hossain, M. and Ahmed, M. (1997) Poverty alleviation, empowerment, and sustainable resource use: experiences in inland fisheries management in Bangladesh. P.141-162. In F. Smith (ed.) *Environmental Sustainability: Practical Global Implications.* St. Lucie Press, Florida.

Chircop, A. and Torell, M. (1997) The legal framework for fisheries co-management in Vietnam. Unpublished Report Prepared for Vietnam-Canada Ocean Policy Project-II, Halifax, Nova Scotia, Canada.

Cholik, F. and Ilyas, S. (1994) The Role of Bottom-up Planning in Poverty Alleviation in Indonesia: The Agriculture Case. Paper Presented at the 7th Biennial Conference of the International Institute of Fisheries Economics and Trade. Taipei, Taiwan, 18-21 July 1994.

Cunningham, P.D. (1998) Extending A Co-management Network to Save the Mekong's Giants. *Mekong Fish Catch and Culture*. Mekong River Commission, Bangkok. **3**(3), 6-7.

Department of Fisheries (1994) A proposed fisheries development and management law. Bangkok, Thailand.

de Sagan, R. (1992) The local government code and its provisions on fisheries. Bureau of Fisheries and Aquatic Resources, Quezon City, Philippines.

Farooque, M. (1997) Regulatory regime on inland fisheries in Bangladesh: issues and remedies. Bangladesh Environmental Lawyers Association, Dhaka.

Hinton, P. (1985) An approach to the study of traditional systems of coastal resource management in Thailand. In K. Ruddle and R.E. Johannes (eds) *The traditional knowledge and management of coastal systems in Asia and the Pacific*. UNESCO, Jakarta.

Hirsh, P. (1998) Holding and Managing Resources in Common Issues of Scale in Mekong Development. Paper Presented at the Seventh Common Property Conference of the International Association for the Study of Common Property. Vancouver, British Columbia. 10-14 June.

Juntarashote, K. (1994) Possible Development of Community-Based Coastal Fisheries Management in Thailand with the Advent of Fishing Rights System. Paper presented at the 7th Biennial Conference of the International Institute of Fisheries Economics and Trade, Taipei, Taiwan. 18-21 July 1994.

Kalagayan, B.N.V. (1991) Philippine initiatives in marine coastal conservation. Lundayan. **II**(1), 2-7.

Kahn, Benyamin and Ahmad Fauzi (2001) Fisheries in the sul sulawesi seas, indonesian country report: assessment of the state of the biophysical, socio-economic and institutional aspects of coastal and pelagic fisheries in the Indonesian part of the Sulu-Sulawesi Seas. WWF. Wallacea Programme-Indonesia.

Lhaopadchan, P. (1993) The pilot project in territorial use rights in fisheries. Department of Fisheries, Bangkok, Thailand.

Lopez, M.D.G. (1983) Notes on traditional fisheries in the Philippines. In K. Ruddle and R.E. Johannes (eds) *The traditional knowledge and management of coastal systems in Asia and Pacific.* UNESCO, Jakarta.

Majid, D.S. (1992) Coastal Fisheries Management in Malaysia. Department of Fisheries. Kuala Lumpur, Malaysia.

Ministry of Fisheries (1995a) Fisheries Sector Strategy. Hanoi.

Ministry of Fisheries (1995b) Co-management of living coastal resources in ASEAN: theory, practice and implications for Vietnam. Hanoi.

Ministry of Fisheries (1997) Master Plan for Fisheries to the Year 2010. Hanoi.

NEDA (National Economic and Development Authority) (1993) Medium- term Philippine Development Plan. draft. Manila, Philippines.

Nikijuluw, V.P.H. and Naamin, N. (1994) Current and future community-based fishery management in Indonesia. Indonesian Agricultural Research and Development Journal. Vol. **16**, No. 2, pp. 19-23.

Piumsombun, S. (1993) The socioeconomic feasibility of introducing fishery right system in the coastal waters of Thailand. Department of Fisheries. Bangkok.

Pomeroy, R.S. and Carlos, M.B. (1997) Community-based coastal resource management in the Philippines: a review and evaluation of programmes and projects, 1984-1994. *Marine Policy* **21**(5), 45-464.

Sajise, P.G. (1995) Community-based resource management in the Philippines: perspectives and experiences. Paper presented during the training course in Co-management of Living Coastal Resources in ASEAN: Theory, Practice and Implications for Vietnam. Ministry of Fisheries, Hanoi, Vietnam.

Sinkham Phomvisay (1997) Policies for Fisheries Management and Research in Laos. Ministry of Agriculture and Forestry, Department of Livestock and Fisheries. Vientiane.

Sulaiman, H. (1994) p. 160. in G. R. South *et al.* (eds) *Traditional Marine Tenure and Sustainable Management of Marine Resources in Asia and the Pacific.* University of the South Pacific, Suva, Fiji.

Tabunda, M.S. and Galang, M.M. (1992) A guide to the local government code of 1991. Mary Jo Educational Supply, Manila.

Thompson, P.M., Islam, M.N. and Kadir, M.M. (1989) Impacts of government-NGO initiatives in community-based fisheries management in Bangladesh. Paper presented at Seventh conference of the International Association for the Study of Common Property, Vancouver 10-14 June 1998.

Thong, H.X. (1995) The problems facing small-scale coastal fishing in Vietnam. Institute of Fisheries Economics and Planning, Ministry of Fisheries, Hanoi.

Thong, H.X. and Thieu, N.D. (1998) Traditional Organizations in Fishing Communities of Small-scale Fisheries in Vietnam. Fisheries Co-management Project Working Paper. ICLARM, Manila.

Tokrisna, R. and Duangsawasdi, M. (1993) Thailand experience in fisheries management in FAO/Japan expert consultation on the development of community-based coastal fishery management systems in Asia and the Pacific. *FAO Fish. Rep.* 474. FAO, Rome.

Tokrisna, R., Boonchuwong, P. and Janekarnkij, P. (1997) A Review on Fisheries and Coastal Community-Based Management Regime in Thailand. Fisheries Co-management Project Working Paper 32. ICLARM, Manila.

Torell, M. (1998) Institutional, Legal and Policy Perspectives on the Management of Aquatic Resources (especially fisheries) and the Aquatic Environment in Wetlands, Floodplains, Lakes and Rivers in the Mekong River Basin. Paper presented at the Seventh Common Property Conference of the International Association for the Study of Common Property. Vancouver, British Columbia. 10-14 June.

Van Zalinge, N., Thouk, N. and Tana, T.S. (1998) Where there is water, there is fish? Fisheries issues in the lower Mekong Basin from a Cambodian perspective. Paper presented at the Seventh Common Property Conference of the International Association for the Study of Common Property. Vancouver, British Columbia. 10-14 June.

Wahyono, U, and Chong, K.C., Suseno and Paklawo, R. (1992) Traditional community-based fisheries management practices in Indonesia. Directorate General of Fisheries, Ministry of Agriculture, Jakarta.

Yahaya, J. and Yamamoto, T. (1988) A socioeconomic study of fisheries management and conservation with particular reference to two artisanal fishing villages in Penang, Peninsular Malaysia. College of Economics, Nihon University, CENU International Publications Series No. 1.

Chapter 7

EXPERIENCES WITH FISHERIES CO-MANAGEMENT IN EUROPE

DAVID SYMES
The University of Hull, Department of Geography, Hull, United Kingdom

NATHALIE STEINS
Dutch Fish Product Board, Rijswijk, The Netherlands

JUAN-LUIS ALEGRET
Universitat de Girona, Girona, Spain

1. INTRODUCTION: THE GEO-POLITICAL COMPLEXITIES OF EUROPEAN FISHERIES MANAGEMENT

Fisheries management in Europe is confronted by a situation of exceptional complexity such as is found nowhere else in the world. Not only is the coastline highly fragmented and deeply indented, with much of the marine space separated off into distinctive semi-enclosed areas like the Baltic, North, Mediterranean and Black seas, but responsibility for fisheries management is also divided among a large and growing number of coastal states.

No fewer than 30 coastal states share Europe's coastline: some have extensive coastlines while others can claim only a few miles. Such is the complexity of land-sea relationships in Europe that only a handful of states fronting directly onto the Atlantic Ocean have relatively uninterrupted 200 mile Exclusive Economic Zones (EEZs). By contrast, for the seven North Sea coastal states and the nine Baltic States median lines define the geographical extent of their EEZs and thus the area of nominally sovereign water is often quite severely restricted. The situation in the Mediterranean, meanwhile, differs markedly from the rest of Europe. Here, the very narrow extent of the continental shelf, averaging barely 25 miles in width, has precluded the development of exclusive fishing zones beyond the 6 or 12-mile territorial limits - and so by far the greater part of the Mediterranean is defined as 'high seas' (Symes, 2000).

To these geo-political complexities must be added the deeply disturbing state of many fish stocks. As Europe has been the global epicentre of industrialization and urbanization, the surrounding seas have been subjected to intensive exploitation by artisanal, commercial and industrial fishing for more than a century. Today, practically all the known commercial stocks are fully exploited and many are being fished outside safe biological limits (ICES, 2001). As a result the immediate priorities for fisheries management are to reduce fishing capacity and effort, prevent further depletion of the resource base and develop long-term strategies for the recovery of commercially important fish stocks.

Even if it were not for the existence of the European Union (EU) and its Common Fisheries Policy, there would be a very strong presumption in favour of international cooperation in the management of the shared but depleted resource base. Paradoxically this same geo-political context would seem to provide a very restrictive environment for the development of co-management, except at the lowest spatial scales. In contrast to many other areas of governance, where devolved management has become the norm, during the final quarter of the 20th century responsibility for fisheries management in Europe has increasingly shifted towards the centre, limiting the scope for national and regional policy initiatives. The underlying reasons are hard to disentangle: globalization of the fishing industry, the increasing interdependence of fishing activities of several neighbouring countries in areas like the North Sea, and the fact that awareness of the need for strong political action to curb the growth in fishing effort coincided with the expansion of the European Community project in the 1970s, are all relevant factors in explaining why fisheries management in Europe becomes a matter of international rather than national political responsibility.

Post-war Europe has been characterised by shifting patterns of political alliance and economic cooperation. Its most recent phase has seen the collapse of the socialist economic bloc and the continuing expansion of the EU, with the anticipated incorporation of a dozen central and east European countries over the next few years. As a result it is likely that only a handful of coastal states will remain outside the EU by 2010 - but these will include three of Europe's most important fishing nations: Norway, Iceland and Russia. Much less certain is the political evolution of the EU towards a more federalist system - and equally uncertain, for the time being at least, is the precise form of the Common Fisheries Policy (CFP) after 2002. What is clear, however, is that enlargement of the EU involves the incorporation of many distinct, not to say divergent, political cultures, which adds still further to the complexities of fisheries management in Europe.

We begin our analysis of co-management in Europe with an exploration of the central institutions of the EU and their failure to date to allow the development of stakeholder participation in defining and elaborating fisheries policy. At present it is left very largely to the implementation process, and to the discretion of individual Member States, for co-management systems to find their full expression. The larger part of this chapter is given over to an analysis of national and local co-management systems in northern and southern Europe. We conclude with a comment on the prospects for remedying some of the deficiencies in relation to stakeholder participation and co-management in the current review of the CFP.

2. NON-PARTICIPATORY GOVERNANCE: THE COMMON FISHERIES POLICY

2.1. Evolution of the CFP

In order to comprehend the role of co-management - or rather the lack of it - in EU fisheries policy, one needs first to understand something of the overall system of governance and the basic nature of the CFP. The CFP had its origins in the early 1970s when, anticipating the accession of four new Member States each of which was an important fishing nation in its own right, the EC6 (Belgium, Denmark, Germany, France, Italy, Luxembourg) seized the initiative in establishing the principle whereby Member State vessels share rights of access to fish 'up to the beaches' of all other Member States and confronting prospective Member States with this *aquis communautaire*. In the event only three of the four candidates elected to join the EC: Norway, potentially the greatest prize, voted against membership. The full significance of the EC6's initiative did not become apparent until several years later when the decision was taken to combine the EEZs of all Member States into a single 'common pond' prior to the development of a unified management framework. Today, the CFP is the largest and most comprehensive regime for international fisheries management anywhere in the world. It embraces not only a resource conservation policy, but also a structural policy aimed at a balanced development of the industry and rational use of resources, a policy for the regulation of markets and a policy for the development of relations with third countries (Holden, 1994).

Although the foundations of a 'common market' for EC fishery products had been laid as early as 1970, the main elements of a resource management policy were not finalised until 1983, following a protracted and sometimes acrimonious period of negotiation lasting some six years. The most contentious area concerned the share of total allowable catches (TACs) to be allocated to Member States in the form of national quotas. This was to form the basic element of 'relative stability', a device intended to guarantee the historical relative size and scope of the national fishing industries of Member States and designed to offset the fundamental principle of non-discrimination which underlies all EU internal policy and which had been secured, in theory at least, over a decade previously through the equal access agreement. The CFP, launched in 1983, was only the second, fully comprehensive common policy, very much in the shadow of the Common Agricultural Policy and created ostensibly for very different reasons. It was hailed as a bold and imaginative, if potentially flawed, experiment in macro-regional governance of fisheries.

While in principle the CFP bound together all Member States, in practice it applied only in its complete form to the waters of the Atlantic and its tributary seas. Mediterranean waters were excluded from much of the resource conservation policy, principally on account of the absence of 200 mile EEZs and the limited presence of the EC in the form of only four out of the twenty coastal states flanking the Mediterranean Sea.

2.2. The decision making process

Some twenty years later, the CFP is regarded by many as a failure and not only by those who are sceptical of the European project in general or by those sectors of the industry disadvantaged by the Policy's inability to halt the depletion of resources, but also by the European Commission itself (Commission, 2001). Identifying any one overriding cause for this failure is difficult: it is possible to lay at least part of the blame at the doors of the administrators, the fishing industry and the scientists. Perhaps a common thread runs through the detailed criticisms, namely the lack of effective communication between these essential actors and, in particular, the failure to engage the main stakeholders - the fishermen - in any meaningful part of the policy process.

There is perhaps a sense of inevitability about the centralist, bureaucratic and authoritarian style of governance to emerge from the EU; in this sense, fisheries policy is no different from any other policy area. Accusations of remoteness and democratic deficit lie at the heart of the eurosceptic argument, alongside the nationalist rhetoric of repatriation of political decision-making. Beyond this, a fundamental structural problem affects decision-making. Tensions exist between the Commission, with sole responsibility for initiating policy proposals, and the Council of Ministers, which acts as the executive and legislative body. Moreover, the policy process seems largely impervious to external influence and immune from the normal processes of parliamentary debate. The final stages of decision-making in the Council of Ministers are preoccupied with issues of distributive justice and ensuring that policies do not infringe the interests of any one national fishing community unduly, in pursuit of the principle of non-discrimination. However, in the sense that non-discrimination in the form of equal access was set aside through derogation in respect of the 12 mile limits, time-limited restrictions imposed through Treaties of Accession and most recently through the fixed quota allocations agreed in the early 1980s - and in the context of progressive but uneven rates of stock depletion - these arguments take on a somewhat surreal nature.

The absence of the usual checks and balances of democratic institutions is compounded by the inadequacy of interest representation at the European level in an industry characterised by heterogeneity of functional and territorial interests (Lequesne, 1999) and in a political environment where lobbying has become a full time, professional occupation denied to working fishermen. National fishermen's organizations are obliged to confront a difficult, expensive and seemingly fruitless task of having to fight on two fronts, first to persuade their own national governments and secondly to campaign for moderation in the Commission's handling of the resource crisis. Dedicated European institutions are unable to lend much effective support. Europêche, the European Federation of Fishermen's Organizations, has proved incapable of melding a robust common position among the Member States and bringing any collective influence to bear on policy making at the European level. Likewise, the Commission's own Advisory Committee on Fisheries and Aquaculture, recently reformed to accommodate other stakeholders including the environmental NGOs, while having the ear of the Commission, does not appear to make much impact on key issues. Indeed, it is arguable that international NGOs - like WWF - with offices in Brussels are better organized and positioned to lobby the Commission effectively. And, whereas the Commission is bound to call for the opinions of the European Parliament (and the non-elected Economic and Social Committee) on all major policy proposals, it is not obliged to take account of their views in its final recommendations to the Council of Ministers. As a result, the European Parliament's views, voiced through its Fisheries Committee and sometimes lacking in consistency and betraying the influence of specific regional interests (Lequesne, 1999), tend to undermine rather than strengthen the legitimacy of the Commission's proposals.

Thus the European Commission sets the agenda for debate and controls the outcomes through its unique position as the initiator of all policy proposals. For its part the Council of Ministers determines the fate of those proposals as much according to the balance of national interests as in an unequivocal concern for the sustainability of the fisheries. In Van Vliet and Dubbink's (1999) three perspectives on the governance of fisheries - hierarchical, market and participative - the EU falls unambiguously into the first category where central institutions claim full responsibility for management, using legal powers and administrative process to enforce necessary rules and regulations on the fishing industry.

2.3. Policy implementation

If one essential element of the co-management concept - partnership in the formulation of policy - has been denied to the fishing industry through the obduracy of the EU's central institutions, the implementation of policy remains largely in the hands of individual Member States. Their task is to incorporate the decisions of the Council of Ministers within the framework of national legislation. Herein lies a further chance for co-management institutions to flourish. Very few decisions by the EU institutions have, in fact, enhanced the opportunities for co-management in the implementation of policy. One decision bucks this trend, namely the setting up of Producers' Organizations (POs) initially introduced under the markets policy as a common measure to help organize the first hand marketing of their members' catches through the operation of a withdrawal price scheme. In most Member States, POs have simply retained this limited function, but in the UK and the Netherlands their remits have been significantly extended to include the management of national catch quotas (see below). With a network of circa 180 POs throughout the EU, a basic organizational framework for co-management is already in place.

It is tempting to hypothesise that traditional forms of devolved and participative management have survived best in those areas relatively untouched by the CFP. The evidence, however, is inconclusive. For example, in the Mediterranean where the CFP's influence is limited to a limited range of technical conservation measures, the persistence of the *cofradía* in Spain and the *prud'homie* in France as local management institutions would seem to lend support to the hypothesis, although both are in different degrees under threat from more modern management systems (Alegret, 1999a; Frangoudes, 1999). However, in other parts of the Mediterranean, most notably in Greece, the underlying social and political cultures offer little positive encouragement to devolved or participative management (Hoefnagel, 1999). Even in the case of coastal waters, where the access derogation allows the Member State considerable scope for introducing additional management measures, the existence of co-management is patchy (see Symes and Phillipson, 2001).

3. CO-MANAGEMENT EXPERIENCES IN EUROPE

3.1. Co-management, decentralization and regionalization

In Europe, several regimes of co-management, decentralised management and regional management are found. It is useful to define these terms. Co-management refers to systems that are based on a division of responsibilities between state and fishing sector, usually involving delegation of responsibilities to non-governmental organizations operating at national, regional or local levels. This shared management can relate to either the decision-making process or the process of implementation or both. The range and type of possible co-management system may vary substantially. In this respect, the distinction between horizontal and vertical models introduced by Pinkerton (1994) is useful. Horizontal models emphasize the balance of initiative, influence and power between the state and the fishing industry; in other words, there is no presumption as to where management powers fundamentally lie. The Spanish *cofradías* can *de facto* be regarded as a horizontal co-management model. In the vertical model, the state holds the legal and moral authority and actual power but voluntarily allows aspects of management to devolve to resource users (Pinkerton, 1994). In Europe, most co-management models in marine fisheries are vertical models. The extent to which co-management responsibilities are devolved differ significantly, with the Dutch quota co-management system arguably being the most fully devolved. In addition, while co-management is essentially a partnership between the state and fishing industry, in the coastal waters we find situations where third parties, such as nature conservation NGOs, are involved in co-management.

Co-management should not be confused with decentralised management, which refers to the transfer of responsibilities and powers from the central government to autonomous regional or local public authorities. The Sea Fisheries Committees in England and Wales, which are the competent authorities for the management of fisheries in the 0-6 mile zone, are an example of decentralised management. But they also assume a degree of co-management in so far as committee membership is divided equally between elected representatives of local government and representation of stakeholder interests appointed by the fisheries ministries. Decentralization is a form of devolved management that can, but not necessarily has to be, part of a co-management system. Regional management systems may be regarded as a special form of devolved management and may relate to: (a) natural maritime regions or regional seas, which are normally shared by more than one nation state and which require some form of joint management to take account of transboundary resources, (b) functional/administrative divisions of space which may be used to differentiate management regimes but which have no connotation of territoriality, and (c) territorial seas, which represent an extension of territorial rights seaward in the form of regional EEZs (Symes *et al.*, 1996). The CFP is such a regional management system covering all of the EU's marine waters, in which the component of territorial management exists by derogation in the 12-mile zone. Thus, regional management and co-management are fundamentally different concepts, yet can be linked.

3.2. The northern Atlantic region

The Dutch *Biesheuvel* system probably ranks as the strongest co-management system for sea fisheries in Europe's northern Atlantic waters. Its introduction was the result of a severe management crisis. In 1976, The Netherlands was one of the first countries in the world to introduce a system of ITQs. Initially, quotas could only be transferred together with the vessel. In 1985, this was changed and quotas became transferable on their own. The introduction of individual quota alone was, however, not enough to prevent overcapacity and overfishing of the national quota (Langstraat, 1999). The government had failed to freeze the total capacity of the fleet. Fishermen began to invest in larger vessels with greater capacity, resulting in ever-increasing pressures on the national quota. A growing number started overfishing their ITQs. Many fishermen consequently found themselves in a classical Prisoner's Dilemma: fear that overfishing by others would inevitably result in closure of the fishery to the detriment of those who intended to spread their fishing activities, compelled them to become involved in the same undesirable race to fish. Market prices decreased as a result of a growing grey market and non-transparent market streams (Langstraat, 1999).

In response to these developments, the government introduced several control and enforcement measures to restrain the growth in fishing capacity and effort including, amongst others, a licensing system and a 'days-at-sea' regime. Despite the introduction of these regulations and the intensification of controls, the aforementioned management problems could not be resolved (Langstraat, 1999). As a consequence, the Minister for Fisheries had to resign and the government asked former Prime Minister Biesheuvel to search for a solution. A steering committee was set up, including representatives of the government and the fishing industry. Its main tasks were to restore confidence between the two parties and to achieve a situation of a controlled fishery together with maximum profit for the fishermen. The steering committee advised the government to form quota management groups within existing POs. In 1993, the so-called *Biesheuvel* system was formally adopted.

The *Biesheuvel* system is a co-management system that is based on the principle of division of responsibilities between the government and the industry. Members are obliged to transfer their right to manage their ITQs to their Group and commit themselves to the joint fishing plan and other rules. They maintain the right to use their own ITQs, and are also allowed to lease quota to or from other members. Transgressors are sanctioned by the Group (see Box 1).

Box 1: The Dutch *Biesheuvel* system for quota co-management

Only Biesheuvel Groups that are formally recognised by the government can perform quota management tasks. Each year, a Group has to apply for renewal of its activities, which will be granted provided it meets a number of criteria. There are eight Groups for the demersal sector and one for the pelagic sector. The Groups operate within the PO system. A total of 97% of the cutter owners joined a Biesheuvel Group. Group members are obliged to transfer their right to manage their ITQs to its executive board and to commit themselves to the rules. However, they maintain rights to use their individual quota under the conditions agreed upon in the fishing plan. Through the administration of the Group, fishermen are allowed to lease (parts of) quota to or from other Group members. If such opportunities are not present, transactions are allowed between Groups or with non-members. Group members are obliged to sell and register their catches through recognised fish auctions. Fines imposed by the executive board have to be paid immediately. These fines are usually higher than those imposed by the public authorities.

To make Group membership more attractive than non-membership, Groups receive an additional days-at-sea allocation of 10%. This means that they have greater flexibility in utilising their catch quota and it also allows fishermen to fish their quota in more remote areas. In addition, while quota transactions between Groups are allowed throughout the year, transactions with non-members are limited to a certain period.

The Groups are obliged to provide all information requested by the relevant authorities. Control takes place through scrutiny of their accounts and through direct controls on landings. In addition, social control plays an important role. In the Biesheuvel system, the executive board is held responsible for overfishing of the quota. Should this happen, the board will not only impose fines on those who are guilty of overfishing, but the offenders also have to pay an indemnity to those who have become the victims of their behaviour. In a scenario where offences by one fisherman have immediate consequences for his colleagues, the offender does not make himself popular. Social control is further strengthened by the threat that overfishing of the Group's quota can lead to a loss of government recognition, which means that the Group is no longer allowed to perform management tasks.

Source: Langstraat (1999)

The *Biesheuvel* co-management system has brought ecological and economic advantages to all parties involved (Langstraat, 1999). Since its introduction, offences against the quota regulations have decreased dramatically and overfishing of the national quota has not taken place. The system made possible the realization of a reliable monitoring system for both the landings and the utilization of the days-at-sea allocation. Landings are now evenly spread over the year. The transparency of the market for leasing and sale of ITQs facilitated the management of the national quota and a more efficient use of the harvesting potential. And, arguably the most important aspect of fisheries management, it restored the fishing industry's confidence in the government and forged a partnership.

At the start of the *Biesheuvel* system, there was much scepticism among the fishing industry. Many doubted that the system would work and that fishermen would be capable of quota self-management. However, it did not take long before fishermen were convinced of the benefits of the system. Fishermen are highly motivated to ensure its continuation and have even been actively promoting the system in light of the CFP review. Recently, the Dutch government and the fishing industry decided to initiate a so-called *Biesheuvel II* project. This project aims on the one hand to fine tune the existing system and to implement regulations that for legal reasons could not be implemented in 1993 and, on the other hand, to delegate additional management responsibilities to the fishing industry. The latter may even go beyond quota management to include, for example, the enforcement of technical resource conservation measures.

In the UK, sectoral quota management by the POs shows some similarities to the Dutch system. As early as 1984, the UK government vested quota management responsibilities with the POs. In 1997, all UK POs were managing quota under the sectoral system accounting for the management of more than 95% of the national quotas (Phillipson, 1999). The POs are obliged to maintain accurate statistical records of their members' catches in order to monitor the national quota uptake. The system is also monitored by the relevant government departments to which the POs regularly report. In general, however, POs are free to manage their own quota in the manner they choose, and quota allocation methods can vary for different species. For example, some POs allocate monthly quota to vessels on a flat-rate basis, while others allocate annual quota to individual vessels according to track record. In 1994, the position of the POs was further extended through their ability to purchase and subsequently 'ring fence' catch track records, which form the basis for the calculation of quota allocations. POs can buy out member vessels whose owners are leaving the industry and share out the vessels' track records. To prevent the purchased track record being lost, if vessels were to leave the PO, the PO has the option of 'ring fencing' the track record in the sense that it is permanently held within the PO. The UK's co-management system does not extend to those fishermen who, for various reasons, have not become PO members. This means that effectively two management systems are in operation: a co-management system for POs and a centralised regulation of non-PO members by the government. Indeed, there are fears among the fishing industry that the sectoral quota management system may break down if all vessels were to become PO members, since any relative advantages currently enjoyed by PO members in terms of their collective quota allocations would in many cases disappear in such a situation (Symes *et al.*, 1996).

In Europe's northern Atlantic region, only the Dutch and British POs have an important co-responsibility in the management of the national quotas. In other northern Member States, the POs mainly fulfil market regulatory functions (eg, application of withdrawal prices). Attempts to involve POs in quota regulation in France in the early 1990s failed mainly because in practice POs were unable to impose restrictions on their members for quota management regulations, particularly in the presence of non-PO members subject to no such controls. Since 1993, responsibility for quota regulations has reverted to the French government (Symes *et al.*, 1996). However, this is not to say that the Dutch and UK systems are the only examples of co-management in Europe's sea fisheries.

In the Scandinavian countries, the involvement of fishermen's associations has had a long tradition and can be traced back to origins of the co-operative movement in the 18th century (Symes *et al.*, 1996). In Denmark, for example, the Regulation Advisory Board includes a broad representation of the harvesting, processing and export sectors. This Board, which also includes scientists and administrators, advises the Fisheries Ministry in respect of structural measures and the regulation of fishing effort. A further link is provided by broad industry representation on the EU Advisory Board, which advises the European Committee of the Danish parliament on matters relating to the development of EU policy. These boards are advisory and do not confer any executive responsibility or exert undue political influence on the fishing industry. In Norway, the system of centralised consultation is one of the most sophisticated models of incorporation of the industry's position within the stream of professional and scientific advice, including the development of sector plans, fisheries regulation, and loan and grant negotiations. The national fishermen's association *(Norges Fiskerlag)* holds the largest single allocation of seats in the Regulation Council, which plays an important advisory role in resource management policy. There are, however, moves to reduce the role of the harvesting sector in favour of the processing sector and a more market-driven approach.

Whereas in sea fisheries, the boundaries for co-management models are set by CFP regulations, Member States have virtually autonomous powers in the coastal waters. This would suggest that opportunities for co-management are greater in the 12-mile zone. Interestingly, while forms of decentralised regional fisheries management are found along Europe's northern coast - the Sea Fisheries Committees in England and Wales being the most sophisticated model - co-management systems are only in force in a few countries. In contrast to co-management in the offshore fisheries, such models often involve partnerships with third parties. In the Netherlands, for instance, a statutory co-management system for shellfish fisheries has been in place since 1993. In this case, co-governance is aimed at balancing the interests of the fishing industry with the demands for nature conservation. In the early 1990s, natural factors in combination with a poorly regulated fishery resulted in very low shellfish stocks and consequent high mortality rates of shellfish eating birds. The relationship between shellfish fishermen and nature conservation groups reached a low point. After heated discussions, agreement was reached to close substantial areas for shellfish fishing and in years with low stocks 70% of the mean average food requirement in shellfish for birds will be reserved. Within these preconditions, the shellfish sector is responsible for implementing and enforcing a fisheries management plan, which has been very successful (Steins, 1999). The nature conservation NGOs were initially closely involved in the establishment of the co-management system resulting in a partnership between former 'enemies'. Recent experiences show, however, that the gap between these sectors is again widening. One of the reasons is the difficulties NGOs have in committing themselves to a co-management system for a fishery, which they believe should be banned in an area of high conservation value (Steins, 2000).

An emerging example of co-management in coastal fisheries is found in Ireland. Traditionally, fisheries management in Ireland has been characterised by a centralised approach in which industry involvement was negligible (Steins, 2001). Following the recommendations of a study into the coastal fisheries sector, the government has recently encouraged the establishment of voluntary Inshore Fisheries Development Committees to promote local involvement in decision-making and management of inshore marine fisheries. Their main objectives are the identification of employment opportunities through diversification into new fisheries, aquaculture and marine leisure/tourism activities; managing fisheries to ensure the sustainability of the resource; enhancing the economic position of the sector; improvement of fisheries-related infrastructure; and providing a forum where issues can be resolved at a local level and where consultation can take place between the state authorities and the coastal sector on future policy initiatives. The Irish Sea Fisheries Board has appointed Inshore Development Officers/Facilitators to co-ordinate the committees. Besides fishermen representatives, the committees generally include representatives from recreational fisheries, the marine tourism sector and nature conservation interest (BIM, 2000). In this light, the successful lobster v-notching programme to protect and enhance the stocks - a fishermen's initiative supported by the government (Steins, 2001) - can be regarded as a catalyst for a changing management ethos based on the devolvement of management responsibilities to the industry.

Probably the most comprehensive and structured system of participative management is to be found in France where the traditions of rural syndicalism have percolated the organizations for the fishing industry - yet even here the close relations between the state and industry do not equate to co-management in the sense of a clear division of responsibility and delegation of powers. Both the administrators (*Affaires Maritimes*) and representation of fishing interests are organized among similar hierarchical lines with interaction - rather than integration - at local, regional and national levels. Each of the 39 *comités locaux des pêches maritimes*, corporate professional organizations comprising vessel owners, crew members, merchants and processors established in 1945, can formulate policy proposals in relation to local issues but have no independent authority to determine local regulations. Instead their proposals are taken forward for consideration by the regional *comités* which occupy a pivotal position in the management system through their relationships with the regional office of *Affaires Maritimes* with powers to implement local regulations (Frangoudes, 2001). Likewise the national *comité*, mainly comprising officials of the constituent unions and associations but usually without the presence of active fishermen, works closely with the central government department in the elaboration of national policy.

3.3. The Spanish Mediterranean

Fisheries management models in the Spanish Mediterranean, as in the French, Italian and Greek fisheries, have traditionally been, and still are, based exclusively on the control of fishing effort (inputs). This is due to the multi-species nature of the fisheries and the large number of vessels dispersed throughout the coast, which makes any attempt to establish management through catch controls (outputs) extremely difficult.

As indicated earlier, the EU's conservation policy does not apply to the Mediterranean. The only element that has been adopted is the harmonization of technical measures (Reg. No. 1626/94), which has proved not to be very successful. The main reason for this failure is the specific character of the Mediterranean fisheries, including an important tradition of self-management, ancient traditions and historical management institutions (*cofradías* in Spain, *prud'homies* in France), small scale fisheries in an area that essentially is defined as 'high seas', the strong regional dependency on fisheries, an important market for undersized fish, a very high per capita consumption of fish and no tradition of strong enforcement in some areas. These characteristics have hindered the extension of the conservation policy into the Mediterranean.

In the past three decades, important political and social changes have taken place in Spain that have led to the consolidation of the democratic system and the creation of a new state model based on the transfer of competencies to 17 autonomous regional governments, some of them with direct responsibilities in fisheries management. Since 1987, the new democratic state has been trying to create a system of POs in the Mediterranean, with the fundamental objectives of adjusting catches to meet demands and allowing producers to benefit from participation in the marketing process, something that the historical *cofradías* did not do (see Alegret, 1996a). Despite these efforts, POs have not been established in the Spanish Mediterranean fisheries. The reasons for this failure lie, *inter alia*, in the social and political costs that would have resulted from replacing the *cofradías* and creating a power vacuum. The reduction in transaction costs that fishermen might have gained from the introduction of POs would have been insufficient to offset the increased social and political costs caused by the demise of the *cofradía*. However, the problems of adjusting catches to demand and of the participation of the fishermen in the process of marketing continue to be significant issues in Spain's coastal fishing sector. At the moment, the *cofradías* and their federations seem incapable of responding to changes imposed by the market: despite numerous attempts, they have been unable to reach agreement on setting maximum catch quotas for small pelagics (Alegret, 1999a); nor have they reached agreement to eliminate some of the privileges that certain local merchants still enjoy. This has brought the *cofradías* to a situation where they cannot represent the global interests of the sector. More specifically they have lost bargaining power in the face of merchants that now dominate the process of determining prices. Also the administration no longer sees the *cofradías* as the true co-management instruments that historically they once were. The consequences of this changing attitude are difficult to foresee.

Historically, the *cofradias* have been managed so as to maintain social order under *Gemeinschaft* conditions although not necessarily with the level of equality they aspired to. The mechanisms used by the *cabildos* (executive bodies of the *cofradías*) to reach consensus in collective action were complex and based on the existence of a network of family alliances, personal connections, clientalism, common values, political balances between fleets and gears and a certain level of local social pressure, possible only in contexts where everyone knows everyone else - all characteristics of community based societies. Membership of a *cofradía* includes both vessel owners and crew members and its regulatory role is defined in terms of a relatively restricted geographical area. In this sense, the *cofradías* reflect a more general range of interests in respect of the fishing industry and community. Since the 1980s, however, the *cofradías* openly started to reclaim areas of responsibility from the state in order to fulfil their own agreements and norms in a real regime of co-management. This development was strengthened throughout the 1990s, thus changing the components of traditional social order, to the point that currently external intervention by the state is deemed necessary to solve questions of internal order in the *cofradías* and demanded as the only possible solution to the multiple crisis situation that the fishing sector is undergoing. One possible explanation for this institutional change is that, due to the economic changes in recent years, the *cofradías* have started to transform their own governance system based on co-management and consensus into a system of top-down governance based on a majority vote (Alegret, 1999b).

One of the characteristics of the *cofradías* as organizations, according to the traditional model, is that they do not have separate governing bodies for formulating and implementing management agreements. The members of the *cofradías* are - or should be - those responsible for the implementation of and compliance with their own decisions, one of the principles of a co-management system. Contrary to popular belief, the *cofradías*, as they have existed over the past fifty years, are not the organization that had earlier existed in the Spanish fishing sector (Alegret, 1999c) The *cofradías* became public law corporations only when this was imposed upon them by the dictatorship in 1943. Before this period, the fishing sector had different organizational structures in tune with the conditions of the time. By granting the *cofradías* the status of public law organizations, the state institutionalised its intervention in the fishing sector, thereby establishing a legal basis for the present co-management system (Alegret, 1996b). This new system is still based on this statutory status of the *cofradías*. Through this state decision, the *cofradías* obtained management co-responsibilities for exclusive territorial areas and administrative tasks, and they became the formal body for consultation with the fishing industry.

Within their jurisdiction and due to their statutory status, each may collaborate with the state in organizing and regulating access to resources. Each *cofradía* undertakes the regulation and control of the access to resources for each of the different fleets operating within their territorial limits. According to the particular fleet - artisanal, trawling or purse seine - each *cofradía* establishes the timetable for port entry and exit, the possible close-season periods, and stipulates the complete set of specific norms for the area under its jurisdiction, including the zones of fishing activity for each type of fleet in accordance with the geographical, ecological, historical and social characteristics of each locality. *Cofradías* then propose to the administration to transform these resolutions into laws.

In addition to resource co-management responsibilities, the state has delegated tasks of an administrative character to the *cofradías*, including the management of catch and sales statistics, vessel registration, the collection of certain taxes, and control of the first stage of the commercialization process.

The juridical status of the sea, the behaviour of the producers, and the complexity of the control of the harvesting process, have been the elements used to explain and justify the existence of mutual collaboration between the state and the fishing organizations in the Spanish coastal fishing sector. This collaboration, which has been presented as a necessary form of co-management, has been consolidated into the kind of corporate institution best suited to the particular historical period. In other words, it is as if this co-management has created different forms of corporatism, capable of surviving as long as the political and economic conditions that generated them continue to exist. A specific kind of corporate co-management exists in the Spanish coastal fishing sector and will surely continue for as long as the model of small scale capitalist exploitation is maintained. It is a system based on the theoretical, formal and legal delegation of certain responsibilities of management and control of the harvesting activities to the fishermen.

The co-management system in the Spanish Mediterranean coast is, however, under threat because of changes resulting from the globalization of the fish markets, the over-capitalization of the fleets and the lack of fishermen's participation in the broader policy making process. At this time of important change, the co-management system represented by the *cofradía* should coexist with a changing market system, which the POs represent. These two kinds of organizations are mutually compatible because they have different structures but complementary goals. The main challenge for the *cofradías* is that they must adapt their ancient structures, procedures and discourses to the context in which new decision processes, new actors, new rules and new policies are present.

4. POLICY REFORM: TOWARDS IMPROVED GOVERNANCE

From the discussion of fisheries management in the northern Atlantic and the Spanish Mediterranean, it becomes clear that - while forms of decentralised and regional management and models that include substantial consultation with the industry can be found all over Europe - the existence of co-management, involving the division of responsibilities between government and industry, is patchy and limited to only a few Member States. The CFP in its present form does not offer a model in which co-management systems could flourish: partnership in the formulation of policy has been denied to the industry, resulting in a lack of essential support for most policies.

In the Green Paper on the reform of the CFP, the European Commission identifies a number of areas of failure, including the fact that 'politically the stakeholders do not feel sufficiently involved in the management of the policy' (Commission, 2001: 4). While the Commission does not anticipate the formal participation of stakeholders in the decision-making process either at Community or regional level, it recognises the value of encouraging new forms of participation in the pre-decision phase of policy making. It recommends the establishment of a network of regional advisory committees (RACs), involving 'national officials, industry representatives, NGOs, fisheries biologists and economists from the Member States with a real interest in the fisheries concerned' (pp 28-29) though it immediately dilutes the case for subsidiarity by suggesting that participation would also be open to industry representatives and officials from other Member States. Such committees, co-financed by the EU, national governments and the stakeholders themselves, would provide an opportunity for the direct involvement of stakeholders in the policy process before proposals were formally put to the Council of Ministers. The new system would, moreover, ensure that 'fisheries governance remains compatible with the legal and institutional framework of the Treaty' (p 29) and so leave the Commission's monopoly over policy proposals unchallenged.

These proposals fall far short of what the English and Scottish fishing organizations had in mind in their proposals for 'zonal management committees' which were to have quasi-autonomous executive powers in a number of important areas (SFF and NFFO, 2000), virtually short-circuiting the Commission's dominant influence in policy formulation. The basic elements of the 'zonal management' proposal were supported by most of Europe's fishing industry. Discussion of the Commission's proposals, however, reveals a significant fault line running through the EU over devolved management. Regionalisation is generally supported by the northern European countries but opposed by Spain not so much for reasons of principle but primarily because of a suspicion that even the modest proposals for RACs would reinforce 'relative stability' and, therefore, the continued exclusion of Spain from effective fishing rights in areas like the North Sea.

The Green Paper stops short of detailing the composition of the regional advisory committees or elaborating their functions. As Grieve (2001: 14) suggests, 'a more productive working environment could be created if the RACs were tasked with pursuing a clear set of objectives and striving for practical and effective management solutions based on open consensus and the best interests of the fishery as a whole'. What is unclear is whether there is any underlying intention to rationalise management policy by allowing a differentiated approach to the regional seas or whether RACs will simply feed comments into what is essentially a continuation of the 'harmonised' approach to management throughout the common pond as a whole. Equally unclear is the ability of RACs, combining several interest groups from several Member States, to function coherently in order to bring forward constructive management advice.

The assumptions implicit in the Commission's proposals are that membership of the RACs will closely reflect national and sectoral 'constituencies'. If, on the other hand, their primary purpose is to provide the Commission with the best possible advice rather than repairing the damage caused by the non-inclusion of fishing interests in the policy process, there might be an argument for preferring the appointment of persons to the RACs on the basis of their individual expertise and standing, as in the case of the Regional Councils in the United States or the Fisheries Resource Conservation Councils in Canada.

A further option for industry involvement lies in the further integration of two components of the CFP, the conservation policy and the markets policy, using the producers' organizations as vehicles for devolved management. Market Regulation 104/2000 imposes a number of obligations on POs, such as the implementation of fishing plans. The regulation also states that the need for sustainable fishing has to be taken into account. This implies that the POs can be asked to assume much of the responsibility for managing the annual catch quota for their combined membership as well as for managing fishing effort. Within such a management system, POs would have to draw up a comprehensive fishing plan, incorporating marketing, monitoring and enforcement plans. Recognition of POs that clearly fail to meet their objectives should be withdrawn, something that the Commission has so far failed to do. The ultimate goal of the system should be that all participants realise that they will punish themselves and their colleagues by violating the rules and that PO membership offers extra advantages, which makes it attractive to become a member (Langstraat, 2001). The attraction of such a system is that it makes use of existing regulatory and organizational frameworks, and is therefore likely to gain acceptance from the Commission, national governments and the fishing industry. The challenge for such a system obviously lies in 'generating collective benefits through the harbouring of individual interest of members in the organization' (Phillipson, 1999: 92), a problem UK and French POs have already encountered. Other reservations may be that POs represent the sectoral interests of only a proportion of the industry and that their marketing function could be eroded. These reservations can, however, be overcome by creating a system of incentives that makes PO membership more attractive than non-membership and by making use of the provisions in the market regulation which offer opportunities to form branch organizations including both fishermen and representation of the downstream sections of the production chain, through which PO activities can be tuned to the needs of the post-harvesting process. Finally, it is essential that POs be appropriately configured to accommodate within their remit broader management functions, including professional leadership and trained staff (Phillipson, 1999, Symes *et al.*, 1996).

At best, the prospects for improved governance of the EU's fisheries seem fairly modest. The Green Paper makes clear that improved governance has to be realised within the existing regulatory framework. In this light, the existing PO system offers opportunities through a further integration of conservation and market policy. Here, the ability of the responsible authorities to implement PO policy consistently is crucial. The Commission itself considers the establishment of RACs as the most viable prospect for improved governance. Much will depend on how well the RACs respond to the limited opportunities offered and how willing the Commission is to incorporate the committees' advice in its formal policy proposals. Thus, despite what is being heralded as the most important shake-up of Europe's fisheries management for a quarter of a century, it seems that the bureaucratic structures of the CFP will remain largely intact and that the real opportunity for co-management will continue to be nurtured at the national level and in relation to coastal waters rather than concerned with the initial tasks of shaping sound and sustainable policies at the level of the EU.

ACKNOWLEDGEMENTS

The authors wish to thank Dick Langstraat of the Dutch Fish Product Board for his valuable comments.

REFERENCES

Alegret, J.L (1996a) Ancient institutions confronting change: the Catalan fishermen's cofradías. In: Crean, K. and Symes, D. (eds), *Fisheries Management in Crisis*, 92-98. Oxford: Blackwell Science.

Alegret, J.L (1996b) Co-management and legitimacy in corporate fishing associations: the case of the confraries de pescadors de Catalunya, Spain. In: Meyer, R.M. *et al.* (eds), *Fisheries Resource Utilization and Policy. Proceedings of the World Fisheries Congress, Theme 2*, 342- 348. New Delhi: Oxford & IBH Publishing Co. Pvt. Ltd.

Alegret, J.L. (1999a) Alternative management models to deal with the purse seiner crisis in Catalonia. In Symes,

D. (ed.) *Alternative Management Systems for Fisheries*, 199-210. Oxford: Blackwell Science.

Alegret, J.L (1999b) Anthropology of fisheries governance: the incipient failure of collective action in Catalan *cofradía*. In: *Europaea* **II**, 45-64.

Alegret, J.L (1999c) Space, resources and history: the social dimension of fisheries in the Northwest Mediterranean. In: Symes, D. (ed.), *Europe's Southern Waters: Management Issues and Practice*, 55-65. Oxford: Blackwell Science.

BIM (An Bord Iascaigh Mhara) (2000) The inshore fisheries sector. At: *http://www.bim.ie*

Commission (2001). *The Future of the Common Fisheries Policy*. Brussels: European Commission.

Frangoudes, K. (1999) Conditions for implementing a licensing system: the French Mediterranean example. In Symes, D. (ed.) *Alternative Management Systems for Fisheries*, 188-198. Oxford: Blackwell Science.

Frangoudes, K. (2001) France. In Symes, D. and Phillipson, J (eds) *Inshore fisheries management in Europe*, 139-155. Dordrecht: Kluwer.

Grieve, C. (2001) *Reviewing the Common Fisheries Policy: EU Fisheries Management for the 21st Century*. London: IEEP.

Hoefnagel, E. (1999) Professional culture, fisheries management and the concept of co-management. In Symes, D. (ed) *Europe's Southern Waters: Management Issues and Practice*, 33-41. Oxford: Blackwell Science.

Holden, M. (1994) *The Common Fisheries Policy*. Oxford: Blackwell Science.

International Council for the Exploration of the Sea (2001) *Report of the ICES Advisory Committee on Fishery Management 2000*, ICES Cooperative Research Report No 242. Copenhagen: ICES.

Langstraat, D. J. (1999) The Dutch co-management system for sea fisheries. In Symes D. (ed.), *Alternative Management Systems for Fisheries*, 73-78. Oxford: Blackwell Science.

Langstraat, D.J. (2001) Fleet policy. *Keynote Speech for the Public Hearing on the Future of the Common Fisheries Policy organized by the European Commission, Brussels 5-7 June 2001.*

Lequesne, C. (1999) Studying European fisheries from a political science perspective: a research agenda. In Symes, D. (ed.) *Multi-disciplinary Research in Fisheries Management*. Hull: European Social Science Fisheries Network.

Phillipson, J. (1999) The fish producers' organizations in the UK: a strategic analysis. In Symes, D. (ed.) *Alternative Management Systems for Fisheries*, 79-92. Oxford: Blackwell Science.

Pinkerton, E.W. (1994) Summary and conclusions. In: Dyer, C.L. and Goodwin, J.R. (eds), *Folk Management in the World's Fisheries: Lessons for Modern Fisheries Management*, 317-337. Niwot: University Press of Colorado.

Scottish Fishermen's Federation and National Federation of Fishermens' Organizations (SFF and NFFO) (2000) *Zonal Management: A New Vision for Europe's Fisheries*. Aberdeen and Grimsby; SFF and NFFO.

Steins, N.A. (1999) *All Hands on Deck: an Interactive Perspective on Complex Common-pool Resource Management based on Case Studies in the coastal Waters of the Isle of Wight, (UK), Connemara, Ireland, and the Dutch Wadden Sea*. Wageningen: Wageningen University.

Steins, N.A. (2000) Co-managing fisheries and nature in the Dutch Wadden Sea. *Paper for Integrating Biodiversity and EU Fisheries Policy, Workshop 1, organized by WWF-UK, Gothenburg 15 March 2000.*

Steins, N.A. (2001) Ireland. In Symes, D. and Phillipson, J (eds) *Inshore Fisheries Management in Europe*, 119-137. Dordrecht: Kluwer.

Symes, D. *et al.* (1996) *Devolved and regional management systems for fisheries. Final report of EU Project AIR-2CT93-1392, European Commission DG XIV*. School of Geography and Earth Resources, University of Hull, UK.

Symes, D. (2000) Rights-based management: a European Union perspective. In Shotton, R. (ed.) *Use of Property Rights in Fisheries Management*. FAO Fisheries Technical Paper 404/1, 276-283. Rome: FAO.

Symes, D. and Phillipson, J. (eds) (2001) *Inshore Fisheries Management in Europe*. Dordrecht: Kluwer.

Van Vliet, M. and Dubbink, W. (1999) Evaluating governance: state, market and participation compared. In: Kooiman, J., van Vliet, M. and Jentoft, S. (eds), *Creative Governance: Opportunities for Fisheries in Europe*, 11-30. Aldershot: Ashgate.

Chapter 8

EXPERIENCES WITH FISHERIES CO-MANAGEMENT IN LATIN AMERICA AND THE CARIBBEAN

ALPINA BEGOSSI
Center of Environmental Studies and Research, State University of Campinas, Brazil

DAVID BROWN
CARICOM (Caribbean Community) Fisheries Unit, Belize

1. INTRODUCTION

Fisheries co-management is a situation where government agencies and fishers share responsibility in managing the fishery (Sunderlin and Gorospe, 1997), or where government agencies and fishers, through their cooperatives organizations, share responsibilities for management (Jentoft, 1989). In this volume it is considered as an arrangement where the responsibility for resource management is shared between government and user groups. The interchange of conserving biodiversity and promoting social and economic development, such as food security and social reproduction (Redford and Stearman, 1993) are the core of ideas for co-management in tropical areas. Such experiences, including interactions of local populations and management, are illustrated in Weber *et al.* (1999), for Latin America and Africa.

This chapter analyses different regimes of interaction fishers-fisheries, illustrating with fisheries from freshwater and marine environments in countries such as Brazil, Belize, Chile, Jamaica and Peru. We take into consideration the type of environment, mechanisms of control and management, as well as conflicts, to analyse why some communities have adopted co-management regimes, whereas others maintain only informal rules useful for management. We then address a number of co-management issues of general relevance to the continent. We conclude with reflections on future possibilities.

2. BACKGROUND

2.1. The Meanings of 'Co-management' in Latin America and the Caribbean

Both fisheries resource management and co-management are relatively new concepts in the Latin American and Caribbean region. The programmes described below as incipient forms of co-management hark back to traditional systems, while actual cooperative management between fishers and governments began to appear on the agenda in some areas in the early 1980s. These ideas came somewhat later to the Caribbean, the concept of co-management became familiar to resource users and policy makers in the mid-1990s in that area. It became part of the lexicon of Fisheries Field Officers doing extension work on the fishing beaches and in the fishing communities, through the Community Involvement sub-project activities under the CARICOM Fisheries Resource Assessment and Management Programme (CFRAMP) project, and through research by professionals and scholars in the general area of fisheries management.

For fishers, the meaning of co-management has largely derived from its contrast with centralized management and possible management by fishers alone, both of which they have consistently rejected in favour of management by co-operative effort by both fishers' groups and government (see Espeut, 1944, Brown, 2001a, 2001b, 2001c). The processes that would lead to the institutionalizing of co-management largely remained obscure, until their introduction to the emerging co-management pilot projects in the region, and the publicizing of these through video and other media channels via educational and awareness building programmes.

To the fisheries administrators and policy makers, the concept is linked to participatory planning and methods and stakeholder consultation for decision-making, as is highlighted by the CFRAMP project. However, the concepts of participation and consultation, depending on the nature of the issue(s) at hand, are applied to both discussions with stakeholders before major decisions are made, or after the major decisions have already been made behind closed doors. That is, when the preferred alternative courses of action have been officially decided upon and implementation is about to begin or has begun.

In the process, mainly due to the increasing popularity of the concept, a tendency has developed in some professional quarters to over-stretch the meaning to apply to even the most casual and transient interactions at the community level between government functionaries and fishers' groups. Projects planned in this vein, are labelled as based on co-management arrangements, thus relegating the concept to the realm of rhetoric.

Clarification of the various shades and levels of the meaning of co-management, and the processes and resultant institutional arrangements, is being addressed through capacity building educational and awareness building programmes, for both resource user groups and fisheries administrators. Through these means, the concept is associated with different points of operation along a continuum, stretching beyond the most centralized, 'top-bottom approach' to complete management control by resource user groups and fishing communities the 'bottom-up approach', with the mid-point, collaborative management decision-making, being currently preferred.

2.2. The Organizational Forms of Co-management

The fisher's organizations of Latin America and the Caribbean present some diversity in terms of their structure and robustness. For example, Costa Rican fisheries show incipient stages of organization compared to other Latin American countries (Breton, 1991). Authoritative states led in some cases to the arrest and prison of cooperative leaders, such as in Paraguay (Turner, 1988) creating obstacles to the success of local organizations. In other cases, such as for the *Colônias de Pesca* in Brazil, state intervention was strong enough to inhibit legitimate forms of fishery organizations. The *Colônias* trace back to colonialism and to state intervention into Brazilian fisheries in 1846, which reported alternately to the Ministry of Marine Affairs or Ministry of Agriculture during Brazilian history; during the military government many *Colônias* were presided over by sergeants or lieutenants (Breton *et al.*, 1996).

In the last ten years, rural organizations have directed priorities towards ecological and economical objectives involving a wide range of natural resources that were related to subsistence and to market exchanges. Beginning with forestry sectors (such as rubber-tappers in Brazil), such organizations transcended specific resource management to more general, multiple, management of natural resources, including fisheries. Some especially well known examples include the Extractive Reserves in Brazil, along with other similar correlates such as the Sustainable Development Reserve (*Reserva de Desenvolvimento Sustentável*) (Table One), and the *ejidos* in Mexico.

In different parts of Latin America co-management has emerged from very different histories. The Brazilian Amazon is one important example. In 1992, following the movement initiated by rubber-tappers, the government created the CNPT, the National Center for the Sustained Development of Traditional Populations. This centre followed the creation of many Extractive Reserves and it was an attempt at the *'Instituto Brasileiro de Recursos Naturais Renováveis'* (IBAMA), the national environment agency, to adopt strategies for co-management in Brazil, specially in the Brazilian Amazon, where local demands were being made explicit by rubber-tappers. This participatory phase included the bilateral programme in which Brazilian and German governments subsidized the IARA project, aiming at the sustainable use of fisheries in Amazon while considering local social factors (Castro, 2000). In the Middle Solimões, the Mamirauá Project represented another attempt at the local sustainable use of resources and it will be analysed below.

Co-management of fisheries in the Amazon is expressed by institutions called locally *Acordos de Pesca* (fishing agreements). These agreements were recognized for some Amazonian communities and lakes, but in other areas the IBAMA insisted in maintaining open access to the lakes (McGrath *et al.*, 1993; McGrath 2000). According to Castro (2000), several populations adopted the fishing accords to have control on the lakes they use, and in spite their illegal status, the fishing accords have spread throughout the Amazon Basin. As stressed by Castro (1992), the fishing accords are becoming the basic institutional unit upon which the co-management systems are expected to be built.

In the Caribbean Community (CARICOM) fisheries, governments faced with serious overfishing and habitat degradation in the inshore fisheries of the region responded in the 1980s by rejecting centralized management which ushered in open access conditions and establishing institutional arrangements that created an enabling environment in which fisheries co-management could thrive. CFRAMP and the Natural Resources Management Unit of the Organization of Eastern Caribbean States both dedicated to the promotion of the sustainable development and management of the fisheries resources of the region. All countries established National Fisheries Advisory Committees for the formulation of advice to governments on fisheries governance, developed Fisheries Management Plans (FMPs) that mapped out the direction of fisheries governance in each country and recommended relevant regulations, and granted powers to ministers responsible for fisheries to declare areas with competent fishers organizations 'Local Fisheries Management Authorities', with responsibility to manage the local fisheries. Progress towards the establishment of co-management institutions has been slow, although some pilot projects have been established. Various impediments and constraints engender significant differences in the rate of progress and the forms of co-management arrangements in different societies.

The Caribbean region has never lost sight of the fact that, 'Co-management of the resources is ... the ultimate goal for the effective involvement and participation of the resource users in the sustainable development and management of the fisheries resources'. It is in the process of creating a new regional institution - the Caribbean Regional Fisheries Mechanism (CRFM) - to take over seamlessly from the CARICOM Fisheries Resource Assessment and Management Programme (CFRAMP) when the latter ends its mandate at the end of September 2001 (CFRAMP 2000). Part of the CRFM's mandate would be:

> '... to ensure that the national fisherfolk organizations are further strengthened, and their capacities enhanced to become co-managers of the fisheries resources. Technical support should be provided for the development of national fisherfolk organizations ... to ensure that organized resource users in all the participating countries are effectively and democratically represented on the National Fisheries Advisory Committees (CFRAMP 2000)'.

The fishers and other stakeholders in the fishing communities in the region, in study after study, have opted for co-management as their preferred model for the management of the region's fisheries resources, overwhelmingly rejecting central management and sole management by the fishers' organizations and communities (Brown & Pomeroy, 1999; Brown, 2001a, 2001b, 2001c; Espeut, 1992, 1994; Mahon & Drayton, 1990). Fishers' organizations, however, were ill prepared for taking on the responsibilities of co-managers of fisheries (Pomeroy, 1998). CFRAMP, through the Community Involvement and Education sub-project, has had capacity building as the core of its strategy. Institutional Strengthening Workshops and Capacity Building Seminars have periodically been organized for Fisheries Field (Extension) Officers and members of the fisher organizations in the participating states. Extension officers have, to a large extent, overcome their lack of desire to go out in the field to interact with fishers, helping in finding solutions to the problems facing the latter and organizing community meetings, training programmes and public awareness programmes for fishers, students and other stakeholders in the fishing communities. Consequently, there has been a general improvement in resource users' awareness of, and responses to conservation and management initiatives. There have been encouraging responses to the consultative and decision-making processes, and most organizations have become more stable and assertive, with new organizations emerging (McConney *et al.*, 1998).

In spite of several obstacles, there are clear signs that the fishers' organizations are realizing their potential to effect changes in management policy, and are moving towards taking up the responsibilities involved in the role of institutional strengthening and further capacity building, to ensure their preparedness for the tasks that lie ahead. Among the regional fishers' organizations, the Belize Fishermen Cooperative Association, discussed below, stands out as a regional classic case of a dynamic partnership between a resource user organization and government functionaries in a form of open-ended co-management arrangement, not based on any formal agreement or legal stipulations, but recognized by all the parties involved.

3. LATIN AMERICAN AND CARIBBEAN EXPERIENCES WITH CO-MANAGEMENT

In this section, Latin America and Caribbean fisheries are approached on three levels: cases in which local rules exist but there is no clear demand for management, cases in which fishers participate in specific regulations, demanding or changing specific laws and legalized co-management. In each case the focus is on the context in which management takes place, the institutional arrangements, the biological aspect associated with subsidies and developmental policies, as well as the conflicts at different levels among fishers and institutions, including the involvement of NGOs as negotiators.

3.1. Incipient forms of co-management

Many different cases can be found in which informal rules govern fishing areas and activities. These rules do not necessarily give a clear direction to conservation, but can have implications for both management and conservation. Informal and formalized fishing areas and territories have been pointed out in the literature on fisheries (Acheson, 1981), and such rules occur in different parts of the world and at different scales, ie, individual, families, groups, clans, villages (Berkes, 1985). These systems include rights, ie a claim protected by a customary law and practice, and rules about how property rights are exercised (Ruddle, 1996). As pointed out by McCay and Acheson (1987), territoriality can be the basis for developing more restrictive common property institutions. Illustrative cases occur in Brazil and Peru.

3.1.1 Artisanal Brazilian Coastal Fishers

Ever since Forman (1967) noticed the secrecy about fishing spots among northeastern fishers in Brazil, scholars, especially Cordell (1974, 1978, 1989), have examined the link between territory and management. Cordell's studies of fishers from Bahia State, Brazil involved the mapping of fishing spots used and the examination of the environment information used in the choice of spots. Rules about the use of these spots were observed by Cordell (1989), especially a very informal rule, *respeito* (respect) of the use of the fishing areas. Fishing decisions depended on environment information, such as the cyclical regularities of the tides affecting the distribution of species in the estuary, along with the fishing methods used (Cordell, 1974).

Other examples include the stable fishing spots found among southeastern Brazilian artisanal fishers. Along the Brazilian southern coast, fishing areas are divided informally into fishing spots and spots used by fishers more than ten years ago in the Atlantic coast are still the same (Begossi, 2001 a,b). Other forms of incipient management are revealed through conflicts among different categories of fishers, such as at Sepetiba bay (Rio de Janeiro State), in which fishers' claims include the exclusion of shrimp and herring trawlers from the bay. The bay is seen as a spawning ground were the tiny nets of shrimp trawlers cause impact on the stocks (Begossi, 1995). These are examples of local knowledge and rules that are already in force, but are not perceived and used by planners and by officials and representatives of Federal or State environmental agencies for conservation measures.

Property rights over fishing spots may be seen as incipient forms of local management, as long as they avoid overlap in spots or areas and the concentration of fishing in one or on a few spots (Begossi, 1995). According to Cordell (1978), fishers learned how to capitalize on patterned behaviours of fish, embodying such knowledge in 'systems of orientation' that enabled them to map the spots through which fish pass periodically. The selecting of fishing spots and temporary territorial claims in the southern coast are decision making processes with implications for local resource management.

3.1.2 Fishers from Lake Titicaca, Peru

Studies from Levieil and Orlove (1990), and Orlove (1991) in 151 fishing communities located at Puno, Lake Titicaca, found local communally controlled territories. Rights over two resources are delineated: the beds of totora reeds, which are plots owned by individuals and with a restriction of totora collection, and the right to fish within the communal territory. The later right is then distributed by open access within the community. The Peruvian State claimed exclusive control of aquatic resources and conflicts resulted as the communal fishing territories continue to operate. This case is almost a case of independent, non-official management in which there were conflicts with the state over attempts at local management.

3.2. Fishers Participation in State Regulations

3.2.1 Ibiraquera Lagoon in Brazil

The active interference of local fishers, demanding and modifying local laws, or even demanding management arrangements directly from the government, is illustrated through the case study by Seixas and Berkes (2000) of the Ibiraquera lagoon, located in the southern coast of Brazil in Santa Catarina State. Shrimps and mullets are the main resources used by about 350 professional fishers. The study by Seixas and Berkes (2000) describes the historical steps towards local fishing management. Until 1960, fishing rules based on respect were decided locally.

The local management system was disrupted during 1960 and 1970, after a shift from the subsistence fishery to a regional-market oriented fishery. A local crisis in the fishery led fishers to prohibit gillnets and cast nets in some sites of the lagoon, but this arrangement did not last long. In the early 1980s, the fisher organization (*Colônia de Pescadores*) demanded regulations from the Federal Government, represented by *Superintendência do Desenvolvimento da Pesca* (SUDEPE). SUDEPE agents worked with local fishers setting up new regulations for Ibiraquera lagoon in 1981 (N-027/81) banning the use of all nets except cast nets and requiring minimum mesh sizes of 2.5 cm for shrimp and 5.0 cm for fin fish. Another regulation in 1986 (N-09/86) banned the use of gas lamps, allowing only kerosene lamps for shrimp fishing. From 1992 to 1998 a shrimp-stocking project was carried out with local fishers and local research institutions, such as the Federal University of Santa *Catarina* and a State research agency (EPAGRI). This research, in agreement with the Colônia proposed establishing a minimum mesh size of 3.0 cm for shrimp (regulation N-115/93). From 1981 to 1994 the regulation worked well as long as there was strong rule enforcement. Since then, another crisis has emerged along with another economic shift from just fishing to including tourism-related activities, and the lagoon system is being pressured by unregulated fishing activities.

Seixas and Berkes (2000) stress that when fisher knowledge is taken into consideration in making official regulations, they are more likely to succeed. However, some recommendations from local fishers of Ibiraquera Lagoon have not been accepted by the government. One such recommendation, from 1984, related the opening of a channel connecting the upper and middle lagoon sites, which are important for fish passage. Similar cases occurred in other coastal sites of Brazil, when local demands from the fishers helped in regulations forbidding nautical sports at Praia Grande Beach (Municipal Law 348/88, Britto, 1999).

Special attention, however, should be taken concerning top and down initiatives of creating Extractive Reserves, as occurred in the Itaipu fishery, Rio de Janeiro. In that community, the CNPT-IBAMA failed in its attempts to create an Extractive Reserve, because fishers had no demand for such a reserve, nor knowledge on the mechanisms, legal status and implications of that reserve for the fishery.

Table 1. Selected cases of legalized management in Latin American fisheries. Sen and Nielsen (1996) categories are IF=informative, CO=Cooperative, AD=Advisory, IS=Instructive, and CS=Consultative.

Author/Locality	Co-management/ Control	Institutions	Management Practices	Conflicts
Almeida and Menezes (1994), Begossi *et al.* (1999) BRAZIL: Upper Juruá Extractive Reserve, Juruá and Tejo rivers, Acre. *[IF]*	Management plan approved in 1994 by IBAMA. There is local participation, through meetings in assemblies with ribeirinhos and researchers.	CNS, ASAREAJ[1], Universities, and NGOs. IBAMA represents the Federal State. Participation of the Catholic Church. The local radio is called Green Forests *(Verdes Florestas)*.	Closed entrance to outsiders. Use of natural resources for subsistence (fishing and hunting). Prohibition to use nets in the mouths of the rivers, among others.	Conflicts occurred with municipalities and with rubber land owners *(seringalistas)* when the reserve was established.
Castilla and Fernandez (1998) CHILE: Management Exploited Areas, Coastal marine areas. *[CO]*	-1976-81, increased landings, which decreased thereafter. -1989-1992: total closure. -1993: Chilean Fishery and Aquaculture Law (FAL) enforced and quotas were used (TAC: Total Allowable Catch)	MEA-FAL, artisanal fishers, small-scale fisher Unions *(sindicatos)*, which operate in small coastal villages *(caletas)*; there are about 190 caletas. Fishers unions, between 1991-93 were active in order to obtain MEA.	Co-management (MEA-FAL): area-specific management plans, where artisanal fishers are allowed to harvest benthic resources; use of TAC; MEA-TAC are rooted into small-scale fishery communities and cannot be transferred.	Conflicts between artisanal and industrial fleets around fishing ground rights; FAL implemented exclusive fishing rights within a five-mile to small-scale vessels.
McDaniel (1997) PERU: Communal Reserve Tamishiyacu-Tahuayo, lakes and rivers. *[AD]*	Communal management in 1984. RCTT established in 1991. Community elects an inspector of the fishery, vigilance system; the fishery is more sustainable than in 1980.	Community leaders, working independently approached Peruvian government. With the cooperation of the Ministry of Fisheries and Agriculture, the community was granted rights to control lakes.	- prohibition of using nets in lakes during low water season; communal regulations to capture paiche *(Arapaima gigas)*, among others.	Conflict with the commercial fleet of Iquitos.
SCM, CNPq, IPAAM (1996)[2] www.pop.tefe.rnp.br /reserva/reserva/htm : BRAZIL: Sustainable Development Reserve of Mamirauá, Solimões and Japurá rivers, Amazonian várzea. *[IS/CS]*	Formed without population consultation (decree 12.836, in 1990). It includes local participation through the SCM (Civil Society Mamirauá). The Management Plan includes entrance restrictions for outsiders, restrictions on boat sizes and lake management.	RDSM (with local participation), and agreements with IPAAM (State Environmental Protection Agency), IBAMA/MMA (Federal Agencies), and CNPq (National Research Council). Support by many NGOs, such as ODA, WCS, WWF, WWF/UK. Participation of the Catholic Church.	Management includes local participation, through meetings in assemblies and a Deliberative Council, with NGOs, governmental agencies, research institutes and universities. There are protected areas and sustainable use areas.	Conflicts with outsiders, such as fishers that demand fishing in reserve lakes. Lakes are closed for fishers from Manaus and from Manacapuru.

[1] CNS is National Council of Rubber Tappers *(Conselho Nacional dos Seringueiros)*, and ASAREAJ is Association of Small Farmers and Rubber Tappers of the Upper Juruá Extractive Reserve *(Associação dos Seringueiros e Agricultores da Reserva Extrativista do Alto Juruá)*

[2] EEM *(Estação Ecológica de Mamirauá)*, SCM *(Sociedade Civil de Mamirauá)*, RDSM: *Reserva de Desenvolvimento Sustentável Mamirauá* (RDSM) IPAAM: Instituto de Proteção Ambiental do Estado do Amazonas, IBAMA: Instituto Brasileiro de Meio Ambiente e Recursos Naturais Renováveis, MMA: Ministério do Meio Ambiente, Recursos Hídricos e Amazonia Legal, CNPq: Conselho Nacional de Desenvolvimento Científico e Tecnológico, ODA (Overseas Development and Administration), WCS (Wildlife Conservation Society), WWF (World Wildlife Foundation).

3.3. Legalized co-management

Compared to other areas of the world, Latin America has few studies on fisheries and management. We find few examples of co-management in the literature on Latin America fisheries, especially if we confine ourselves to legitimate co-management, in which local organizations really participate in management programmes (Table 1).

Local organizations have often adapted to new global and environmental requirements as they set up environmental regulations for fishing activities. Legalized management is tricky, because it is often difficult to be sure that is a bottom up approach, or that there is real participation by the state, rather than simply a local demand for the state's participation. The examples of legalized co-management described in Table One include the Extractive Reserve of the Upper Juruá , in Brazil (Almeida and Menezes, 1994; Begossi *et al.*, 1999), located in a riverine environment, the Management Exploited Areas in Chile, covering coastal marine areas (Castilla and Fernandez, 1998), the Reserva Communal Tamishiyacu-Tahuayo in Peru (McDaniel, 1997), and the Sustainable Development Reserve of Mamirauá (SCM, 1996). In three of these cases there is conflict between artisanal and commercial fisheries, and in the Brazilian case, with other institutions including state representatives. The four cases include specific regulations concerning the use of natural resources, especially fish, including exclusion of outsiders, areas defined for use, and specifications concerning the use of technology or the exploitation of certain species. The State is represented by environmental agencies or resource specific agencies. The Peruvian and Chilean cases deal only with fisheries while two Brazilian cases include multiple resources (Table 1).

3.3.1 Extractive Reserves in Brazil

Extractive Reserves, considered forest areas in which long term usufruct is given to local communities, were legalized in Brazil in 1990, with the creation of the Upper Juruá Extractive Reserve (Begossi, 1998; Begossi *et al.,* 1999). The land of an Extractive Reserve is from the Nation, directed by IBAMA (Almeida and Menezes, 1994). The first plan of the reserve was done by the National Council of Rubber-Tappers in 1988, and the first management plan was approved in 1991 (*Plano de Utilização*) (Almeida and Menezes, 1994).

An important aspect of co-management in Extractive Reserves is its informative (Sen and Nielsen, 1996) kind of management. Management is explicitly carried out by local families, which are locally organized and tied to the country and to the world through the CNS (*Conselho Nacional dos Seringueiros*). They have the chance to elaborate management plans, associating the economic necessity of extraction and cultivation to the ecological necessity of conservation of natural resources. One reason of the global ties and for the national success of the Extractive Reserves was the movement organized by Chico Mendes. After his death, the movement continued and brought more alliances through NGOs and Brazilian Universities. There is a strong tie between Extractive Reserves and research, as a result of rubber-tapper demands. The national and international pressure on the Brazilian government in order to create Extractive Reserves was strong enough to bring about the first legalized Extractive Reserve in 1990, the Upper Juruá Extractive Reserve. The impact of the rubber-tapper movement at international levels provided a strong push legalizing the rubber-tapper claims for extractive reserves (Begossi, 1999).

3.3.2 The Portland Bight Sustainable Development Area

The PBSDA is an example of NGO sponsored co-management. The Caribbean Coastal Area Management Foundation (C-CAM), formerly known as the South Coast Conservation Foundation, became the first NGO dedicated to the sustainable management of coastal resources, to respond to the serious stock depletion situation in Jamaica. It established a complex Community-based Coastal Zone Resource Co-Management project in the Portland Bight area of southern Jamaica. This is the largest embayment in Jamaica, hitherto very rich in fishery resources, with a complex eco-system, including nursery grounds for fish, crustaceans and mollusks. The project is a multi-coastal resource project, going beyond fisheries management to the management of wetlands, forestry, coral reefs and other rare fauna.

The project was established through the provision of technical support for mobilization and formation of Fishers Associations embodying active fishers and other stakeholder groups in the communities, the building of the capacities of these organizations for co-management through education and awareness building programmes, the zoning of marine space among multiple user groups, the establishment of fish sanctuaries, and the establishment of a professional administration with officers responsible for capacity building and surveillance and enforcement of resource use rules and regulations.

The project instituted a local resource management council - the Portland Bight Fisheries Management Council (PBFMC) of 32 members drawn from all stakeholder groups, including representatives of fishers' organizations, the local branches of the security forces, the Natural Resource Conservation Authority (NRCA) and C-CAM. Among other things, this supreme body is charged with the responsibility of making and enforcing local resource use rules and regulations. The legal backing was provided by the NRCA, the public agency charged with the responsibility of declaring and managing protected areas on behalf of the government, and to transfer the authority to manage particular areas to competent NGOs. By this means, the C-CAM became a facilitator of resource co-management in Portland Bight. Management sustenance of the fisheries resources is invested in the PBFM, government's recognition of C-CAM's management responsibilities, and fiscal sustenance of the project, through resource user fees and nature tourism activities in the area.

This co-management arrangement enhances community democracy and empowerment, while acknowledging the duty of the state to be involved in the process. C-CAM therefore functions as a facilitator. Ultimately, as institutional strengthening and capacity building take firm hold, C-CAM would withdraw from the project to give way to a direct co-management between government and the communities.

3.3.3 The Belize Cooperatives

The Belize Fishermen's Cooperatives are an example of a fisher organization model for co-management. The Cooperatives emerged in the 1960s, some two decades before Belize achieved political independence, as a mass movement that had wrestled lobster processing and exporting rights from foreign monopolists, accused of exploiting the local producers through exploitative producer prices. Each cooperative is owned and managed through an elected management committee. Its umbrella organization, the Belize Fishermen Cooperative Association (BFCA) formed by representatives of the member organizations, uses the defence of this privilege-turned right of monopoly over the production, processing and exporting of lobsters, and later on, of conch, as a rallying point for unity against any imagined or real threat from both internal and external sources.

The organizational strength of the movement is primarily dependent on the defence of this monopolistic right; their economic strength is derived from the lucrative trade in lobster and conch products. This significantly enhances their independence and reduces their dependence on government largesse. The BFCA operates from a position of both political and economic strength, and has been successful in warding off any attempts to deprive them of their hard won rights. The movement embraces the conservation ethically, and regularly organizes educational and community awareness programmes for its members and stakeholders in the fishing communities, thus building their capacities for the responsibilities involved in co-management. The practice of fiscal responsibility and accountability by the leadership makes them above reproach, which contributes to group cohesion.

The Cooperatives have the ability of bargaining for concessions from governments, and are able to influence decision-making through dialogue, lobbying, negotiations and effective use of their membership on the National Fisheries Advisory Board. The BFCA remains the only appropriator organization in the region that consistently accepts, or invites itself to decision-making forums. It is one of a few that vehemently protest, when decisions are made without their involvement.

Elements of co-management built into the process include supportive surveillance of the fishery, and participatory decision-making in the formulation and application of conservation regulatory measures. This is a classic case of a dynamic partnership between a resource appropriation organization and government functionaries in management relationships, in which the scale of strength and influence seems to weigh in favour of the former. This is a form of open-ended co-management arrangement, not based on any formal agreement, or legal statutes, but recognized and respected by all parties involved in the relationship.

4. LESSONS FROM LATIN AMERICAN AND CARIBBEAN CO-MANAGEMENT EXPERIENCES

4.1. The Co-management Spectrum

The four cases of co-management described in Table 1 are classified, according to Sen and Nielsen (1996) and Sverdrup-Jensen and Nielsen (1997) co-management spectrum, from the government stand point to the user group action as: a) Sustainable Reserve of Mamirauá, Amazon, Brazil, as a case between *instructive to consultative*, because the decision about the reserve was a government decision, but there are mechanisms for consulting users; b) Management Exploitation areas, in the coast of Chile, as a *cooperative* co-management, because there are organized fisher unions and an active government participation; c) Reserva Communal Tamishiyacu-Tahuayo, Amazon, Peru, as *advisory*, because community leaders approached the government, and granted rights to use lakes (a case of users advising government that endorsed the decisions), and d) Extractive Reserve of the Upper Juruá, Acre, Brazil, as *informative* because the government delegates authority to local decisions. There are slight differences among those categories, and we may find difficulties in finding if some form of co-management is advisory or informative, for example, also because we do not have enough information in order to classify such cases. But Sen and Nielsen (1996) categories are certainly very useful to identify legitimate forms of co-management. Considering the importance and authority of the local organizations for the success of co-management, the first case, of Mamirauá in Brazil is probably more a case of consultative arrangements (Jentoft, 1989) rather than of co-management.

An alternative classification used for the CARICOM area (Figure 1), combines ideas from Sen & Raakjær (1996), Pomeroy & Williams (1994), Borrini-Feyerabend (1997) and Pomeroy (1998). This approach conceptualizes co-management arrangement as a process becoming more consolidated along a linear continuum with co-management institutionalized somewhere at mid-point as discussed above at the end of section 2.1.

According to this scheme, the fishing practices that are heavily based on traditional TURF management systems, in which there is minimal information exchange and occasional consultation between government functionaries and fishers, and by which fisheries regulations and their enforcement fall within the functional areas of the fisheries administrations and final decisions are made by government, are classified as falling between Instructive and Consultative. A second group falling between Consultative and Cooperative Negotiation covers Sea Urchin production in St. Lucia and Barbados, and the Lobster and Conch fisheries of Belize. Under this there are more regular consultation and information sharing. The co-management arrangement is mainly limited to collaborative rule making, monitoring and surveillance; organized resource users are strongly represented on the National Fisheries Advisory bodies, thus fostering an avenue for the resource user groups to influence decision-making, but there are no formalized and legalized co-management institutional structures in place. A third group falls between Cooperative Negotiation and Institutionalized Co-Management. Under this government regularly involves resource user and other stakeholder groups in the decision-making process; co-management is formally and legally institutionalized, with localized sustainable management structures in place, and user groups and stakeholders operate as partners with government fisheries administrators and NGOs, the latter as catalysts of the process.

Comparing the classifications used in Table 1 and Figure 1 is instructive. The first group on Figure 1 might be equated with the Sustainable Reserve of Mamirauá, Amazon, Brazil; and the second could be the equivalent of the Management Exploitation Areas on the coast of Chile; and the third group would come under the Informative with the Extractive Reserve of the Upper Juruá, Acre, Brazil as an example. None would be exactly equivalent to the Advisory Reserve Communal.

4.2. Environmental Factors

Environmental features are difficult to track and to relate to forms of co-management. Three of the co-management cases in Table 1 are localized in river-lake systems. The possibility of defined territorial sites might contribute to local control and management. However, other factors might also be responsible to observe co-management in some sites and in other not. We may suggest that historical patterns might influence the capacity for local management to evolve. For Brazilian sites, the high level of local organizations, including fisheries, found in the Amazon, compared to the south, might be explained by local history, political alliances and international interactions. Considering the high standard level of infrastructure in the south (highways, communication), the north developed creative ways of communication, including the radios that communicate families along the isolated Amazonian rivers, such as the Green Forest radio (*Verdes Florestas*) of Cruzeiro do Sul, Acre (Begossi, 1999). Aswani (1999) suggested for the Pacific Region, that not only exogenous agencies, but that also a complex of historical autochthonous intraregional processes resulted in different sea tenure systems in the Pacific.

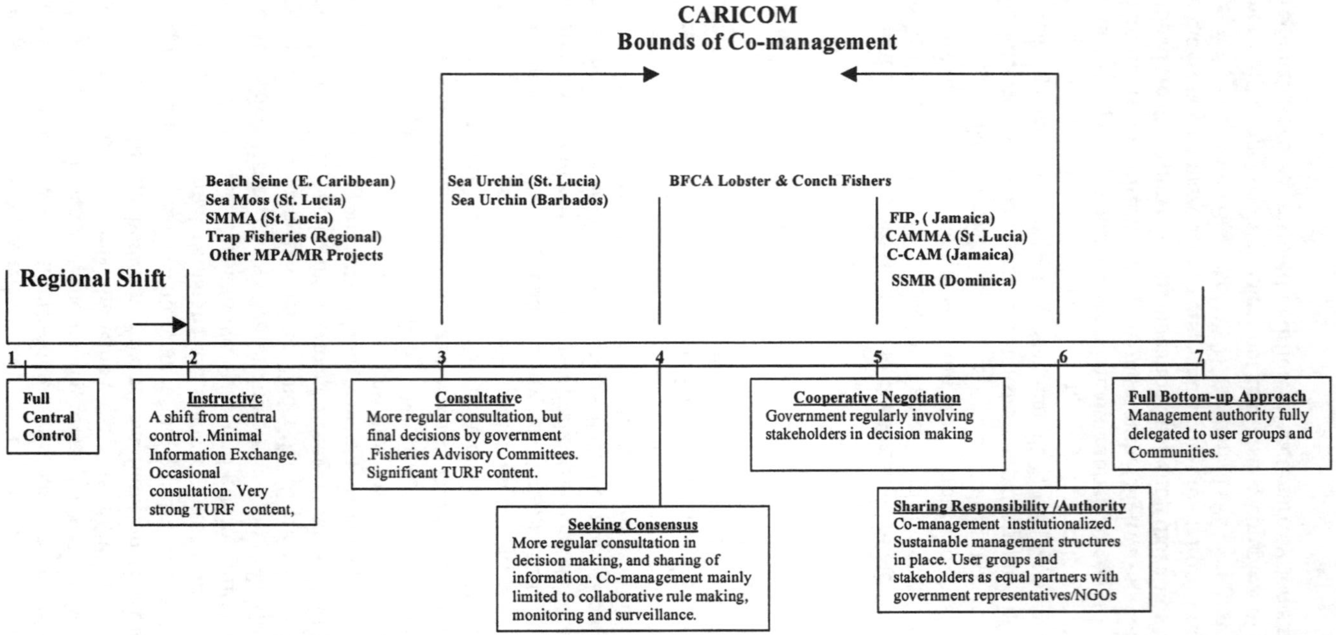

Source

- Pomeroy &Williams (1994)
- Sen & Rackjaer (1996)
- Borrini-Feyerabend (1997)
- Pomeroy (1998)

Acronyms (Pilot Projects)

SMMA: Soufriere Marine Management Area

CAMMA: Canaries-Anse La Raye Marine Management Area, St. Lucia

CCAM: Caribbean Coastal Area Management Foundation, Jamaica.
SSMR: Soufriere- Scotts head Marine Reserve
FIP: Discovery Bay Fisheries Improvement Project, Jamaica

Figure 1. Spectrum of Co-management Arrangements in the CARICOM Region

4.3. Forms of Organization

Co-management seems to work when fishers have responsibility for distributing their share and for determining rules of access to fishing grounds (Sinclair, 1990). Due to the authoritative aspects of the Latin America States, exemplified by so many military governments in a variety of countries, it is hard to be sure if an appropriate amount of independence should be guarantee in a co-management regime. Substantial independence for local organizations might be a condition for co-management to succeed in Latin America. Extractive reserves seem to be in such a position, and maybe fishers cooperatives might sustain some co-management. Fisher's cooperatives organizations are in a position to make more equitable regulations than are governments, and they are more capable to respond to special needs (Jentoft, 1989). McGoodwin (1989a) illustrated the importance the cooperative in Sinaloa, Mexico regarding the rights on the shrimp fisheries. He pointed out that cooperatives should be encouraged as independent corporate entities. Mexican fisheries have demonstrated their effectiveness in a variety of studies (Dyer and McGoodwin1994, McGoodwin, 1989b) . However, there can be a great distance between legal structures and the actual practice of co-management (Pinkerton, 1992) and there are many cases in the literature that are interpreted simplistically as community-based success (McCay and Jentoft, 1998). NGOs, as discussed below, can be an important factor in how 'community' are effectively mobilized.

The larger contexts in which co-management takes place is not static. They undergo changes over time. The exclusionist principle that enabled the communities adjacent to the particular fishing grounds to claim exclusive access and communal property rights to the resources facilitated the traditional process of resource use control. This principle has, however, come under attack in post-colonial CARICOM. Demographic, economic and rapid technological changes in the fishing industry, have created spheres of multiple use conflict in the inshore fisheries.

In Latin America, the exclusion of potential users has been a difficult task. These conflicts are particularly true in the implementation of the fishing accords in Amazonian lakes, and in the process of exclusion of commercial trawlers in the SE coast of Brazil, in waters bordering areas of the Atlantic Forest coast. Institutional arrangements made up by the government in Brazil, such as the CNPT (IBAMA), created to deal with the local demands of rural dwellers and fishers (named 'traditional') are still far from reaching an approach that might deal with local demands, conservation and scientific knowledge. On the other hand, these institutions are probably the first government initiative that open a channel for co-management, through the legalization of Extractive Reserves and through the support of some of the Fishing Accords in the Amazon.

Changes in political regimes following general elections are not necessarily accompanied by commitment by the in-coming governments to implement policies enunciated by the out-going governments. Moreover, the diminutive stature of fisheries, compared with other economic sectors in the national economies, do not make it politically prudent and urgent to pay much attention to fisheries-related issues. The institutional arrangements made by the governments, such as the FACs that could have institutionalized macro-level co-management in the countries, and FMPs, that could have ushered in progressive management measures to curb the slide toward the collapse of the inshore fisheries, have not lived up to expectations in many countries, due to inconsistencies in policy implementation and lack of political courage to implement politically risky policies.

4.4. Subsidies and Incentives

The technological and developmental policies of the immediate post-colonial era, was supported by an elaborate incentive scheme. Government subsidies were offered on inputs for productive purposes at a time when it was politically expedient to cite job creation and reduction of abject poverty among the small-scale fishers as justification. Fishing effort increased substantially but as long as most of the fisheries had not reached the Maximum Sustainable Yield (MSY) levels, the subsidies acted as justifiable incentives, in line with the declared policies of the governments. However, the biological context of the fisheries has changed substantially since the 1980s, as the evidence of stock depletion and habitat degradation confirm. Maintaining these subsidies at this time makes it a perverse incentive to the inculcation of the conservation ethic.

Boat-owning fishers such as the lobster and conch fishers of Belize and the Bahamas are notably enjoying middle class to lower upper class lifestyles, and those targeting migratory pelagics in Guyana and some of the East Caribbean island states, no longer exhibit the abject poverty reminiscent of the pre-Independence and immediate post-Independence eras. The justification for maintaining the subsidies on poverty grounds has been severely undermined. However, the possible negative socio-political fallout of large-scale unemployment and likely increase in crime and other social problems following their displacement of the poorest groups from the industry has left the policy situation in a limbo.

4.5. Stakeholder Relations and Conflict

Multiple user conflicts are becoming increasingly evident. The Soufriere Marine Management Area in St. Lucia, the Canaries, the Anse La Raye Marine Management Area also on the western coast of St. Lucia, and the Soufriere-Scotts Head Marine Reserve in Dominica, are examples of community-based co-management structures established to reduce and contain conflict between fishers' groups and other competing resource user groups in the tourism sector.

At the centre of the web of stakeholders at the national level are the fisheries administrators and the fishers. The fisheries administrations in the region wield enormous power over the control of the resources on behalf of the governments. Fishers complain about bureaucratic impediments that frustrate their efforts to communicate and transact business with government officials. The process of making decisions on rules and regulations and their enforcement are usually devoid of the initial consultation with resource users, and where they consult, it mainly takes the form of information dissemination on critical decisions already made behind closed doors. In most cases, these decisions are not reversed as a result of the belated consultative process. The difficult relationship of government officials, researchers and locals, as exemplified by the difficulty in getting the fishing accords demand attended in the Amazon illustrates this point.

The fisheries administrations lack the resources, both human and material, to efficiently and effectively manage the fisheries on behalf of the governments, on their own. This is more glaring in the area of fisheries surveillance and enforcement of management rules and regulations. The evidence in the region shows resource users who have divergent objectives and conflicting interests, and who are in constant conflict over incompatible gears, multiple use of sea space, setting artisanal fishers against industrial fishers, and artisanal fishers using set nets and lines against those using drift or drag nets; all compounded by the intrusion of other sea space users in the tourism and industrial sectors.

4.6. Gender

One important characteristic of the fisheries in the region is its male-dominance. Cultural, traditional and stereotypical biases are reinforced through the socialization process to skew the conventions, norms, rules of resource use and rights of access against women. Hence, with perhaps the singular exception of the Pearl Lagoon area of Nicaragua, women are virtually absent from the productive process, even though some have made some headway as boat owners. As stakeholders, the majority is in the distributive sub-sector as vendors, traditional processors and financiers, with a few holding the usual administrative positions in the fishermen's Cooperatives and Associations. Though having fewer alternatives than men, many women have proved their mettle as small-scale business managers in the private sectors of the national economies. Lack of managerial capabilities is one of the major weaknesses that render the male-dominated fisher folk organizations weak and unstable. However, acceptance of females as members of the Cooperative Societies and Associations has been one of the progressive ideas being promoted by CFRAMP but strongly resisted by their male counterparts.

4.7. NGOs as Potential Facilitators

Non-Governmental Organizations (NGOs) play a leading role in the institutionalizing of Community-based Co-management of coastal resources in other parts of the world. The NGOs in the CARICOM region have not found operating in the capture fishery sector attractive as they have, in general environmental issues and other land-based sectors. Both fisheries administrators and the NGOs tend to cite the so-called independent nature of fishers, stemming from the nature of their occupation, as making it difficult to mobilize and organize them for management purposes. Extractive Reserves seem an interesting example of the interchange among local dwellers or fishers, NGOs, and researchers from Universities, with the support of the state agencies.

The fisheries administrators cite lack of qualified staff and other resources as their main handicap, yet they would not collaborate with NGOs with expertise in grassroots mobilization and organization, community education and rural development, to complement their work. There is a sceptical attitude towards the real intentions of some of the radical and dynamic NGOs, hence the fisheries administrations prefer to seek occasional assistance from them, rather than forging long-term working relationships. There seems to be a subtle power struggle between the two groups, with the fisheries administrators preferring to guard their decision making powers, rather than sharing it with NGOs. There is still a need to build the capacities of NGOs for dealing with marine resource management issues, and the building of constructive working relationships between NGOs and the fisheries administrations in the participating countries. The outstanding example of an NGO-driven resource co-management initiative is that of the Caribbean Coastal Area Management Foundation (CCAM) managing the Portland Bight Sustainable Development Area (PBSDA) of South Jamaica.

5. FUTURE PROSPECTS

Management problems in fisheries have common features in different parts of the world, and Caribbean and Latin American fisheries show similarities that go back to colonialism and authoritarian governments. Solutions are local, tied to resources and users, and may deal with single or with multiple resources. In the case of Brazil and Mexico, Extractive

Reserves, Sustainable Reserves and *ejidos* seem to be the future for co-management of multiple resources. The Fishing Accords in the Brazilian Amazon have strong chances to be an alternative for managing lake resources, dealing with single resources and minimizing user conflicts, but they still depend upon the support of the IBAMA. Actually, Table 1, which includes examples of co-management in Chile, through the Management Exploited Areas, and in Peru, through the Communal Reserves, shows the first stages of what could be followed to deal with co-management in fisheries. Especially in tropical areas, where the conservation of hot spots obligates a multiple resource approach to management, extractive reserves, sustainable reserves, *ejidos*, or communal reserves seem to give future directions. Forms of co-management, such as using local rules already under course, and local demands, such as the Fishing Accords, will depend upon a higher flexibility of the government in accepting local demands and scientific knowledge.

ACKNOWLEDGEMENTS

We thank Doug Wilson for the valuable comments and suggestions, FAPESP and CNPq for research grants to AB, F. de Castro for providing material and for suggestions.

REFERENCES

Acheson, J. (1981) Anthropology of fishing. *Annual Review of Anthropology* **10**, 275-316.

Alcorn, J. B. and Toledo, V. M. (1998) Resilient resource management in Mexico's forest ecosystems: the contribution of property rights. In Berkes, F. and Folke, C. (eds), *Linking social and ecological systems*, Cambridge U. Press, pp. 216-249.

Almeida, M. W. B. and Menezes, M. A. (1994) Reserva Extrativista do Alto Juruá . In: *O Destino da Floresta*, Dumará Distribuidora de Publicações Ltda, Rio de Janeiro, pp. 165-225.

Aswani, S. (1999) Common property models of sea tenure: a case study from the Roviana and Vonavona lagoons, New Georgia, Solomon Islands. *Human Ecology* **27**(3), 417-453.

Begossi, A. (1995) Fishing spots and sea tenure: incipient forms of local management in the Atlantic Forest coastal communities. *Human Ecology* **23**(3), 387-406.

Begossi, A. (1998) Resilience and neo-traditional populations: the cai‡aras (Atlantic Forest) and caboclos (Amazon). In Berkes, F. and Folke, C. (eds), *Linking social and ecological systems*, Cambridge U. Press, pp. 129-157.

Begossi, A. (1999) Scale of interactions of Brazilian populations (caiçaras and caboclos) with resources and institutions. *Human Ecology Review* **6**(1), 1-7.

Begossi, A. (2001a) Mapping spots: fishing areas and territories in the Atlantic Forest coast, Brazil. *Regional Environmental Change*, **2**, 1-12 .

Begossi, A (2001b) Cooperative and territorial resources: Brazilian artisanal fisheries. In: *The Commons Revisited: an Americas Perspective*, J. Burger, R. Norgaard, E. Ostrom, D. Policansky and B. Goldstein (eds), Island Press, Ch. 5, pp. 109-130.

Begossi, A., Silvano, R.A.M, Amaral, B.D. and Oyakawa, O. (1999) Uses of fish and game by inhabitants of an extractive reserve (Upper Juruá , Acre, Brazil). *Environment, Development and Sustainability* **1**, 1-21.

Berkes, F. (1985) Fishermen and the 'tragedy of the commons'. *Environmental Conservation* **12**, 199-206.

Breton, Y. (1991) Economic Anthropology and interdisciplinarity in Costa Rica fishing. *Maritime Anthropological Studies*, **4**(1), 1-12.

Breton, Y., Benazera, C., Plante, S. and Cavanagh, J. (1996) Fisheries management and the colonias in Brazil: a case study of top-down producer's organization. *Society and Natural Resources* **9** (3), 307-315.

Britto, R. C. C. (1999) Modernidade e tradição: construção da identidade social dos pescadores de Arraial do Cabo, RJ. EDUFF, Niterói, RJ.

Brown, D. N. (1996) 'Fishermen as Co-Managers of Communal Property in the CARICOM Region.' Paper presented at the 6th Annual Conference of the International Association for the Study of Common Property, University of California- Berkeley, USA, June.

Brown, D. N. and Pomeroy, R.S. (1999) Co-Management of Caribbean Community (CARICOM) Fisheries, *Marine Policy*, vol. **23** (6), 1999.

Brown, D. N. (2001a) Report of the Multidisciplinary Survey of the Fisheries of Suriname, CFU, Belize City, Belize. January.

Brown, D. N. (2001b) Report of the Multidisciplinary Survey of the Fisheries of the Dominican Republic, CFU, Belize City, Belize.

Brown, D. N. (2001c) Report of the Multidisciplinary Survey of the Fisheries of the Bahamas, CFU, Belize City, Belize.

Brown, D. N. (2001 d) Structures and Processes of Fisheries Co-Management in the Caribbean. Paper presented at the Inaugural Conference, Centre for Maritime Research (MARE), Amsterdam, the Netherlands. 30 August - 1 September.

Castilla, J. C. and Fernandez, M. (1998) Small-scale benthic fisheries in Chile: on co-management and sustainable use of benthic invertebrates. *Ecological Applications*, **8**(1), S124-S132.

Castro, F. (2000) Fishing Accords: the political ecology of fishing intensification in the Amazon. CIPEC, Dissertation Series no. 4, Indiana University, Bloomington.

CARICOM Fisheries Resource Assessment and Management Programme (CFRAMP) (2000) Proposal for the Establishment of the Caribbean Regional Fisheries Mechanism (CRFM), Report of the Working Group. Revision 2, May 2000:12pp.

Cordell, J. (1974) The lunar-tide fishing cycle in Northeastern Brazil. *Ethnology* **13**, 379-392.

Cordell, J. (1978) Carrying capacity analysis of fixed territorial fishing. *Ethnol.* **17**(1), 1-24.

Cordell, J. (1989) A sea of small boats. Cambridge, MA: Cultural Survival, Inc. 418pp.

Dyer, C. L. and McGoodwin, J. R. (1994) Folk management in the world fisheries. University Press of Colorado, Niwot.

Espeut, P. (1994) 'A Socioeconomic Survey of Thirty Fishing Communities and Twelve Countries in the CARICOM Region.' CFRAMP, Belize.

Espeut, P. (1996) 'Provisional Management Plan for the Portland Bight Sustainable Development Area, 1997-2000.' Jamaica, November.

Finlay, J. A. (1993) 'Rights-based Fishing in the Beach Seine Fishery of St. Vincent and the Grenadines, St. Lucia and Dominica.' Grenada Fisheries Division.

Ford, A. and Montes, J. A. (1999) Medio Ambiente, uso de la terra y desarrollo sostenible: la Reserva Arqueológica de El Pilar para la flora y fauna Mayas de Belice y Guatemala. *Mesoamérica* **37**, 31-50.

Forman, S. (1967) Cognition of the catch: the location of fishing spots in a Brazilian coastal village. *Ethnology* **6**, 417-426.

George, S. (1994) 'Coastal Conflict Resolution: A Case Study of Soufriere, St. Lucia.'Gulf and Caribbean Fisheries Institute Inc.

Haughton, M. O. (1998) 'Draft Outline of Vision & Strategy for Sustainable Fisheries Development in The CARICOM Region.,' CARICOM Fisheries Officers Meeting, CFRAMP, Belize City, Belize.

Jentoft, S. (1989) Fisheries co-management: delegating government responsibility to fishermen's organization. *Marine Policy* **13**(2), 137-154.

Lawrence, N. *et al.* (1997) 'SMMR Management Plan.' Fisheries Division, Ministry of Agriculture and Environment, Roseau, Dominica.

Levieil, D. and Orlove, B. S. (1990) Local control of aquatic resources: community and
ecology in Lake Titicaca, Peru. *American Anthropologist* **92**(2), 362-382.

Mahon, R. and Drayton, N. (1990) 'Trap Fishery Management in Barbados: The Fisherman's Perspective', *Caribbean Marine Sciences*, pp. 1-15.

McCay, B. J. and Acheson, J. (1987) Human Ecology of the commons. In: B. J. McCay and J. Acheson (eds), *Question of the Commons*, University of Arizona Press, Tucson.

McCay, B. J. and Jentoft, S. (1998) Market of community failure? critical perspectives on common property research. *Human Organization*, **57**(1), 21-29.

McConney, P. (1998) 'Current Issues in Fisheries Governance in the Caribbean Community (CARICOM), *Marine Policy*, vol. **22**, No. 1.

McGoodwin, J. R. (1989a) Conflicts over shrimp rights in a Mexican fishery. Pages 177-198 in J. Cordell (ed.), *A sea of small boats*. Cambridge, MA: Cultural Survival, Inc.

McGoodwin, J. R. (1989b) Do randommizing devices aid marine hunters? shark fishermen in Pacific Mexico. *Maritime Anthropological Studies* **2**(2), 134-153.

McDaniel, J. (1997) Communal fisheries management in the Peruvian Amazon. *Human Organization* **56**(2), 147-152.

McGrath, D. G. (2000) Avoiding a tragedy of the commons: recent developments in the management of Amazonian fisheries. Chapter Tem in Anthony Hall (ed.), *Amazonia at the crossroads*, Institute of Latin American Studies, London.

McGrath, D. G.; Castro, F.; Futemma, C.; Amaral, B. D. and Calabria, J. (1993) Fisheries and the evolution of resource management on the lower Amazon floodplain. *Human Ecology* **21**(2), 167-195.

Orlove, B. S. (1991) Mapping reeds and reading maps: the politics of representation in Lake Titicaca. *American Ethnologist* **18**(1), 3-38.

Pinkerton, E. W. (1992) Translating legal rights into management practice: overcoming Barriers to the exercise of co-management. *Human Organization* **51**(4), 330-341.

Pomeroy, R. S. and Williams, M. J. (1994) Fisheries co-management and small-scale fisheries: a policy brief. ICLARM, Manila.

Redford, K. H. and A. M. (1993) Forest-dwelling native Amazonians and the conservation of biodiversity: interests in common or in collision? *Conservation Biology*, 7(2), 248-255. www.pop.tefe.rnp.br/reserva/reserva/htm.

Renard, Yves (1994) Community Participation in St. Lucia. *Community and the Environment Series*, No. 2. Panos Inst. & CANARI, St. Lucia.

Renard, Yves (1997) 'Collaborative Management for Conservation,' in B. Feyerabend ed., *Beyond Seeking Social Sustainability in Conservation*, IUCN, Gland, Switzerland.

Ruddle, K. (1996) Traditional management of reef fishing. Pages 315-335 in V. C. Polunin and C. M. Roberts (eds), *Reef fisheries*, Chapman & Hall, London.

Salafasky, N., Dugelby, B. L., and Terborgh, J. W. (1993) Can extractive reserves save the rain forest? An ecological and socioeconomic comparison of nontimber forest product extraction systems in Petén, Guatemala, and West Kalimantan, Indonesia. *Conservation Biology*, **7**(1), 39-53.

Sandersen, H. T. (1995) Co-Management in Caribbean Fisheries: The Case of St. Lucia, Paper Presented at the 5th Annual Conference of the International Association for the Study of Common Property, Bodo, Norway, 24-28 May.

SCM, CNPq, MCT, IPAAM (1996) Mamirauá: Plano de Manejo. Brasília, SCM, CNPq, MCT.

Seixas, C. S. and Berkes, F. (2000) Learning from fishers: incorporating local knowledge in policy design and assessment. In: Paper presented at the International Workshop on Conservation and Development of Coastal Zones on the Eastern Coast of South America, Florianópolis, SC. Brazil, December 4-10, 2000 (in press).

Sen, S. and Nielsen, J. R. (1996) Fisheries co-management: a comparative analysis. *Marine Policy* **20**(5), 405-418.

Sunderlin, W. D. and Gorospe, M.L.G. (1997) Fisher's organizations and modes of co-management: the case of San Miguel Bay, Phillipines. *Human Organization*, **56**(3), 333-343.

Sverdrup-Jensen and Nielsen, J.R. (1997) Co-management in small-scale fisheries: a synthesis of southern and west African experiences. Pages 303-321 in Normann, A K., Nielsen, J.R., and Severdrup-Jensen, S. (eds), *Fisheries co-management in Africa*, Research Report 12, ICLARM, IFM, Denmark.

Weber, R., Butler, J., Larson, P. (eds) (2000). Indigenous peoples and conservation organizations. WWF, Washington, D.C.

Chapter 9

EXPERIENCES WITH FISHERIES CO-MANAGEMENT IN NORTH AMERICA

LAURA LOUCKS
Simon Fraser University, British Colombia, Canada

JAMES A. WILSON
School of Marine Sciences, University of Maine, Orono, ME 04469-5782, USA

JAY J.C. GINTER
National Marine Fisheries Service, Alaska Region, P.O. Box 21668, Juneau, Alaska

1. INTRODUCTION

The authors of this chapter focussed first on the question of who was using the term co-management in North America and what was meant by it. We used the definition proposed by the editors of this book: 'an arrangement where responsibility for resource management is shared between the government and user groups' (Sen and Nielsen, 1996). However, much of our discussion soon shifted into a comparison of power sharing in Canada and the United States, particularly who has decision-making authority over fisheries resources and how this changes over time. In discussing the ideal conditions for government user cooperation and a mutual willingness to support local governance, the idea of Jeffersonian Democracy came to mind. Jefferson envisioned a national government limited purely to national concerns and subsequent divisions into state, county and township governments such that '... each might do for itself what concerns itself directly, and what it can so much better do than a distant authority' (McHugh, 1972).

Although the Jeffersonian ideal may well fit the co-management model, co-management is neither a panacea for sustainable fisheries management nor can it be implemented in the same way in every situation. Rather, it is an evolving process for building collaborative institutions and better government-resource user relationships in which management rights and responsibilities are negotiated to best fit the circumstances. In its highest form, co-management opens up new possibilities for cooperation between governments, First Nations, resource users, and communities; providing new forums for dialogue, improved communication, reduced over-investment, increased user participation in allocation decisions and new means for resolving complex resource management problems (Pinkerton,

1989). Our own process of dialogue, debate and discussion in writing this chapter, is a testimony to the fruits of collaboration. Perhaps this is a good reminder that the concept of co-management originates from the idea of democracy: the practice of social equality by vesting the power of government in the people being governed.

2. ORIGINS, CONTEXTS AND MEANINGS OF CO-MANAGEMENT IN NORTH AMERICA

In North America, the idea of 'co-management' originally emerged from the assertion of aboriginal treaty rights, both in Canada and the United States, as First Nations people began challenging federal government authority over aboriginal title to their traditional territories. In 1974, Judge Boldt, senior federal judge for Western Washington, affirmed the first fisheries co-management principle of shared resource use. In the famous trial *U.S v. Washington*, also known as the 'Boldt decision', the Washington tribes' 'right to fish' was interpreted to mean 'sharing equally' with non-native citizens (Cohen, 1986). The effect of this ruling was to re-define property rights for First Nations, yet it wasn't until the second ruling in 1980, 'a right to environmental protection of fish habitat' that the tribes could begin to co-manage the pacific salmon fishery (Singleton, 1998). This decision provided the impetus and financial means for Washington State tribes to strengthen their organizations, define membership criteria, hire technical support, collect their own data, incorporate local knowledge in decision-making and improve communication both internally and externally (Pinkerton and Keitlah, 1990). Today, the tribal-state co-management system in Washington State is arguably the most sophisticated hybrid governance model that combines state control with local decentralized decision-making and accountability (see Pinkerton, Chapter 4).

In Canada, the Supreme court case *Calder v. Attorney General of B.C.*, in which the Nisga'a people were recognized as having aboriginal title to their territorial lands, catalyzed the first Canadian Federal Land Claims policy (Berkes, 1989). In 1975, shortly after this decision, the Cree and Inuit signed the first co-management agreement in Canada. Nearly a decade later, aboriginal rights were 'recognized and affirmed' in section 35(1) of the 1982 Canadian Constitution Act. As a result, several Federal Land Claim co-management agreements have now been signed, formalizing the principles of self-governance over aboriginal fishing and hunting. Subsequently, First Nations are gaining ground in asserting their aboriginal fishing rights and demonstrating new models for shared governance. However, while these processes have strengthened the self-determination of several First Nation governments, there remains an ongoing struggle for power and rights over resources, especially the right to protect habitat. In many cases, outside development interests supported by the Canadian government, conflict with the interests of the Aboriginal people and their communities. For example, the Cree in James Bay are still fighting for authority over their traditional hunting and fishing territories, despite the 1975 agreement (Feit, 1988). Evidently, co-management agreements are ineffective if the government is not accountable to their commitments and gives priority to another set of interests. Underlying this dilemma is a larger set of questions: how does government authority get transferred and under what conditions? Who benefits from co-management agreements and are these benefits always sustainable?

The issues of government authority, accountability and resource sustainability relate to the constitutional framework that authorizes government decision-making as well as the prevailing ideological paradigm that influences how the government views its role in protecting the 'public interest' in public resources. Tony Charles, in his recent book *Sustainable Fisheries Systems*, suggests three paradigms or 'world views' in which policy objectives generally fall: rationalization , conservation and social/community (Charles, 2001 p. 251). The rationalization paradigm focuses on wealth maximization and economic efficiency goals, whereas the conservation paradigm is based on long term fish stock protection and scientific assessment. In contrast, the social/community paradigm revolves around the belief that social and cultural institutions play a critical role in maintaining sustainable harvesting practices, including habitat protection. Significant tension exists between these perceptions of what constitutes a sustainable fishery, particularly since each paradigm has different implications for ownership rights, responsibilities and the distribution of benefits (Charles, 2001). Consequently, as fishery resources decline, conflicts increase within the fishing industry as prevailing policies exclude various user groups.

In Canada, the federal Fisheries Act gives the Minister of the Department of Fisheries and Oceans the exclusive authority over ocean fisheries allocations and conservation decisions, while the provinces have jurisdiction over near-shore shellfish fisheries, aquaculture and fish processing. Following this, the provinces are typically more concerned with social/community objectives such as rural employment, while the federal government's more recent policies are generally more concerned with rationalization and conservation goals such as wealth maximization and financing scientific stock assessments. Yet at the same time, the federal Minister of Fisheries is also accountable to the 'public interest' by way of the public trust doctrine, on which the Fisheries Act is founded. Adopted from the English public right of navigation and fishing, both Canada and the United States have recognized a responsibility to hold fisheries in trust for the benefit of the people (Smallwood, 1993). In this view, the Minister of Fisheries must carefully weigh the outcomes of various policies to ensure that the interests of the Canadian public are truly being served. However, the exclusive authority of the Minister is often mediated by the fact that it resides in a highly political and democratic regime. Strong and sometimes determinative political influences can affect the Minister's decisions, often swaying policy objectives according to powerful political lobbying efforts. As a result, the more political fisheries decision-making becomes, the more user conflicts arise, often causing sizeable public demonstrations as various groups seek more participation in decision-making processes.

Co-management plays an important role in providing alternative public entry into the highly politicized fisheries management arena. Government-user agreements can stabilize the uncertain political climate by including users in a process that explicitly defines mutually beneficial rights and responsibilities. Collaborative agreements between the federal Minister of Fisheries and various stakeholders, on matters concerning coastal zone management, are possible with the 1997 Oceans Act. However, fisheries co-management agreements vary according to their dominant underlying ideological paradigm. Moreover, the question of whether these agreements always serve the greater public interest depends on the prevailing view on sustainability and how rights, responsibilities and benefits are distributed accordingly.

In the United States, the government's obligation to protect public interests is conceived in a way that is similar to Canada's, but the implementation has taken a very different turn for two principal reasons. First, in the early development of the public trust doctrine, as applied to fisheries, was exercised initially at the state rather than the federal level. The English Crown's public trust authority during colonial times was assumed by the individual state governments after US independence as a sovereign responsibility. US federal authority is limited to those powers explicitly enumerated in the US Constitution. Powers not delegated to the federal government nor prohibited to the states by the Constitution are reserved to the respective states. This has become a basic tenet of federalism in the US Hence, state governments have exclusive authority to manage fisheries within their boarders. Marine fisheries often extend beyond the seaward boundary of coastal states, however, and those and interstate commercial fisheries arguably may be regulated by the federal government under application of the public trust doctrine at the federal level.

Second, US citizens have legal standing to vindicate public trust interests in the US Subsequently, marine fisheries management regulations, for example, must comply with the limitations and procedures set out in a body of federal statutes and executive orders. Currently, these include 11 statutes and six executive orders, many of which are not specific to fisheries (such as the Administrative Procedures Act) but nevertheless constrain the federal government. Unlike Canada, in which the Minister of Fisheries has ultimate authority, many opportunities for public participation are mandated for fisheries and environmental decision making, such as broadly constituted advisory committees and regional councils. Consequently, the operational rules governing the public process in the US appear to be practically written with co-management in mind. Someone reading the legal requirements for US federal regulation and strictly interpreting the Sen and Nielson definition of co-management (above) might be tempted to conclude that little difference exists between normal fishery management procedures in the US and co-management. This might be why defining the term is so confusing especially to people fully involved in the current process.

Yet there remains a serious disconnect between the kind of large-scale, high transaction cost politics of the regional level decision making and the small-scale, local concerns that characterize interest in co-management in the US The large scale and correlated high transactions costs of the US decision making process disadvantages small-scale interests. They face high initial organization costs as well as on-going costs of coordination. In addition, the decentralized or regionalized decision making in the US may be superceded by actions of Congress (eg the American Fisheries Act) or by appeals of those who feel disenfranchised at the local level. As American historian Arthur Schlesinger, Jr. has observed, '... local government is characteristically the government of the locally powerful, not of the locally powerless; and the best way the locally powerless have found to sustain their rights against the locally powerful is through resort to the national government' (Schlesinger, 1971). Hence, whether appealing to a higher more centralized authority or to a regional authority, small-scale local concerns face high transaction costs. These costs may be reduced and the ability to achieve an effective voice in the process is enhanced, however, when there is a certain homogeneity of interests among small-scale operators. Ecological or socio-economic differences consequently make effective representation less probable and tend to deprive the system of a full sense of the diversity of local situations. As a result, the scale of implementation and the high costs of participation appear to be a major problem that has strongly biased the process against small-scale operations.

Both methods of implementing the public trust doctrine have led to impaired results for co-management. In the US, the scale of implementation and the high costs of participation, with few exceptions, appear to be a major problem that has strongly biased the process against small-scale operations. Similarly in Canada, a bias favouring large-scale operations appears to have arisen in support of those groups that can afford to pay for scientific stock assessments as well as monitoring and enforcement costs. Yet, the legal constraints on the actions of the Canadian government tend to be much less than in the US because of the manner of implementing the public trust doctrine. In the US, citizens can challenge a state or federal government's actions in court. Consequently, the government follows procedures to augment its credibility and legitimacy in maintaining its public trust obligation. But in Canada, the Government can choose its own definition of what constitutes the 'public interest' and take actions essentially constrained only by its need or desire to be re-elected. Therefore, when the Canadian Government does engage in the sharing of management responsibility, this is a meaningful departure from its legally mandated role. Yet despite the government's commitment to shared responsibility, lacking any legal obligation, its accountability depends on which prevailing 'world view' the Minister or political advisors subscribe to.

For these reasons, the history and emphasis of co-management in Canada and the US differ. Even the connotations and meaningfulness of the term vary because of the legal context in which fisheries are placed. We are continually challenged to clarify what we mean when we talk about fisheries co-management, especially in the context of the underlying 'world view' that determines how the public interest could benefit from the transfer of fisheries management rights and responsibilities. We present the following case studies to describe how authority changes in Canada and the US, and specifically how rights and responsibilities are exchanged and benefits are distributed within different contexts of fisheries co-management. We chose examples from eastern and western regions of Canada and the US that provide insight into the consequences of contrasting paradigms that underlay the different meanings for co-management and the conditions under which the federal government is motivated to transfer decision-making rights and responsibilities. These cases illustrate, however, that while rights and responsibilities are more equally shared, the highly participatory processes generate additional transaction costs. Taken together, these cases provide us with a better understanding of the meaning and practice of co-management in North America and, more importantly, address the question: who benefits from shared government-user responsibility?

3. SOME CURRENT EXAMPLES OF CO-MANAGEMENT APPROACHES IN NORTH AMERICA

3.1. Corporate Fisheries Co-Management: The Canadian Atlantic Sea Scallop Fishery

Atlantic Sea Scallops are harvested on the Banks of the Scotian Shelf by deep-sea dredges that are towed along the seabed floor. In the past, these fishing areas were the cause of great dispute between Canada and the United States. However in 1984, an International Court of Justice established a boundary, known as the Hague Line, separating the scallop fishery into two exclusive zones (Repetto, 2001). Shortly after this decision, the Canadian scallop fishing industry adopted an enterprise allocation (EA) system and consolidated their quota into nine corporations. In support of this approach, the federal Department of Fisheries and Oceans further divided the Canadian scallop fishery into two distinct offshore and inshore fleets, separating the fishing grounds between them. As a result, the offshore fleet gained exclusive access to the Canadian side of the Georges Bank, as well as Brown's Bank, German Bank, and St. Pierre Bank off the Eastern Scotian Shelf.

Today, the offshore sea scallop fishery generates in excess of $90 million dollars in annual landed value. Quota-holders collaborate closely with government managers and scientists through an informal co-management agreement with the purpose of stabilizing harvesting strategies according to the scallop's fluctuating recruitment patterns. Industry not only finances most of the standard science, monitoring and enforcement costs of this fishery, it invests millions of dollars in a government research programme to determine an accurate and precise estimate of recruitment abundance. Using sophisticated benthic sonar mapping technology and sample surveys, together with daily harvesting data, scientists are able to determine the location and abundance of specific year classes of the sea scallops and recommend a harvesting strategy that will even out the market supply of scallops in later years (Repetto, 2001).

The willingness of industry to maintain conservative quotas in favour of stable landings from year to year is believed to be related to the organizational structure of the enterprise allocation system. Each enterprise, or company, receives their own license with a specific percentage of quota attached. In exchange, the company is responsible for specific costs such as dockside monitoring, electronic at-sea monitoring boxes, port sampling, and scientific research. With each license and enterprise allocation, the company has the management right to decide how, when and where it will harvest its quota. By carefully following scientific advice, the companies have created a harvesting strategy that has smoothed out the variations in recruitment patterns, resulting in higher abundance and overall market stability. As a result, the companies can carefully control their economic inputs and outputs, capitalizing on larger and more efficient technology. For example, the number of vessels have been reduced in the fishery from 68 vessels in 1986, to 28 vessels in 1999, in favour of larger factory freezers that can process product at sea (Fisheries and Oceans Canada, 2000). In the future, the offshore sea scallop industry hopes to reduce their number of vessels by half, as they continue to invest in larger trawlers. As well, the changes in quota holdings demonstrate a trend in consolidation from nine corporations down to six, with one company currently holding nearly 50% of the quota.

The overall objectives of economic efficiency, user-pay monitoring and enforcement, stable year to year harvesting patterns, and technological efficiency are based on the underlying 'world view' of a rationalization paradigm. Similarly, the strong emphasis on scientific research and precision recruitment estimates supports the values attributed to a conservation paradigm. Yet neither of these views recognize the significance of habitat protection or the equitable distribution of resource benefits characteristic of the social/community paradigm. While some analysts give the prognosis that this fishery is 'prosperous and largely content with its rights based system' (Repetto, 2001), these issues challenge the success of this model.

For example, it is unclear that the offshore sea scallop industry is entirely responsible for increasing the abundance of scallop stocks on George's Bank, and that they will be able to sustain adequate abundance in the future. In 1994, three areas of George's Bank were closed to all fishing as a result of the collapse of Atlantic groundfish stocks. This moratorium not only reduced fishing pressure, it also prevented the seabed habitat from being disturbed by high impact technology such as scallop dredging and groundfish dragging. Research has demonstrated that scallop abundance rises dramatically in areas closed completely to these fisheries, suggesting that the habitat protection is critical for stock recovery (NE Fisheries Management Council, 2000). The co-management agreement between industry and government does not include habitat protection explicitly in their conservation strategy, nor does it address the possible conservation costs of increased habitat damage caused by the wider dredging tows of larger vessels. Furthermore, while the economic efficiency of the offshore fishery is generating prosperity for the corporations and covering the costs of co-management, it is difficult to assess whether or not the total level of fishing effort has been reduced. For example, as the number of vessels in the offshore fleet decrease, their harvesting capacity has shifted to the inshore fleet. Consequently, as the effort of sea scallop dredging increases within the inshore boundary, negative impacts such as habitat damage and gear conflicts increase.

3.2. Community-Based Fisheries Management: The Fundy Fixed Gear Council

In April 1996, after several groundfish closures and steady declines in Atlantic cod landings, the Department of Fisheries and Oceans (DFO) implemented a plan to divide the Nova Scotia groundfish fishery into seven geographic management areas. Accordingly, each area was required to organize themselves into Community Management Boards and manage a groundfish quota that was based on a previous catch history in the area. The Fundy Fixed Gear Council (FFGC), the management board for the fixed gear groundfish fishery on the Nova Scotia side of the Bay of Fundy, represents more than 200 inshore fishermen, including longline, handline and gillnet fishermen. Unlike other areas, in which the shift to Community Management Boards created serious allocation conflicts as various groups vied for management authority, the Fundy Fixed Gear Council adopted a highly democratic community-based co-management model. While the economic success of the model is difficult to evaluate at this early stage, given that Atlantic cod landings remain low and the groundfish fishery remains a fraction of what it once was, there is little question that the vision for community-based co-management is driven by a deep desire for shared responsibility for resource management.

The events leading to this co-management arrangement are significant in understanding what motivated the exchange of rights and responsibilities between fishermen and DFO. Traditionally, fishermen in Nova Scotia have never been very eager to participate in membership organizations. However, in 1996 a large coalition of fishermen joined together to protest DFO's proposal that handline fishermen should not be considered as 'bona fide' professionals due to their lack of economic viability. This government action, taken together with the 1993 cod fishery closures and proposed regulations to increase user fees, created a profound sense of mistrust and anger amongst small boat inshore fishermen. Their greatest fear was that the government was privatizing and corporatizing the inshore groundfish fishery by pushing rationalization policy instruments such as ITQs. As a result, fishermen throughout Nova Scotia and New Brunswick joined together and occupied several DFO buildings across Nova Scotia for several weeks, demanding a meeting with the federal Minister of Fisheries. After significant media attention, a meeting was held and fishermen met with senior DFO staff for three days to discuss many of the fishermen's concerns. While several issues were put on the table for discussion, the most frequently heard call for change focussed on the lack of democratic decision making processes and the need for locally developed polices and rules. However, despite these concerns and discussions, DFO implemented their new system, taking a 'one size fits all' approach to the whole province.

Community Management Boards, implemented within each of the seven geographic areas, were given the financial responsibility for dockside monitoring and enforcement in exchange for the management rights attributed to having a community quota. The quota, by its nature, transferred the right and responsibility to determine when and how much to fish. However, the way in which these rights, responsibilities and benefits were distributed depended largely on how the Community Management Board made decisions about dividing the community quota.

In this sense, the Fundy Fixed Gear Council had an advantage over other regions. They had already taken a leadership role in building alliances throughout the Bay of Fundy area and had developed a vision for grassroots democratic self-governance. However, as part of its new approach, DFO imposed a quota system that divided the groundfish Total Allowable Catch among 'communities' based largely on provincial county lines. Using the aggregated catch histories for the period 1986-1993 as a basis for determining the 'community share' of overall Scotia-Fundy groundfish quota, the FFGC received the lowest community quota of all the regions. Their fishing history was widely spread outside their designated county lines and a large percent of their groundfish fishermen were small-scale handliners with catch histories often not recorded.

As it turns out, the low quota had some unexpected advantages. Some community leaders have suggested that because the Bay of Fundy Fishermen had so little to lose, they were unanimously willing to try something completely new. Even so, several other factors helped set the new plan for community-based management in motion (Bull, 1998). First, the Nova Scotia side of the Bay of Fundy is a well defined geographic area. Fishermen have an extensive understanding of the ecological systems inherent to the Bay of Fundy as well as a strong personal identity as 'Fundy fishermen' that provides a common basis for social cohesion. Second, one other Community Management Board model with a similar social/ community 'world view', had already been developed, thereby providing a learning opportunity (Loucks, 1998). Third, a great deal of organizational groundwork had already been done in the area by the Maritime Fishermen's Union and the Coastal Community Network, particularly around the concept of fisheries co-management. Fourth, two fishing organizations already existed with a relatively unified membership and good working links between them. Consequently, the Council could build a Community Management Board from the existing institutional capacity within the Fundy area. For example, the Bay of Fundy Inshore Fishermen's Association and the Maritime Fishermen's Union, Local 9, were essential for forming the decision-making structure of the Fundy Fixed Gear Council. Today, each group elects three representatives, appointing one as a co-chairperson. In addition, the Council itself appoints three community representatives. All decisions concerning inshore cod, haddock and pollock are made entirely by consensus. A number of committees provide advice to the Council, including three Gear Committees (made up of elected representatives from each gear sector organization within the Bay of Fundy area) a Research and Advisory Committee, an Infractions Committee and a Licensing and Access Committee.

As a highly democratic model, the underlying social/community 'world view' of the Fundy fixed gear council contrasts significantly with the prevailing government 'world view' represented in both the rationalization and conservation paradigms. Consequently, the definition of co-management that the Council strives to achieve is largely ignored by the federal government. The objectives of the Council are: (1)community involvement; (2) accountable and democratic decision making structures; (3) a system of self-regulation and compliance; (4) community-based research; (5) policy and planning; (6) an ecological approach based on interactions between community, economics, fishing fleets, habitat, and fish stocks; and (7) conflict resolution.

The movement towards community-based management in the Bay of Fundy region has emerged from a collaborative participatory process of informal 'kitchen table meetings', designed to generate new ideas around sustaining rural communities and their resource base. In 2000, the Fundy Fixed Gear Council, in cooperation with the New Brunswick Conservation Council, the Bay of Fundy Marine Resource Centre and the Bay of Fundy Fishermen's Council participated in developing a more ecological approach to fishing, based on regulating how, when and where fishing should occur. A set of Fundamental Principles for good fisheries management were identified together with nine Operational Principles including social, economic and ecological values. The Fundy Fixed Gear Council has been a leader in the field of community-based fisheries management. In this role, it has developed a grassroots democratic self-governance structure, which includes decision-making processes relating to all aspects of local management. In only four years since it was first organized, it has undertaken a wide range of research on fish stocks, habitat and fishing patterns, has made strong partnerships with organizations locally, regionally, nationally and internationally, and has been involved a number of training and professionalization initiatives. In spite of these positive steps forward, the FFGC still faces many challenges.

The underlying social/community 'world view' of the FFGC is compromised somewhat by the prevailing paradigms underlying the Community Quota system. The transfer of rights and responsibilities of the quota system is largely characteristic of a rationalization paradigm. Consequently, this significantly limits the flexibility for harvesting options, making it difficult to implement effort control approaches that take habitat and life history cycles into consideration. Similarly, the Council lacks any formal decision-making authority over the broader set of management rights that community-based management requires. Therefore, they are reliant on DFO approval for all aspects of management decision-making. In addition, the FFGC only governs fixed gear groundfish in the Bay of Fundy, while other organizations are responsible for mobile gear groundfish and inshore species such as herring, lobster, sea urchins and clams. The mobile fleet operates with an ITQ system and relies on controversial fishing technology known as 'draggers'. These large boat dragger fleets drag trawl nets across the ocean bottom and have long been suspect of causing significant habitat damage. Consequently, due to these limitations of the FFGC manadate, the principles of multi-species management, habitat conservation, and effort controls are seriously compromised.

3.3. Local Self-Governance and Small-Scale User Group Organizations in USA: The Maine Lobster Fishery

Co-management in the Maine lobster fishery grew out of a long history of fishermen's frustration and sense of alienation from both a state legislative process and a federal administrative process. State legislators were generally ill informed and had short tenure on the Marine Resources Committee. With each new crop of legislators the discussion tended to begin back at the beginning. The different needs and interests arising from a diverse coast and fishery usually led to legislative impasse. Only when all areas of the coast could agree on a measure - which was not often - would the legislature pass meaningful rule changes. The federal process, conducted through an advisory body called the New England Fisheries Management Council, was on paper, an open, transparent process. In practice, however, it was one in which fishermen had almost no meaningful input and had little understanding of why any given action might have been taken. Typically, Maine fishermen had to drive four to six hours to attend a two day Council meeting where they might be able to make a two or three minute speech before bored, but polite, Council members. Like the legislature, the Council was faced with a lack of information among its members and a range of differing circumstances and interests that tended to immobilize decision-making.

Alienation from the process was deep and pervasive among fishermen. Discussion about some form of decentralization, or co-management, started in the mid-1980s at the annual meeting of the Maine Fishermen's Forum. During the early nineties these ideas became more concrete and constructive, finding their expression in ideas about democratic, local control of some aspects of the fishery. It was argued that the alienation from state and federal processes undermined any sense of stewardship and that this carried with it high enforcement and conservation costs. There was a sense that many local ecological issues - habitat and gear impacts for example - could not be dealt with effectively at the state or federal level. There was a strong sense that effective science needed to be carried out at a more local scale. It was argued that local areas, given appropriate flexibility, were in a better position to adapt and evolve new policies than were the large scale state and federal processes.

In 1995, the Maine legislature, at the initiative of the Commissioner for Marine Resources, approved a bill that provided for local zones, each with elected councils and the ability to change certain rules through referendum. Within a year seven local zones were established, each with approximately 1000 fishermen. Initially the Councils were given control over rules that were deemed purely local in character, including: (1) the ability to limit the number of traps fished by each boat, (2) the ability to set rules about how those traps were deployed (rigged) and (3) the ability to set the days and times during which fishing would be permitted. Authority for rules whose impact was deemed wider than local (eg, rules related to the reproduction of lobsters) was retained at the statewide level. After about three years the Councils were given the right to limit entry in each zone. Any change in rules is initiated by the Council but requires a two-thirds approval in a referendum of license holders.

The zones have developed and implemented new restraining rules with lightning speed. In their first year of operation all seven zones approved trap limits and by their third year of operation five of the seven zones had approved limitations on entry. The ease of rule change in the local zones was facilitated by the relative homogeneity and, where homogeneity did not exist, by the continuity of relations and improved ability to communicate and negotiate at the local level. Rules limiting trap numbers and license holders are central rules that challenge centuries old traditions of open access and, in some rural communities, threaten the loss of already limited economic opportunities. The discontinuity and ill-informed decision making that characterize the legislative and federal advisory processes basically dissolved under the changed conditions of local decision making.

An entirely new, and generally younger, population of 'activists' emerged. Local meetings, while time consuming, tended to be relatively free of game playing and seriously addressed issues facing the local area. In some instances - eg, potentially crippling rules for dealing with endangered right whales - the zone councils became the pivotal organizations around which the industry mobilized. There has been a marked change of attitude towards the role of science in the fishery. Previously science was almost viewed as an arbitrary (and off-the-wall) tool of government. With the zones (and a group of scientists interested in working with fishermen) there was a remarkable surge of interest in science, with a statewide committee of scientists and lobstermen representing each zone being formed in 2001.

Implementation of the zone councils was not, of course, without its problems. Given the numbers of fishermen involved (approximately 7000), simply establishing voting districts, jurisdictional boundaries and learning how to hold regular and orderly meetings took an extraordinary amount of effort on the part of several state employees and an enthusiastic group of volunteers in each zone. In the first year, a problem arose that highlighted the potential conflict between the democratic processes within the zone councils and their legal status as advisory committees. One of the zones voted (with 92% approval) to create a trap cap of 1200 and then to reduce it to 1000 and then to 800 over the following two years. The heavy majority in favour led its proponents to argue that the strong majority indicated a need for a lower limit or a faster reduction. They immediately went to referendum a second time asking for an 800 trap limit in the first year. This second referendum received 77% approval but was fiercely opposed by fishermen who had been fishing large numbers of traps (well over 1200) who thought they had negotiated a gradual build-down and had supported the first referendum. These fishermen also pointed out that the local council had adopted operating rules that stipulated no more than one referendum per year on a given issue. They set out to register their complaint but had no where to go within the council system. Rules and procedures had been kept to a minimum on the theory that it was good to 'keep it simple', especially at the beginning. No appeals or dispute resolution procedure had been built into the system and the complaining fishermen had to go to the legislature, the governor and, eventually, the courts asking for invalidation of the second referendum.

Other fishermen saw this as an attempt to end-run a democratic process and felt that their authority would be undermined if this succeeded. They loudly pointed out that they would lose their incentive to put in the hours of work necessary to bring about consensus at the local level, if it were possible for a small minority to overturn that work by appeals to higher authorities, or for that matter, if it were possible for higher authorities to simply erase their work at the local level if they (the higher authorities) disliked the results (legally a possible outcome).

Eventually, the referendum issue was resolved (the second referendum was disallowed by the courts) and a valuable lesson was learned - the boundaries of authority given to local governance units have to be defined and there have to be provisions for dispute resolution. In other words, the normal appurtenances of self-governance have to be in place. If they are not, the threat of arbitrary actions by higher authorities seriously weakens the willingness of local users to participate in the management process. The incentive to participate does not disappear entirely because it may still be possible to influence the process, but the prerogatives of governance are transferred to higher authorities.

Legally, the Maine system is still part of a top-down administrative process where the elected councils of fishermen are only advisors to the state's Commissioner of Marine Resources. In practice, however, the broad democratic foundations of the councils mean the Commissioner has the political freedom to overturn or modify council decisions only in extraordinary circumstances. Additionally, as the authority of the councils becomes better defined through legislative, administrative and court actions, the Commissioner's discretionary authority tends to be further restricted.

3.4. Cooperative Management in Federal Fisheries off Alaska, USA

Commercial fisheries in Alaska are big business. The fishing industry is the largest employer within the state and, although not the largest revenue producer, it is among the big three state industries of oil, tourism and fishing. By volume, commercial fisheries in Alaska annually account for about 4.5 billion pounds of landed fish worth about one billion US dollars. These landings account for roughly one half and one third, respectively, of the total volume and value of commercial fisheries in the US (DOC, 2001). Subsistence/personal use and recreational harvests of fish add small but socially and economically significant amounts to the overall total of marine species harvested in Alaska.

Management of these fisheries, not surprisingly, is also 'big business' for government. But for a few minor exceptions, marine fisheries management is done entirely by either the

State of Alaska Department of Fish and Game (ADF&G) or the federal government through the National Marine Fisheries Service (NMFS). The fishery management authority of NMFS is limited to implementing five fishery management plans pertaining to groundfish fisheries in the Bering Sea and Aleutian Islands area and Gulf of Alaska, as well as certain crab, scallop and salmon fisheries. Federal authority, limited to fishing activities seaward of the state's three-mile territorial sea, is also responsible for implementing allocation measures in the Pacific halibut fisheries inside and outside of the territorial sea off Alaska and for controlling the harvests of certain marine mammal species by Alaska Natives. Whereas the state is responsible for managing the harvests of all other marine species, including large wild salmon fisheries and herring fisheries.

Managing human behaviour in what are, in some cases, large industrial-strength fisheries has all the potential for producing an over bearing, centrally controlled and intrusive regime of governing. The federal government experience in Alaska during the last quarter of the 20th century, however, has been one of partnership in the development of fishery management policies. This partnership has involved a range of regional to local interests, advisory and consultative mechanisms, fishermen, processing firms and community representatives, and non-governmental organizations (NGOs) all of which bring biological conservation and social equity concerns to the table. In light of a significant national interest in the marine resources off Alaska, this experience is unique for an agency of the federal government. The experience is firmly rooted in the first half of the 20th century when the federal authority was the only authority over fisheries in the Territory of Alaska. This exclusive and highly centralized authority was heavily influenced by the salmon canning industry, and it arguably lead to the virtual collapse of the salmon resource and declaration of Alaska as a 'disaster area' in 1953 by President Eisenhower (Cooley, 1963). This event, added to other failures of the federal management regime, became the *cause celebre* for Alaskan statehood, which was achieved in 1959. Hence, the pre-statehood political history of Alaska is characterized by an absence of sharing the governance of marine fisheries with the people of the territory. The post-statehood history, however, reflects a sensitivity toward local involvement that has been a hallmark of marine fishery management at the state level for over 40 years, and has influenced latter-day federal authority.

In 1976, the Fishery Conservation and Management Act was introduced (now the Magnuson-Stevens Fishery Conservation and Management Act (MSA)) to remove, or at least diminish, the effects of foreign fishing within 200 miles of the US coastline, most notably off New England, the Northwestern states and Alaska. Of equal and more enduring importance, however, was the MSA's establishment of a domestic fishery management regime, the centrepiece of which is public participation through a system of eight Regional Fishery Management Councils (MSA sec. 302). The North Pacific Fishery Management Council was charged primarily with developing and recommending fishery management policies for fisheries off Alaska to the U.S Secretary of Commerce. Once approved by the Secretary, the Council-generated policies are implemented by federal regulation.

The North Pacific Fishery Management Council, comprised of 11 voting members, is an interesting form of cooperative governance because it is neither a state nor federal government entity. The Governor of Alaska nominates five of the seven voting seats appointed by the Secretary, and the Secretary is officially represented by one voting seat on the Council. The other voting members represent the interests of the states of Washington (2) and Oregon (1). Although the Council functions officially as an advisory body to the Secretary, its recommended policies can be only approved or disapproved or partially approved by the Secretary. Consequently, the Secretary has no authority to choose a policy different from the one recommended by the Council. Furthermore, if the Secretary decides to reject the Council's recommended policy, the disapproval must be for reasons of inconsistency with the MSA or other applicable law; not simply because the Secretary prefers a different policy. Hence, in crafting federal fishery management policy off Alaska, the Council plays more than simply a consultative role. However, the Council is not vulnerable to legal challenge of its recommended policies. Because the final decision on whether to approve a Council-recommended policy rests with the Secretary, only the Secretary may be challenged in court to defend the policy and the manner in which it is implemented.

Another facet of Council cooperation in the development of federal fishery management policy off Alaska is the Council's substantial reliance on the advice of its two principal advisory bodies, the Advisory Panel (AP) and the Scientific and Statistical Committee (SSC). The AP is comprised of representatives of all sectors of the fishing industry including various gear groups and processing modes, Alaska Native interests, environmental, and recreational fishing concerns. Likewise, the SSC includes members of the social and biological sciences from academic and government institutions. Council policy on any particular fishery management issue is formulated and analysed primarily through the oversight of the AP and SSC. Rarely does the Council ignore these panel and committee recommendations completely. Consequently, this process enhances public participation in the development of what eventually becomes federal fishery management rules because the affected public may testify at all three levels, the SSC, AP and Council, in addition to commenting at the Secretarial review stage. The result of this lengthy and sometimes tedious public process is a fishery management policy that fairly represents the majority views of the affected public but leavened with biological, economic and social equity wisdom. As such, this system represents a partnership or shared responsibility among all fisheries stakeholders, fishermen, NGOs, fishing communities, academic institutions, and state, local and federal governments in the formulation of fishery policy which closely fits a definition of fisheries co-management (Berkes, *et al.*, 2001).

Several programmes have been introduced through the Council that may not otherwise have succeeded as grassroots proposals at the federal level. For example the Community Development Quota (CDQ) programme was created by the Council in 1991, and approved by the Secretary in March of 1992 (NMFS, 1992). Originally conceived as a means for starting and supporting commercial seafood activities in Western Alaska - an economically deprived and isolated region - the CDQ programme emerged as a political artifact of a contentious debate over a large and highly developed industrial fishery for pollock in the Bering Sea. Advocates of a special allocation of pollock for economic development purposes successfully linked the Community Quota proposal to a larger issue the Council was attempting to resolve which was an allocation of the pollock resource - involving a Total Allowable Catch (TAC) of roughly one million metric tons per year - between inshore and offshore processing interests (Ginter, 1995).

The CDQ programme functions by annually allocating a portion of the TAC of a species to a CDQ reserve. This is subsequently allocated among six CDQ non-profit corporate entities or groups, which are then challenged to derive maximum economic and social benefits from their respective allocations for their member communities. Currently, 65 villages on the Bering Sea coast are eligible to participate in the CDQ programme. Residents benefit either through directly fishing for their CDQ group's allocation, or, more commonly, through revenues derived from leasing the group's allocation to large fishing firms. The economic benefits of the CDQ programme have been considerable. During the early, pollock-only period of 1992 through 1994, CDQ royalties totalled about $53 million (Ginter, 1995). With the expansion of the programme to other species the total value grew to about $1.2 billion in 1996 (National Research Council, 1999). These revenues are used to provide vocational and academic education opportunities for village residents, employment in the fishing and processing industry, and investments in vessels and seafood related businesses. The CDQ programme now provides an estimated 1,000 jobs and wages in excess of $8 million annually for an area of Alaska that experiences chronic high unemployment.

Although the CDQ programme ultimately is a federal programme, it is also very much a State of Alaska community development programme. It is like a three-legged stool supported by the federal and state governments and the CDQ groups. While each CDQ group is responsible for fishing its allocation, the challenge to a CDQ group is to derive the most benefit for its member communities. A CDQ group will not be able to harvest every last pound of every species TAC it is allocated. Hence, it must organize or contract its fishing effort to optimize harvests of the most valuable species within the constraints of the incidental harvest of lower valued and prohibited species. This constraint has been eased in recent years so that incidental catches do not unduly prevent prosecution of a CDQ fishery. A CDQ group also must develop its community development plan (CDP), basically a business plan, which serves as its application to the State of Alaska for a biannual allocation of the CDQ reserve. Each CDQ group competes with the other groups through their CDPs for a portion of the CDQ reserve. Not surprisingly, the total of all groups' requested allocations far exceeds the total CDQ reserve for each species.

The state provides routine administrative oversight of the CDQ programme, reviews and recommends CDQ allocations based on each CDP, and assures that the social and economic development goals of the state for the Western Alaska region are being met. Federal oversight of the programme is provided by NMFS which also develops regulatory reforms in consultation with the state and the Council. NMFS reviews and approves Council and state recommendations on CDQ allocations and CDP/amendments and generally assures that the programme is operating fairly and equitably in accordance with applicable law.

4. CONCLUSIONS

In summary, there is a significant difference in the legal context between the Canadian federal government authority and accountability for fisheries policy and that of the United States. In Canada, the federal Fisheries Act gives the Minister of Fisheries the exclusive and centralized authority for ocean fisheries, whereas in the United States, this authority is shared between the federal and state governments at the regional scale and further tempered by various avenues for public participation. Moreover, the general public has a legal right to challenge the state or federal government in court. The United States has invested heavily in expensive and highly democratic processes for public input into fisheries policy decisions.

One outcome of this investment, as evident in these case studies, is that fishery policies in the United States are more closely aligned with the social/community paradigm than in Canada. The Magnusen Stevens Act not only provided the legal basis for regional councils and public participation, it also made the government accountable to the majority 'world view'. Consequently, a balance of paradigms and policy priorities has evolved. Sometimes this approach has worked very well; at other times it has led to highly impaired results.

Innovative programmes, such as the Community Development Quota programme described in section 3.4, have emerged. These programmes not only re-invest in the health of the fisheries through training and employment initiatives, they are also big business. In generating substantial wealth they reflect some of the positive elements of the rationalization and conservation paradigms that Canadian government policies so strongly support. In other instances, such as the Maine lobster fishery, only by opting out of the federal process has it been possible to move toward co-management.

In contrast, collapses of Canadian fisheries have threatened government credibility. As a variety of groups have called for new decision making structures and processes, government managers have mainly responded by delegating management rights and responsibilities to narrowly defined user groups that meet objectives consistent with the rationalization and conservation paradigms. They have largely adopted the paradigm that the public interest is best served if government ignores equity or ecosystem issues in favour of trading access to resources for reduced government spending on science, monitoring and enforcement costs. In cases such as the Atlantic Offshore Scallop fishery, this approach to co-management has largely reflect the views of corporations. The high level of wealth generated in these corporate fisheries is believed to generate economic rents for the public as companies pay high sums for their licenses relative to other fisheries. In exchange for these benefits, the rights transferred to the industry include the right to own and sell a proportion of the Total Allowable Catch, as well as the right to decide how, when and where to fish. While this regime may achieve improved science and economic efficiency for corporate resource users, the benefits for public interests may be less than those for private interests. When these positive outcomes are measured against a range of social and conservation costs such as habitat damage, by-catch, fishing effort displacement, over capitalization, unemployment, and unresolved First Nations treaties, it is unclear whether this approach produces sustainable and broad public benefits.

These mixed experiences may be attributable to a mismatch between the scale of the relevant fisheries ecology and the scale of management, as discussed in Chapter 2. Biological processes occurs on a variety of scales. When rule-making authority is assigned such that their impact corresponds to the physical jurisdiction of the management unit, much tighter feedback, and faster learning, can be expected occur. This suggests that management institutions, to be truly sustainable, must complement the complexity of their ecosystems. In other words, an ecosystem approach is critical. From the economic perspective this tends to reduce, as much as possible, spatially occurring externalities and all the degenerative incentives that accompany them (Wilson, 2002). As the Maine lobster case study illustrates, complex fisheries require complex management institutions. Yet even when management institutions are designed according to the ecological patterns of the fishery, highly participatory systems are costly, particularly when conflicts arise. Accordingly, the benefits from these systems must be weighed against the full range of costs. In our analysis of co-management, we must deepen our understanding of who benefits and who pays.

Like the Maine Lobster fishery, the Community Quota system in Nova Scotia, is slightly more aligned with the social/community 'world-view' than the corporate co-management system, as the benefits from the resource are distributed more broadly. However, the level of equity on which these benefits, rights and responsibilities are distributed, depends largely on the capacity and willingness of the Community Management board to implement democratic decision-making. In the case of the Fundy Fixed Gear Council, the decision-making process has been extremely successful at building self-determination, yet this model is largely unsupported by the federal Department of Fisheries and Oceans (DFO). This is problematic because democratic institutions have high transaction costs, particularly when they take on additional responsibilities such as local research and data collection. The ability of the Fundy Fixed Gear Council to implement its 'principles for good fisheries management' is constrained by lack of government support and adequate allocations.

Yet, as demonstrated by the Alaska Community Development Programme, local communities and their organizations can generate large revenues when they are given access to the resource. Consequently, as DFO rationalization strategies reduce the resource access of coastal communities, a significant opportunity for building and financing more locally driven democratic institutions is lost. Furthermore, the benefits of local ecological knowledge and a local scale of scientific understanding, necessary for sustainable ecosystem-based fishing practices, are also lost.

Fortunately, Canadian government adherence to the rationalization paradigm may be changing. The 1996 *Oceans Act* mandates more inclusive, ecosystem approaches to fisheries management, and cannot be ignored for much longer. Furthermore, as First Nations continue to assert their traditional rights and negotiate unresolved treaties, they advance the social/community paradigm. As described by Pinkerton in Chapter 4, First Nations collaboration with other coastal communities, aquatic resource interests, and federal and provincial governments, creates a forum where different world views and paradigms can be balanced. Key to this is the ability for the organization to take an ecosystem-based approach in which diverse interests work towards common principles. It is not clear whether such an approach to co-management can survive if the federal government does not depart further from its current rationalization paradigm and address issues such as equitable access to resources and habitat degradation.

Co-management has evolved differently in Canada than in the United States, as evidenced by the various ways in which rights and responsibilities are exchanged and benefits are distributed. It seems that the highly participatory processes in the United States have evolved from a regional and local scale of management decision-making, that is accountable to the public trust and balances a range of 'world views', representing elements of each of the rationalization, conservation and social/community paradigms. While these processes often include significant transaction costs, they must be understood within the context of full cost accounting, rather than ignoring the possible social and environmental externalities incurred without democratic decision-making. Issues such as social equity and aquatic ecosystem functions must be addressed in any sustainable co-management regime. This approach needs to be supported by:

a equitable and adequate access to resources in order to provide the kinds of benefits required to support its expanded goals of economic efficiency, conservation, equity and habitat protection;
b clear dispute resolution procedures;
c common guiding principles and;
d management authority at the appropriate ecosystem scale.

Finally, we must be wary of defining co-management regimes so narrowly that we ignore the trade-offs of social and ecological externalities when calculating economic benefits. Rather, we must continuously ask the question 'who benefits from co-management?' and in so doing, we must command a deeper understanding of the trade-offs between private and public costs and benefits, particularly as various rights and responsibilities are exchanged.

REFERENCES

Berkes, Fikret (1989) 'Co-Management and the James Bay Agreement.' Pp. 181-208 in *Co-operative Management of Local Fisheries: New Directions for Improved Management and Community Development*, edited by E. Pinkerton. Vancouver: University of British Columbia Press.

Berkes, Fikret; Mahon, Robin; McConney, Patrick; Pollnac, Richard; and Pomeroy, Robert (2001) Managing Small-Scale Fisheries, Alternative Directions and Methods, International Development Research Centre, Ottawa, Canada.

Bull, Arthur (1998) The Fundy Fixed Gear Council: Implementing a Community Quota. Pp 59-61 in *Managing Our Fisheries, Managing Ourselves*, edited by Loucks, L., Charles, T. and Butler, M. Halifax, N.S.: Gorsebrook Research Institute for Atlantic Canada Studies.

CCN (1999) 'Creative Solutions for Coastal Economic Development.' in *1999 Conference of Coastal Communities*, edited by E. E. Tamm. Richmond, British Columbia.

Charles, Anthony (2001) *Sustainable Fisheries Systems*, vol. Series 5, Edited by T. J. Pitcher. Osney Mead, Oxford: Blackwell Science Ltd.

Cohen, Fay G. (1986) *Treaties on Trial: The Continuing Controversy over Northwest Indian Fishing Rights*. Seattle: University of Washington Press.

Cooley, Richard A. (1963) Politics and Conservation, The Decline of the Alaska Salmon, Harper & Row, New York, NY.

DOC (2001) Fisheries of the United States, 2000, Current Fisheries Statistics No. 2000, US Dept. of Commerce, NOAA, NMFS, Silver Spring, MD.

Feit, Harvey (1988) 'Self-management and State-management: Forms of Knowing and Managing Northern Wildlife.' Pp. 72-91 in *Traditional Knowledge and Renewable Resources Management in Northern Regions*, edited by M. M. R. Freeman and L. N. Carbyn. Edmonton: Boreal Institute for Northern Studies.

Fisheries and Oceans Canada (2000) 'Scotia-Fundy Offshore Scallop Integrated Fisheries Management Plan, Maritime Region'.

Ginter, Jay J. C. (1995) 'The Alaska community development quota fisheries management programme,' in *Ocean and Coastal Management*, Vol. **28**, nos. 1-3, pp147-164.

Loucks, L. (1998) 'Sambro Community Quota Fisheries Management: A Case of Innovative Community Based Decision Making.' Pp. 54-58 in *Managing our Fisheries, Managing Ourselves*, edited by L. Loucks, T. Charles, and M. Butler. Halifax, N.S.: Gorsebrook Research Institute for Atlantic Canada Studies.

McHugh, J. L. (1972) 'Jeffersonian Democracy and the Fisheries,' in: *World Fisheries Policy, Multidisciplinary Views*, Brian J. Rothschild, Editor, University of Washington Press, Seattle, pp 134-155.

NE Fisheries Management Council (2000) 'Scallop FIshery Management Plan SAFE Report (Stock Assessment and Fishery Evaluation)', Gloucester, Mass.

NMFS (1992) Federal Register notice published in vol. 57, page 54936, on 23 November 1992.

National Research Council (1999) The Community Development Quota Programme in Alaska, National Academy Press, Washington, DC.

Pinkerton, Evelyn (1989) 'Introduction: Attaining Better Fisheries Management Through Co-Management: Prospects, Problems, and Propositions.' Pp. 3-33 in *Co-operative Management of Local Fisheries: New Directions for Improved Management and Community Development*, edited by E. Pinkerton. Vancouver: University of British Columbia Press.

Pinkerton, Evelyn and Keitlah, Nelson (1990) 'The Point No Point Council: Innovations by an Inter-Tribal Fisheries Management Cooperative.' School of Community and Regional Planning, University of British Columbia, Vancouver.

Repetto, Robert (2001) 'The Atlantic Sea Scallop Fishery in the USA and Canada: A Natural Experiment in Fisheries Management Regimes.' Marine Policy Centre of the Woods Hole Oceanographic Institute.

Schlesinger, Arthur, Jr. (1971) 'Is it Jeffersonian?' in: New York Times, Saturday, January 30, 1971.

Sen, Sevaly and Nielsen, Jesper Raakjær (1997) 'Commentary on 'Fisheries Co-Management: A Comparative Analysis'.' *Marine Policy* **21**, 545-546.

Singleton, Sara (1998) *Constructing Cooperation: The Evolution of Institutions of Co-management*. Ann Arbor: The University of Michigan Press.

Smallwood, Kate (1993) *Coming Out of Hibernation: The Canadian Public Trust Doctrine*, unpublished Master's Thesis, University of British Columbia, Faculty of Law.

Wilson, James A.(2002) 'Scientific uncertainty, complex systems and the design of common pool institutions'. In *The Drama of the Commons*. Committee on the Human Dimensions of Global Change, E. Ostrom, T. Dietz, N. Dolsak, P.C. Stern, S. Stonich, and E. Weber, eds. Division of Behavioral and Social Sciences and Education. Washington, DC: National Academy Press.

Chapter 10

EXPERIENCES WITH FISHERIES CO-MANAGEMENT IN AUSTRALIA AND NEW ZEALAND

REBECCA METZNER
Food and Agriculture Organization of the United Nations (FAO), Rome, Italy

MICHAEL HARTE
Falkland Islands Government, Falkland Islands, South Atlantic

DUNCAN LEADBITTER
Marine Stewardship Council, New South Wales, Australia

1. THE CONTINENTAL REGION

Efforts to describe the Oceania countries of Australia[1] and New Zealand as a whole are problematic, at best, because these are two truly unique and different countries. However, it can be said of both that they lie in the South Pacific Ocean; they enjoy exclusive economic zones with valuable and diverse fisheries - including rock lobster, abalone, tuna, prawns, scallops, orange-roughy and Antarctic toothfish caught in the Antarctic's Ross Sea; and that aquaculture is an increasingly important sector of the seafood industry.

Australia has a population of approximately 19.1 million people who occupy the world's sixth largest country of 7,686,850 square kilometres. The world's smallest continent, it nonetheless has a coastline some 25,760 km long and an exclusive economic zone of some 11 million square kilometres (5.7 million square miles) which is the third largest in the world. A federal-state system, there are two territories - the Australia Capitol Territory (ACT) where the capitol, Canberra, is located and the Northern Territory[2] - and the six genuinely distinct states of Western Australia (WA), South Australia (SA), Victoria (VIC), Tasmania (TAS), New South Wales (NSW), and Queensland (QLD).

In contrast, the approximately 3.8 million people of New Zealand occupy an area of 268,676 square kilometres (103,736 square miles) or nearly 3.5% the size of Australia. Despite this size difference, New Zealand has à coastline of 15,134 km or nearly 59% that

[1] Formally, Australia is the Commonwealth of Australia.
[2] Although not technically a state, for purposes of this discussion of fisheries management in Australia the Northern Territory can be considered in the same light as the other states.

of Australia, and it has the world's fourth largest exclusive economic zone.

2. FISHERIES MANAGEMENT IN THE REGION

The physical differences of these two countries aside, there are two basic fisheries management issues in this region that are similar, namely:

- the fisheries resources of these countries are finite and, thus, there is an inevitable need to allocate them amongst various user groups (Metzner, 1997); and
- the space in which these resources occur is finite and, thus, there is an inevitable need to allocate amongst different stakeholder groups the space in which fisheries-focussed user groups find (or raise) and harvest fish resources.

As a result of these allocation issues, there are concomitant emerging or existing allocation battles - and it is these battles that are increasing the need for both cooperative management strategies such as ascribed to 'co-management' and, more narrowly, conflict management mechanisms and procedures.

2.1. Fisheries management in Australia

In Australia, state fisheries jurisdiction extends offshore from 0 to 3 nautical miles, and the Commonwealth's jurisdiction extends from 3 to 200 nautical miles.

At the Commonwealth level, two Acts form the basis of fisheries management - the Fisheries Administration Act 1991 and the Fisheries Management Act 1991, and the Australian Fisheries Management Authority (AFMA) has the responsibility for managing Commonwealth fisheries in what is called the AFMA 'partnership approach'. This approach explicitly recognizes the experience available within industry and the need for government and industry to work together to achieve desired fisheries management outcomes.

The respective states have their own enabling fisheries legislation, much of which was rewritten and enacted in the mid-1990s after the revision and enunciation of the Commonwealth's new policy directions in *New Directions for Commonwealth Fisheries Management in the 1990s* (Commonwealth of Australia, 1989).

For interjurisdictional fisheries spanning state and Commonwealth waters, there are four management scenarios that are possible under the Fisheries Amendment Act 1980 and the Offshore Constitutional Settlement:

- Joint authority and management, where the Commonwealth and a state or states form a single legal entity with legal management powers;
- state management of the entire fishery;
- Commonwealth management of the entire fishery; or
- status quo management, where responsibility is split between the state(s) and the Commonwealth at the 3 mile boundary (Commonwealth of Australia, 1989).

In terms of fisheries management, limited access rules and input controls (gear, spatial, and temporal restrictions) are the predominant forms of regulatory tools in use, although there are now some moves towards greater use of transferable input controls such as transferable traps or other gear-based systems. There are also a few fisheries managed under individual quotas (IQs) or ITQs, but these are by far in the minority.

2.2. Fisheries management in New Zealand

The Fisheries Act 1996 forms the statutory basis for all of New Zealand's fisheries management systems, all of which are managed by the State, with separate management systems for recreational, customary Maori, and commercial fisheries. The purpose of the Fisheries Act 1996 is to enable the utilization of fisheries resources while ensuring sustainability, and it includes provisions for:

- Environmental protection;
- customary fishing regulations;
- recreational fishing regulations;
- bringing new species into the quota management system;
- resolving disputes between fishers over access; and
- consultation on fisheries management.

Recreational marine fisheries operate as open access fisheries and, as such, are either non-exclusive or excludable at only very high cost, and the rights to the fisheries are held in common. The regulations for these fisheries are arguably lightly enforced and include regulations such as daily bag limits, minimum fish sizes, method and gear restrictions, closed areas and closed seasons.

The management of customary Maori fisheries is based on a territorial use rights system where harvesting rights are restricted to specific groups or communities. Shares are allocated within the group through a variety of administrative or negotiated processes such as:

- Rahui - a ban on taking of kai moana (seafood);
- Mataitai - an area of seashore that is managed as a traditional subsistence fishery by iwi or hapu (tribe or sub-tribe);
- Taiapure - an area of coast that is managed by an iwi committee that has customary authority to make rules regarding access and exploitation rates that are legally binding on all fishers. In most cases there is no exclusivity to the stock, though spatial exclusivity is guaranteed in the case of mataitai, so they are affected by (and in turn affect) extractions from the same stock by commercial and recreational fishers.

The main method for managing commercial fisheries is the quota management system (QMS) that is based on ITQs or rights to shares of controlled levels of commercial catch. Catch limits are set for each fish stock, and these rights to harvest fish are acquired by using or purchasing annual catch entitlements that are a function of the quota shares held by individuals and businesses. Non-QMS fisheries are managed through restricted entry licensing, catch limits and gear restrictions, with the exceptions of the tuna fisheries (excluding southern blue-fin tuna) that are effectively open access commercial fisheries.

Quite comprehensive in its application, 44 species or groups of species - representing over 85 per cent of the total known fish catch in the EEZ - are currently managed as 284 separate fish stocks under the QMS. Subject to a statutory requirement to have regard to the costs and benefits of introduction, the New Zealand's Ministry of Fisheries is working to bring more of the commercially caught species and fish stocks under the control of the QMS over the next few years. Significantly, the introduction of new stocks into the QMS is also the mechanism by which the Crown's Treaty of Waitangi (Fisheries Claims) Settlement Act 1992 obligations are met. Te Ohu Kai Moana, on behalf of Maori, receives 20 per cent of the quota for all new stocks introduced into the QMS. Maori customary fishery rights are also protected by the Settlement Act.

The strengths and weaknesses of the New Zealand QMS are well documented in fisheries management literature (Bess and Harte, 2000; Bautzen and Sharp, 1999; Clark *et al.*, 1988; Dewees, 1989; Harte, 2001; Memon and Cullen, 1992; Sissenwine and Mace, 1992). What is evident is that owners of ITQs have a large incentive to invest resources into the sustainability of the fishery simply because any lowering of catch limits reduces the value of their investment in the fishery (Harte, 2001). As Jentoft *et al.* (1998) suggest, the private nature of QMS rights has given a more accurate indication of who the users are than under previous management regimes. Importantly, this clearly defined set of holders of exclusive rights makes it easier to assign responsibility for devolved and/or decentralised management of a given fishery.

3. SO-CALLED FISHERIES 'CO-MANAGEMENT'

Attempting to interpret whether fisheries 'co-management' exists in Oceania is relatively straightforward because, using broad definitions of co-management, it can be said that it does exist.

However, attempts to interpret the nature and extent of the fisheries 'co-management' in Oceania is, at best, subjective, controversial, and a function of the critic's perspective. Even geographical location can play a part in the analysis, because what may be deemed to be relatively unsuccessful co-management activities in one location may be described as wildly successful when compared with those co-management activities occurring in other places.

These difficulties are further highlighted if one uses a broad definition where 'co-management' is used to describe shared responsibilities for resource management between the government and user groups, because the analysis is still subject to interpreting just what constitutes the notion of 'sharing'. Even trying to notionally agree that co-management is somehow 'truly cooperative' is fraught with the difficulties of different opinions and perspectives, and the result may well involve including virtually all cooperative arrangements along the spectrum from 'instructive' to 'informative' - with perhaps the exceptions of community/traditional management on one end and pure top-down government management on the other.

Thus, any sort of analysis of co-management in fisheries begins with the recognition that:

- A plethora of co-management situations currently exists in Oceania;
- these arrangements range along the spectrum of types of co-management; and
- the process of assigning the real world situation to points along this spectrum will be intensely subjective.

3.1. Fisheries 'co-management' in Australia

3.1.1 Defining 'co-management'

In Australia, the use of the phrase fisheries 'co-management' can mean just about anything, depending on the stakeholder(s) with whom one is speaking and the fisheries which are being discussed. There is no 'national' or legislated definition, and the use of the phrase in the context of fisheries management has emerged from and is used by different sectors at different times and in different ways, depending on the agendas, politics, and participants involved.

Informally, and from a historical perspective, one could say that the very personalized, individual-to-individual management that initially occurred in some fisheries has constituted co-management. Quite simply, some state fisheries have historically involved so few participants that management arrangements between fisheries managers and the commercial fishermen have been quite personalized and not particularly bureaucratic or otherwise formalized, and there are early examples of the use of both formal and informal participatory processes in the form of various committees composed of commercial fishers and fishery managers for directed purposes such as designing structural adjustment or buyout programmes. However, as the numbers of participants in particular fisheries have grown - and as allocation issues, both within and between fisheries, have emerged - these informal approaches and concept of co-management have become more formalized.

The legal notion of fisheries 'co-management' was legislatively introduced in the early to mid-1990s in Australia at the time when most of the fisheries legislation in Australia was being substantially revised, and this new legislation defined a variety of advisory and management committees intended to provide advice to fisheries agencies and their Ministers. For example, in 1991 the co-management concept of the 'partnership approach' was codified in the Commonwealth's fisheries legislation. Similarly, the various management advisory committees (MACs) and ministerial advisory committees (also MACs) which were established in the respective states during the 1990s - either legislatively or informally - reflect some formal recognition of the concept of co-management (Table 1).

3.1.2 Who is using the term co-management now?

It is only relatively recently, however, that the term 'co-management' has come to be used to cover the range of legislative and informal arrangements found in Australia, reflecting a growing awareness of both alternative dispute resolution techniques and academic literature on the subject.

The use of the term 'co-management' by fishermen's groups tends to vary by fishery and by state and covers the range from covering informal consultations between fishermen and the managers all the way to participation in formally legislated advisory groups. In some states, the commercial sector is increasingly using 'co-management' as the phrase both describing (or, at least, calling for) greater participation by fishermen or commercial fishing organizations in management decision-making activities and, secondarily, but increasingly, as a phrase to describe the desire to have actual fisheries management decisions devolved to industry.

In the recreational sectors around the country, 'co-management' seems to be taking on a meaning that reflects that sector's growing political influence and lobbying power - and the seriousness of the sector's intent to wield such power. In short, co-management is being used to describe efforts by that sector to have greater control over allocation decisions between the recreational fishing sector and other sectors such as commercial fishing, tourism, and so forth.

Table 1. Legislative examples of 'co-management' structures in Australia[3]

Jurisdiction	Legislation	Co-management Components
Commonwealth	1. Fisheries Administration 1991 2. Fisheries Management Act 1991	AFMA's statutory functions include: (c) to consult, and co-operate, with the industry and members of the public generally in relation to the activities of the Authority;
New South Wales (NSW)	Fisheries Management Act 1994	NSW Commercial Fishing Advisory Council (CFAC) Regional Advisory Committees
Northern Territory	Fisheries Act 1985	Potential committees
Queensland	Fisheries Act 1994	Queensland Fisheries Management Authority (QFMA)
South Australia	South Australia Fisheries Act 1982 (and subsequent amendments)	South Australian Fishing Industry Council (SAFIC) South Australian Recreational Fishing Advisory Council (SARFAC) Fisheries Management Committees (FMCs)
Tasmania	Living Marine Resources Management Bill 1995	Schedule 1(c) to encourage public involvement in resource management and planning
Victoria	Fisheries Act 1995	Objective 3(f) To encourage the participation of resource users and the community in fisheries management Part 9, Schedule 1 Membership and Procedure of Bodies
Western Australia	Fish Resources Management Act 1994	Advisory Committees including: • Rock Lobster Industry Advisory Committee (RLIAC) • Recreational Fishing Advisory Committee (RFAC) • Aquaculture Development Council (ADC) • Fishery Management Advisory Committees, and • Other Committees Part 19 Miscellaneous including Schedule 1, Constitution and Proceedings of Advisory Committees

Use of the term 'co-management' involving Aboriginal and the Torres Strait Islander people varies both by location within Australia and by the groups involved. In general, most of such 'co-management' efforts are in response to requests by Aboriginal people

[3] Adapted by Metzner from Metzner and Rawlinson (1998)

throughout Australia asking fisheries management agencies to include and recognise Aboriginal people's use of fish for food, economic development and cultural reasons while also involving them in decision-making processes regarding commercial fishing, aquaculture, recreational fishing, and ecotourism.

For legal reasons, government efforts to respond tend to be decoupled from the legal issues of Native Title and thus are somewhat grey or unofficial in nature, and this is especially true in the states. Nonetheless, there are examples around the country where fisheries managers are making genuine efforts to develop better relations and inclusive fisheries management strategies. However, given the difficult legal nature of Aboriginal claims and as can be expected when relatively immediate commercial interests collide with the pace of national and international law regarding indigenous peoples, there is also a quietly emerging world of 'co-management' arrangements or, more basically, private contracts between specific communities and various commercial fishing operations or industries.

Finally, at the Commonwealth level, 'co-management' can be used to describe AFMA's participatory process which primarily involves industry and state agency participants, although, it can be argued that in Environment Australia, the phrase tends to be used primarily to suggest greater participation by non-governmental organizations (NGOs) in management decision-making activities than it does participation by either recreational or commercial fishing interests.

3.2. Fisheries 'co-management' in New Zealand

3.2.1 Defining 'co-management'

Until recently, fisheries co-management has not been a specific objective either of Government or of fisheries stakeholders in New Zealand. It is only with the advent of an extensive international literature on the subject of co-management that the term has come to be used for a range of legislative and institutional arrangements that long been a feature of fisheries management in New Zealand.

In fact, the term 'co-management' was first used in New Zealand as a key recommendation of an independent review of the Fisheries Act 1996 carried out in 1998 (PricewaterhouseCoopers, 1998) that had been instigated because of operational difficulties in implementing the Act and stakeholder dissatisfaction with key aspects of the legislation. There, the reviewer called for the co-management of fisheries resources in New Zealand. Drawing on the work of Jentoft *et al.* (1998) a broad interpretation was given to the term co-management defining it as '... an ongoing collaborative and communicative process, where resource users, together with government representatives and other actors, are in an entrepreneurial and creative role' (Jentoft *et al.*, 1998, p. 426).

The review recommended that the Fisheries Act 1996 be amended to allow the Minister responsible for fisheries to devolve some management functions carried out to recreational, customary and commercial fishery stakeholders organizations, but that such devolution should only occur if stringent standards and specifications - as demonstrated via a comprehensive fish stock management plan - could be met by the particular stakeholder organization in question. Moreover, the intent was that these plans would be a joint product of Government and fishers and would be developed in with the participation of other stakeholder groups such as environmentalists.

The outcome would be decentralised management for some fisheries where fishers are empowered to take collective responsibility for their actions. This model is already working for South Island customary fisheries, the rock lobster fishery, and the southern scallop fishery.

3.2.2 Who is using the term 'co-management' now?

The term co-management has failed to gain popular currency in New Zealand. Both fishers and fisheries managers deliberately avoid the term - thereby leaving only authors writing in academic publications to use it - because both recreational fishing and environmental stakeholders groups tend to interpret the term as being synonymous with the abdication of state responsibility for fisheries management.

There are two key reasons why this has happened:

- The terms of reference of the independent review of the Fisheries Act 1996 referred only to the management of commercial fisheries. The review made reference to the need for inclusive co-management and recreational, customary Maori fishers were recognised as being important partners in management. Nevertheless the nature of their involvement was addressed superficially.
- The Ministry of Fisheries was caught by surprise by the independent review's focus on institutional arrangements for fisheries management as well as the expected technical fixes to the Fisheries Act 1996. The Ministry was not ready to respond to these recommendations, and because the recommendations went to core of the Ministry's own role in fisheries management, it may be that the Ministry had difficulty in fully addressing the implications the recommendations had regarding their operations at that time (Bess and Harte, 2000).

This is not so say that co-management does not exist in New Zealand. Despite the absence of an explicit policy of co-management, it has been an implicit part of fisheries management in New Zealand since at least the mid-1980s. Commercial, recreational and customary fishers, environmental organizations, and Government agencies engage in several interrelated consultative processes that are part of fisheries management in New Zealand. These are:

- research planning;
- stock assessment; and
- cost recovery of fishery management costs.

Over time, these processes have evolved from being instructive in nature where there was only minimal exchange of information between government and stakeholders regarding management decisions available to the situation where government and stakeholders co-operate as partners in decision making. In some matters the relationship as developed to the extent to which stakeholders advise government of the management decisions to be taken and government, as appropriate, endorses such decisions (Harte, 1999, 2001).

Concurrently with the evolution of these consultative processes and as a consequence of the positive incentive structures of the quota management system the commercial fishing industry evolved a set of institutions to co-ordinate their own activities. These management associations are based on functional (by species such as hoki, squid or orange roughy) and/or territorial (such as the inshore fisheries of the North of New Zealand's North Island) communities. Depending on the fishery, these associations have a number of purposes including (Bess and Harte, 2000; Harte, 2001):

- To facilitate the collection of funds to finance fisheries management activities such as research or reseeding and to manage the delivery of such services;
- to make fisheries management rules and to impose sanctions on non-compliance of

company shareholders;
- to represent the interests of shareholders in government processes that involve consultation - such as determining (government required) fisheries management services and the setting of total allowable commercial catches; and,
- to defend against erosion of harvesting rights and to promote the expansion and development of management rights.

4. POSITIVE MODELS OF FISHERIES CO-MANAGEMENT

4.1. Successful models from Western Australia

In Western Australia, the choice of a successful example or model of some form of co-management in fisheries depends on how one wishes to describe or otherwise define 'co-management'.

For example, here are various advisory committees established as required under the West Australian Fish Resources Management Act 1994:

- Rock Lobster Industry Advisory Committee (RLIAC);
- Recreational Fishing Advisory Committee (RFAC);
- Aquaculture Development Council (ADC);
- Fishery Management Advisory Committees; and
- provisions for the establishment of other committees for various purposes.

Membership, while only broadly defined in the legislation, is gradually being interpreted by the agency to include other stakeholders outside the scope of the particular fisheries involved. That said, the actual process of forming and selecting membership on these committees is not fully transparent and is still subject to considerable discretion of both the Department of Fisheries and the Minister for Agriculture, Forests and Fisheries, and the roles of the committees are, technically, only advisory. That these committees can be successful in their work is as much a function of the issues facing them as it is of their composition, chairmanship, and other leadership.

A sharing of power that puts stakeholders (including the Department of Fisheries) on a more equal basis serves as the basis for the *Guidelines for Voluntary Resource Sharing* described in Metzner (1999). (See also Wright, Metzner and Chevis, 2000) (Metzner, 1999). Cooperatively developed by a small team of people from the commercial, recreational, and government sectors - some of whom had taken profession alternative dispute resolution training courses, the Guidelines process was crafted on interests-based facilitative mediation principles. It was intentionally designed to provide a mechanism for capturing differences of opinion about fisheries-related issues, cooperatively developing arrangements to solve such problems, and to then deliver the agreed arrangements to the Minister and the Department for approval and implementation.

The Guidelines process has been used in a handful of cases, with the most successful outcome to date being a set of arrangements for reducing both commercial capacity and catch in a small crab fishery while also reallocating a portion of the catch to the recreational fishing sector.

In this instance, the success of the process was a function of the participants involved - two of the creators of the Guidelines process were involved in the negotiations of the arrangements - and the vision of the other participants. Knowing that the future would hold increasing numbers of allocation battles for sharing limited fish resources, the commercial participants were driven in large part by their sheer determination to demonstrate that the then fledgling process could provide a vehicle for resolving user conflicts and by the looming need for commercial capacity reduction in that particular fishery. Similarly, the government's support for the process was carried by a well-intentioned manager and one of the creators of the Guidelines process. And, in the absence of a polarized constituency seeking an all-or nothing outcome, the recreational sector's representative was both supportive of the process and in a position which allowed for genuine negotiation.

4.2. A successful model from New Zealand

Fortunately for proponents of co-management in New Zealand's fisheries, there are many successful examples of co-management to draw on (Harte, 2000, 2001). Among the most successful is the co-management of New Zealand's Rock Lobster fisheries (Harte, 2001).

The fishery is managed through the multi-stakeholder National Rock Lobster Management Group (NRLMG). Membership of the NRLMG comprises government agencies, commercial, recreational and indigenous fisher representatives, environmental non-governmental representatives and science advisers. Recognized as a primary source of advice to Ministers on all matters pertaining to rock lobster fisheries, the NRLMG is funded by industry by way of provision of an independent chairman, meeting venues, catering, and an administrative support role shared with the Ministry of Fisheries. The marriage of the practical working knowledge of rock lobster fishers, the research and management experience of government agencies, and expectations of other sector groups has been a successful and productive one.

Key to the success of the NRLMG has been a commercial sector committed to the sustainable and inclusive management of the rock lobster resource. The New Zealand Rock Lobster Industry Council (RLIC) is a successful example of the potential that commercial stakeholder organizations have to succeed in a number of fields of fisheries management including research.

RLIC is an umbrella organization for nine commercial stakeholder organizations operating in each of the rock lobster management areas of New Zealand (Sykes, 2000). These organizations have been established as incorporated societies or limited liability companies and are known as CRAMACs.

Membership of CRAMACs comprises quota owners, processors, exporters, and fishermen (owner-operators and lease holders) in each region. Governance is based on a two-tiered voting procedure that gives priority to quota ownership on issues affecting total allowable commercial catch decisions, levy setting, and certain government consultation processes. All nine CRAMACs hold a majority mandate of crayfish quota holders in the regions. CRAMACs are shareholders in RLIC and appoint the nine person board of directors, one from each CRAMAC.

In 1997 RLIC became an accredited research provider to the Minister of Fisheries, and since then has successfully tendered for, and executed, four rock lobster stock assessment contracts. Research contracts are undertaken in collaboration with national science providers and internationally recognized stock assessment consultants contracted to RLIC. RLIC also uses accredited technicians employed by science providers and by CRAMACs to undertake an extensive stock-monitoring programme.

5. NEGATIVE MODELS/EXAMPLES OF FISHERIES CO-MANAGEMENT

5.1. Unsuccessful examples in Australia

5.1.1 Circumventing co-management structures

One of the most common categories of unsuccessful co-management throughout Australia is when management decisions are extracted by Ministers/agencies outside of management structures and rationale (Chatterton & Chatterton, 1981). An example of this occurred in New South Wales[4] where commercial estuarine seining was labelled as environmentally damaging - despite departmental research documenting both the minimal impact on seagrasses and the success of bycatch reduction devices - and commercial fishing operations were shut down. To make the decisions even more clearly political and outside the realm of the management process, the fisheries department was also making a number of pronouncements quarantining prawn trawling, both offshore and in sensitive marginal electorates, from the same sorts of environmental judgements.

Commercial fishermen have also become (quite rightly) sceptical about the role and value of the so-called co-management bodies created by fisheries departments, primarily because it seems as though the existence of these bodies depends on whether or not the department that created them receives advice which concurs with the department's view. Although this can be described using an example from New South Wales, such perceptions are not limited to that state or even to commercial fishermen. The fisheries department established a series of fishery-based management advisory committees (MACs) which contained representatives from commercial and recreational fishing groups as well as representatives from environmental groups. It was not the composition of these MACs, but rather the administration of them that became problematic both in terms of the participants' roles, responsibilities, and outputs and in terms of the support for the MACs. Quite simply, the commercial fishermen quickly became scathingly critical of these co-management structures because the department tried to manipulate the MACs' decisions, especially in inshore commercial fisheries which were being closed down and the fish were being reallocated to the recreational fishing sector in the form of recreational fishing parks.

The failures of the existing co-management structures and processes are also evidenced by the increasing resort to litigation - a pathway now being utilized by individuals, groups and sectors - and to legislative tools outside the fisheries arena. There is a rapidly escalating body of legislation that applies to fisheries which is not administered by the fisheries agencies themselves. Specifically, environmental legislation regarding wildlife, marine conservation, and environmental impact assessments is increasingly being applied to (commercial) fisheries, yet it is generally implemented by environment or conservation agencies which do not have either the co-management structures or the co-management ethos. Indeed, such agencies sometimes even appear to cultivate a distance from the commercial fishing industry.

Ironically, however, environmental legislation commonly has third party appeal rights that tend to be broader than those available under fisheries legislation, and this can be used by parties dissatisfied by a decision of a MAC to seek another decision. Thus, in 1979 a recreational fishing group brought a case under the NSW Environmental Planning Assessment Act which successfully argued that the issuing of a commercial fishing licence required an environmental impact statement (EIS). The case was an ultimately pivotal point

---

[4] Similar examples of politically motivated reallocation decisions being made outside existing co-management processes have also occurred in other states around Australia.

in a series of moves by the recreational sector to remove commercial fishing from estuaries. It also reflected the view by the recreational sector that the management structures put in place by NSW Fisheries were not addressing some of the issues that they had thought were important. (Moreover, it flushed out the fact that NSW Fisheries had been avoiding the law for nearly 20 years, despite having been internally advised of that fact for over ten years.) And, demonstrating just how rapidly the use of litigation can escalate, there is now a case being brought before the same court by a commercial fishing group which is arguing that if commercial fishing should have EISs then so should recreational fishing.

Another example is the application of changes to Schedule 4 of the Commonwealth Wildlife Protection Act, 1994, whereby the export of seafood products, once exempt from this legislation, now have to be evaluated by the federal environment department, Environment Australia (EA), in addition to the existing required export permits. This change allowed EA - which is not a part of any of the existing fisheries co-management structures - to make potentially enormous changes to the management of the fishery without having to participate in the established fisheries management processes that all other sectors were obliged to follow and, in theory, it could overturn hard won agreements within and amongst various fisheries sectors without any co-management sorts of activities at all.

5.1.2 Co-management structures without substance

In 1987 the NSW fisheries agency of the day had become frustrated at an inability to get a coherent view from the commercial fishing sector. Years of open access management had created an excess of small scale fishermen with few resources (and little incentive) to form effective fishing associations. The agency created a Commercial Fishing Advisory Council to which all licenced fishermen had to become members. The Council had a staff of 3 and provided advice to the Minister for Fisheries. It was funded by a levy on each commercial fishing licence.

In 1995 following a change of government and a period of dispute with the recreational sector over a rights based management system the agency changed its director of fisheries and dissolved the Commercial Fishing Advisory Council. The agency reduced (but did not abolish) the levy on fishermen and created a new, interim representative body. However, no funds were ever provided and the levy was appropriated by the agency. This body eventually folded, and a voluntary body was created by some fishermen.

Several years later, the fisheries department established an Advisory Council on Commercial Fishing, and it remains unfunded in the same way that the Commercial Fishing Advisory Council was.

5.2. An unsuccessful example from New Zealand

Just as there are examples of successful co-management models there are unsuccessful models and examples of co-management failures. Unfortunately, these examples tend to destroy the legitimacy of the co-management model not just in their particular context but also for co-management in general because most stakeholders enter into co-management situations expecting success.

Failure to achieve the expected outcome is often perceived by the participants not just as a failure of process but a personal failure. Most people are willing to repeat a successful experience but few people are prepared to repeat an exercise that has failed previously. Moreover, the media and stakeholder interests antagonistic to co-management give disproportionate attention to failure compared to successes.

Tauranga Harbour is an enclosed harbour in the Bay of Plenty located on the east coast of New Zealand's North Island. It provides for a valuable small-scale commercial beach seine fishery and is also a sheltered recreation fishery within easy reach of a third of New Zealand's recreational fishers. There has been a long history of conflict between recreational fisheries and commercial fishers in the harbour that had become personalised and entrenched as part of the local culture.

Ongoing intense lobbying by recreational fisheries and local politicians led the then Minister of Fisheries to invoke the dispute resolution provisions of the Fishery Act 1996 and to appoint a Dispute Resolution Commissioner to report on the dispute and make recommendations for the Minister's consideration. Many groups submitted to the Dispute Commissioner, and local commercial fishers, supported by national and regional seafood industry associations proposed that:

- In the short term commercial fishing should occur unchanged in Tauranga Harbour because there are no sustainability grounds for banning commercial fishing and because a ban would have a significant economic effect on commercial fishers;
- commercial fishers would immediately develop an industry enforced management plan for commercial fishing in the Harbour that involved quota owners, fish receivers and fishers; and,
- in the medium to long term a territorial use rights for fishing (TURF) agreement between commercial, recreational and customary fishers and marine farming interests in Tauranga Harbour should be developed.

The Commissioner duly produced a long and considered report that found that Tauranga Harbour had historically low catches compared to outside the Harbour and that high catch rates for commercial and recreational fishers were positively not negatively correlated. In other words, in seasons when fish came into the Harbour everyone benefited, and in seasons when few came everyone caught fewer fish. The Commissar further concluded that the commercial sector was already heavily regulated with closed seasons and other measures to avoid conflict with the main recreational fishing demand in summer and that by and large these were effective. The biggest future threat was unregulated recreational fishing and loss of coastal habitat from pollution and urbanization

The report also endorsed proposals to develop a joint management plan for the co-management of the Harbour between central and local government, recreational, customary Maori and commercial fishers, and environmental groups. In particular the Commissioner emphasized the need for all stakeholders to work together to maintain and enhance the health of the harbour. The job of developing the co-management regime was tasked to the Ministry of Fisheries whose officials were based some 200 kilometres from the Harbour.

Despite the best attempts of the Ministry of Fisheries, commercial, customary Maori and some recreational fishers, the local boating club purporting to represent the majority of recreational fishers refused to accept any proposals for co-management that did not include a complete ban on commercial fishing in the Harbour. Attempts to develop a management plan were subsequently abandoned by the Ministry of Fisheries on the grounds that stakeholders were unwilling to participate.

Under the Fisheries Act 1996, the Minister of Fisheries was still bound to make a determination in the dispute. Ignoring the basic points of the Dispute Commissioner report, the Minister did four things:

- Banned any new commercial fishers from fishing in the harbour;
- allowed existing commercial fishers to continue fishing in the Harbour, but if they did

not wish to fish their beach seine permit they could not transfer it to another fisher (contrary to the fundamental principles of New Zealand's QMS);

- imposed further restrictions on commercial fishing including making formerly voluntary agreements statutory regulations on the grounds that they had been so successful it was necessary to formalize them as statutory regulations; and
- imposed no restrictions nor made any comment about recreational fishing activities.

At best, the Tauranga Harbour example demonstrated that senior politicians and the Ministry of Fisheries lacked an understanding of the potential for co-management to be broadly applied to fisheries management in New Zealand. At worst, the Tauranga Harbour case study suggests that there is no significant institutional or political support for co-management in New Zealand. At a practical level commercial fishers in the North of the North Island have become cynical about future involvement in dispute resolution processes and/or proposals for new co-management arrangements from the few in the region who still believe in their potential.

6. THE PROSPECTS FOR FISHERIES 'CO-MANAGEMENT'

6.1. The prospects for fisheries 'co-management' in Australia

Legally, there are no impediments to the co-management of fisheries in Australia, and the physical infrastructure - in the form of ministerial advisory committees, management advisory committees, stakeholder organizations, and the like - currently exists.

Despite this, the perceptions about whether 'co-management' sorts of processes and participatory fisheries management actually work, vary throughout Australia, depending on where one is and who one asks[5]:

> 'Generally, a number of the [co-management] groups are not working particularly well. They are an advisory committee to the Minister ... however the main problem in a number of cases is that the participants actually consider that they have the right to do whatever they want ... In other cases the problem seems to be one of not understanding the issues ... Additionally, co-management needs to be seen as an opportunity for stakeholders to do something constructive as opposed to bashing government'.

Even within the same state, stakeholders from two different groups can, and do, offer differing perceptions about the relative success of co-management activities, with comments such as:

> 'The [co-management process] is alive and well. We have consultations going on, and they have been a success in that people have had the opportunity to express their views in a non-confrontational arena ... and the interests of stakeholders are being appropriately represented in the process. The new government is committed to supporting the notion of resolving resource issues; however ... funding is going to be tight ... all signals are that funding is going to be significantly less than before'.

contrasting with comments such as:

> 'The general feeling is that the [co-management] process is struggling to survive. We

---

[5] These comments have been paraphrased from personal communications from stakeholders around Australia.

have yet to reach closure, and funding is non-existent'.

And, there are examples where members from different stakeholder groups will even comment that:

> 'We never really have had anything like fisheries co-management. The closest we have come was with an [advisory committee], even with all its faults. However, rather than being seen as a mechanism for better strategic fisheries management by getting away from the tyranny of small decision making, it became seen as a threat and strangled at birth ... We, too, have had to resort to the courts ... Our only hope is a change of government, but the possibility of that is slim and some time away'.

As the comments reflect, Australian perceptions of fisheries 'co-management' vary. Indeed, it may be most accurate to describe co-management in Australia as a fragile concept that is exceedingly subject to interpretation. Furthermore, it is probably accurate to say that the future success of fisheries co-management will, in large part, depend on the extent that stakeholders can work to minimize the very factors that put the implementation of co-management at risk; namely:

- The ability to circumvent the existing co-management structures by increasing resort to courts, by increasing the amount of legislation that indirectly applies to fisheries (such as environmental or trade-related legislation), or by decisions being extracted by Ministers/agencies outside of normal management structures and processes;
- the creation of structures without financial or other forms of support;
- the poor definition of roles, responsibilities, and deliverables; and,
- the lack of commitment, integrity and trust that is essential to co-management.

In short, the potential of co-management and its future prospects of being used for effecting fisheries management has been, and very clearly will continue to be, a function of the characters involved and their goodwill, energy, and genuine personal investment in these processes. And, while the concept of 'co-management' will undoubtedly persist, it remains to be seen as to whether it persists in a positive, active way or whether 'co-management' becomes a term relegated to academic analyses.

6.2. The prospects for fisheries 'co-management' in New Zealand

New Zealand has no legal impediments to co-management. As far as achieving sound fisheries management outcomes, the Fisheries Act 1996 is enabling rather than prescriptive. Additionally, although the Minister retains the responsibility for fisheries management outcomes, these outcomes can be achieved using 'top down' ministerial direction, 'bottom up' collective action by commercial and recreational fishers or various combinations of these approaches. It should also be noted that customary Maori fishers have a number of legislated territorial use right mechanisms that provide for devolved management responsibility to iwi and hapu organizations.

It is institutional and political factors, rather than legal barriers, which proscribe the potential for co-management in New Zealand. Furthermore, the prospects for co-management can be starkly divided into two categories:

- *Relatively poor* - in fisheries that are shared by commercial, recreational and customary Maori fishers (shared fisheries); and
- *relatively bright* - in fisheries that are predominantly harvested by commercial fishers (commercial-only fisheries).

6.2.1 Shared fisheries

The prospects for the inclusive co-management of mixed fisheries shared by commercial, recreational and customary Maori fishers are poor in the short to medium term. A number of barriers exist to the full involvement of recreational and customary Maori fisheries in the co-management of fisheries resources in New Zealand (Harte *et al.*, 1998). These barriers include the following:

- Poorly defined the rights and responsibilities of some fisheries participants, especially recreational fishers;
- failures by government to consistently empower fishers to collectively manage fisheries to which they have harvesting rights;
- lack of capacity among recreational and some customary fisher groups to effectively participate in co-management;
- an absence of incentives to encourage commercial, recreational and customary fishers to work together to find win-win solutions to conflicts over access to resources; and
- an absence of non-regulatory mechanisms for binding fishers to agreements reached between commercial fishers and/or between commercial, customary and recreational fishers.

These barriers have created a far from level playing field for moving to a fisheries management system based on co-management. In particular, recreational and customary fishers lack the financial, technical and human resources to effectively participate in co-management. Without the resources critical co-management roles for recreational and customary stakeholders such as policy formulation, developing harvesting rules, monitoring and enforcement will automatically default to the Government.

It is notable that New Zealand's successful examples of co-management of shared fisheries are in high value, single species fisheries such as rock lobster and scallops. Although speculative, some of the reasons for this may be that:

- The commercial fishers tending be to residents of the same communities as recreational and customary Maori fishers;
- the species having high marginal value both as a recreational and commercial species;
- there are similar catching methods for recreational, customary and commercial fishers that depend on high catch per unit efforts such as potting and dredging in shared fishing grounds; and,
- the fisheries have experienced past collapses and/or localised depletion that has required cooperation between sectors to rebuild the stocks and support from the Ministry of Fisheries in establishing durable local institutions and processes for the management of the resource.

Long-term prospects for the co-management of shared fisheries are brighter. For the potential of co-management to become more than a rare exception though, commitment to a stable policy and political environment for fisheries management is required from the state. This environment must also empower and support all fisheries stakeholders.

6.2.2 Commercial only fisheries

Prospects for the co-management of commercial fisheries that are not shared with other sectors are brighter in the short to medium term than for shared fisheries. In these instances, there are two extremely key characteristics which make the concept of co-management possible:

- The Government and commercial fishers have a similar concern with the sustainability of the fishery; and,
- the Ministry is not placed in the position of being an arbitrator of allocative decisions between commercial and non-commercial fishers.

It is in these fisheries that both commercial fishers and the Ministry of Fisheries see commercial fisher organizations becoming increasingly responsible for developing fisheries plans that set the management objectives and performance measures and also specify the rules for management and governance. Using standards and specifications developed by the Ministry of Fisheries as the framework, the functions of defining and implementing what are considered necessary management activities (including research, administration, monitoring and compliance, and establishing funding arrangements) are increasingly shared by fisher organizations and government.

7. POLICY RECOMMENDATIONS

Essentially, the lessons learned from the diverse set of co-management efforts in Australian and New Zealand fisheries all link back to the fact that co-management is a tool and, as such, is a tool that can be used when managing people who fish. The following recommendations are offered as part of advancing the use of co-management because it is inherently appealing as a cooperative approach to fisheries management.

- The fragility of co-management is not only due to the design and operational support of co-management structures or processes, but also due to the commitment, integrity, and trust amongst the participants. Fledgling co-management systems are open to being undermined or otherwise compromised. As a result, it is all the more important that new co-management systems be exceedingly carefully designed and initially applied in instances where they are most likely to succeed until they have become part of the cultural landscape of fisheries management and are more or less generally accepted as a way of doing business.
- The design of co-management structures and processes needs to minimize opportunities to circumvent the co-management process such as resorting to the courts or resorting to legislation administered by agencies other than fisheries agencies.
- There must be genuine support for co-management structures in terms of their funding and operational support. Goodwill and trust alone may result in short-term achievements, but participants may eventually abandon unfunded co-management activities and resort of other forums such as the courts to seek resolution of issues.
- Fisheries agencies should consider options for arranging funds for stakeholder groups to cope with any increased levels of co-management activities such as facilitative interest-based mediation. Although the near-term costs of doing so are not necessarily low, the durability and support for the cooperatively determined outcomes of co-management processes is likely to reduce future costs, both in terms of crisis management and ongoing management.
- Co-management structures need to designed to cover the full range of legitimate user groups. Doing so helps to increase the legitimacy of the co-management process itself by creating the opportunity for equitable representation and even balances of power. It also internalizes sources of potential dissent that could otherwise circumvent the co-management process.

- Engendering support for co-management will involve cultural change and education on the part of all participants/stakeholders. Thus, training needs to be extended to both the senior people in agencies and stakeholder groups to engendering support and cultural change and to the operational participants (staff) from *all* stakeholder groups - as a way of training and equally empowering the people who will actually participate in co-management activities.
- Particular styles of co-management - such as conflict management and interest-based facilitative strategies such as mediation - can be very efficient and effective ways to address management issues in fisheries. This is in large part due to the fact that they provide clarity about participation, responsibilities, process, and outcomes. As such, they provide explicit and useful models for other types of co-management activities. Furthermore, because such techniques are already applied and used in other settings and, thus, do not need to be re-invented - agencies and stakeholders should seriously consider developing and extending the use of such techniques.
- The respective responsibilities of participants in co-management structures need to be clearly stated and clearly understood. Participants must be clearly recognized by and have clear guidance from their organizations as to the extent of their delegated responsibilities and ability to make commitments on behalf of their organizations. Participants also need to be clearly supported by the respective delegating authorities. Thus, even top level political and agency support for co-management strategies needs to be explicitly enunciated prior to *and* during the implementation of such strategies.

REFERENCES

Bautzen, C.J. and Sharp, B.M.H. (1999) New Zealand's quota management system: the first ten years, *Marine Policy* **23**, 177-190.

Bess, R. and Harte, M.J. (2000) The Role of Property Rights in the Management of New Zealand's Fisheries, *Marine Policy* **24**, 331-339.

Chatterton , L. and Chatterton, B. (1981) How much political compromise can fisheries management stand? Premiums and politics in closed coastal fisheries, *Marine Policy* (April),114-134.

Clark, I., Major, P. and Mollet, N. (1988) Development and implementation of New Zealand's ITQ Management System, *Marine Resource Economics* **5**, 325-349.

Commonwealth of Australia (1989) *New Directions for Commonwealth Fisheries Management in the 1990s*, AGPS: Sydney.

Dewees, C. (1989) Assessment of the Implementation of Individual Transferable Quotas in New Zealand's Inshore Fisheries, *North American Journal of Fisheries Management* **9**(2), 131-139.

Harte, M.J. (1999) Guarding the Consensus: Stakeholder Participation in the Management of New Zealand's Fisheries Resources, *Public Sector* **21**(6), 2-9.

Harte, M.J. (2000) Industry Perspectives: Taking the Initiative for the Management of New Zealand's Commercial Fisheries, In Use of Property Rights in Fisheries Management, Proceedings of the FishRights99 Conference, Fremantle, Western Australia, November 1999, *FAO Fisheries Technical Paper 404/1*, Rome: FAO, pp. 270-273.

Harte M.J. (2001) Opportunities and Barriers for Industry-led Fisheries Research, *Marine Policy* **25**, 159 167.

Harte, M.J., Arbuckle, M., and McClurg, T. (forthcoming) Property rights and the Evolution of Fisheries Management in New Zealand, in *Private Rights and Public Benefits: Proceedings of the Environment and Property Rights Conference*, Lincoln University, Canterbury New Zealand, November 1998.

Jentoft, S., McCay, B.J. and Wilson, D.C. (1998) Social theory and fisheries co-management, *Marine Policy* **22**(4-5), 423-436.

Memon. A.P. and Cullen, R. (1992) Fisheries Policies and their impact on the New Zealand Maori, *Marine Resource Economics* **7**, 153-167.

Metzner, R. (1997) *The Myth of Unlimited Fish Resources In Western Fisheries, WA's Journal of Fishing and the Aquatic Environment*, Fisheries Western Australia, Perth, Spring 1997.

Metzner, R. and Rawlinson, P. (1998) *Fisheries Structural Adjustment: Towards a National Framework.* Canberra: Commonwealth Department of Primary Industries and Energy.

Metzner, R. (1999) Resolving User Conflicts: A Process and Tools for Effecting Voluntary Resource Sharing. In Eide and Vassdal (eds), *Proceedings of the 9th International Conference of the International Institute of Fisheries Economics & Trade, Tromso, 8-11 July, 1998*, vol. **2**, IIFET, Corvallis.

PricewaterhouseCoopers (1998) *Fishing for the Future: Review of the Fisheries Act 1996*. Ministry of Fisheries, Wellington.

Sissenwine, M.P. and Mace, P.M. (1992) ITQs in New Zealand: The era of fixed quota in perpetuity, *Fisheries Bulletin* **90**, 147-160.

Sykes, D. (2000) Are you ready for this? *Seafood New Zealand* **8**(2), 12-17.

Wright, G., Metzner, R., and Chevis, H. (2000) Using Mediation to Solve Fisheries Issues in the *Guidelines for Voluntary Resource Sharing Process*. Perth: Fisheries Western Australia. http://www.fish.wa.gov.au/corp/broc/mediation.

Section Three

MULTIPLE STAKEHOLDERS IN FISHERIES CO-MANAGEMENT

Chapter 11

CONFLICT AND SCALE: *A Defence of Community Approaches in Fisheries Management*

DOUGLAS CLYDE WILSON
Institute for Fisheries Management and Coastal Community Development
Hirtshals, Denmark

The idea of fisheries co-management is that 'communities' and the 'state' should work together to manage fisheries. The role of this chapter is to introduce the section on co-management and multiple stakeholders. This section is about multiple and competing uses of the fisheries resources, in other words, about co-management's relationship to existing and potential *conflicts*. This chapter, however, is about *scale*. The argument is that understanding fisheries co-management, or more specifically understanding the relationship between the state and the community in fisheries management, requires understanding the intersection of conflict and scale.

1. SOME CURRENT CRITICISMS OF THE COMMUNITY APPROACH

Those of us who advocate a strong emphasis on the participation of fishing communities in management are being called upon to respond to a new wave of criticism. One that is, as described in Chapter 1, very similar to earlier waves. This is all to the good. Such waves are helpful and necessary, especially as we are currently in a phase where the 'community' position is the orthodoxy that young scholars must set their sites on, as expressed in the title of a recent conference on the 'Tyranny of Participation.' These days this tyranny is often revealed in careless and uncritical assumptions about the nature of community participation and its potential role in resource management (Agrawal and Gibson, 2001). This chapter presents a theory of scale that I believe can help identify the chaff in community participation while allowing us to keep the wheat.

Two recent works do a good job of articulating the newest embodiment of the critique of community approaches. Allison and Ellis (2001) argue that current approaches rely on local administrations that may not have any interest in conserving fish stocks. They also suggest that community approaches assume that territorial use rights based on communities is an approach that almost always fits both the resource and the people exploiting those resources. While Allison and Ellis' (2001) criticism does tend to overgeneralize from just a few of the many approaches to community involvement in management, these problems should be taken seriously. They also argue that recent approaches to community involvement in fisheries management reflect static understandings of community based on a false assumption that community can be defined. These criticisms are harder to give credence to. Static images of community may appear in journalistic works or early project documents and proposals, but it is doubtful that either serious scholars, or anyone else who has been working in an actual community, can continue very long treating it as static. The question of definition is a straw man, communities are no more or less definable than any other social group, ie they can't be defined outside of some context that gives rise to a particular definition.

Allison and Ellis (2001) also point to the same three false assumptions presumed to underlie community approaches to management as appear in Agrawal and Gibson (2001): community approaches assume that communities have a homogeneous social structure; that they reflect common interests and shared norms; and, that the benefits to management stem from communities being a small spatial unit. The first two, as discussed in Chapter 1, comprise the classic criticism of community approaches that have been heard over the last 30 years. In fact, most sociologists have long since dropped the classical concept of a 'norm' as a *sui generis* rule that everyone in some group accepts and follows. When community approaches do, in fact, make these assumptions they are wrong in doing so. The only adequate defence from the critique is to point to the fact that many, if not most, efforts to involve communities in natural resource management do attempt, after 30 years of repetition, to incorporate these lessons. Of course, such incorporation is not easy, as management programmes must work with someone and considerable legitimacy can be derived from calling that someone 'the community.'

The third criticism about scale is newer and more interesting. Agrawal and Gibson (2001) point out that there is no necessary relationship between the scale on which community managers operate and the effectiveness of their management. They point to cases where small-scale groups dealing with small-scale resources have done a poor job of management, while mobile groups dealing with larger resources have done well. Being a small-scale group may, in fact, be a hindrance if the dynamics of the resource operate at a larger scale.

If scale is a reason for advocating community involvement in management then they have made a strong case that it may be as empty an assumption as that of homogeneity and shared norms. Nevertheless, I would like to argue that scale is the main reason for advocating community involvement in management. There is much more to scale than just 'smaller is better' versus 'that ain't necessarily so.' A systematic understanding of the reason for scale's importance, however, requires an expanded understanding of institutions. Agrawal and Gibson (2001) call for examining the interactions between multiple actors in communities and how these interactions are expressed in processes and institutions. Their beginning point, like much strong, recent scholarship in management and development, is an 'actor-oriented' approaches from which they build up an analysis of institutions, sometimes using the 'rules of the game' concept of institutions from New Institutional Economics (NIE) (North, 1990). Approaches that begin from actors' behaviour, however, are not able to sufficiently appreciate the relationship between scale and institutions.

The central argument of this paper is that a clearer understanding of scale can be achieved by supplementing these actor-oriented approaches with a new approach, focussed not on the individual actors but on the properties of institutions themselves. The approach is based on how institutions coordinate social action. The key motivation of actors in this theory is mobilizing social power in response to *conflict* situations. And the key variable in this theory is *scale*.

The chapter begins with an outline of the reasoning behind these properties of institutions. Once the overall theoretical approach is described, co-management becomes the focus of attention. The second section discusses the motivations that both the state and the community bring to co-management. This is followed by a discussion of co-management and the social construction of the fisheries resource. The last section, before a brief conclusion, examines the question of institutional interactions between scales.

2. THE PROPERTIES OF INSTITUTIONS

Actor-based theories of institutions used by NIE and most institutional political science and sociology are powerful tools. They have given rise to design principles based on raw empirical comparisons (eg Ostrom, 1990) that have proven very robust and are perhaps the most important contribution thus far to our understanding of institutions. However, they are a list of observed regularities, not properties of institutions that allow a systematic explanation of their operation.

The main weakness of the NIE-influenced approaches is their reliance on a reduced definition of institution as "rules of the game, " often citing North (1990). This definition has been extensively critiqued elsewhere (Jentoft *et al.*, 1998; Leach *et al.*, 1997). The bottom line of this critique is that a functioning institution involves more than just rules. Functioning institutions pattern actual behaviour, indeed, this patterned behaviour is how they are expressed and reproduced. In addition to the "rules" dimension, which is simply the way that behaviour patterns are described, institutions also entail a cognitive dimension and a normative dimension (Scott, 1995). The cognitive dimension defines institutional realities in the thought processes of actors, it defines "what counts as what" (Searle, 1995), for example, the ball going through the posts counts as a point. The normative dimension points to the fact that an institutionalized behaviour pattern is experienced as normal behaviour. In all real, as opposed to "paper," institutions all three of these things are going on at the same time in all aspects of the institution. Everything about a functioning institution can be described as a rule. Everything about a functioning institution depends on the way the institution defines the reality that people are experiencing. Everything about a functioning institution is experienced as normal, including people violating the institution and therefore deserving the normal consequences of such violation.

A stronger actor-based way to understand institutions, which is the one drawn on by Agrawal and Gibson (2001), is the "actor oriented approach" pioneered by Long and Long (1992), which emphasizes a dynamic concept of institutions. They avoid much of the reductionism found in the "rules of the game" definition. Their work, which draws heavily on such theorists of action, power, and social structure as Giddens and Clegg, focusses on collective actors as well as individuals. It emphasizes the importance of actors drawing on discourses, their embeddedness in cultures and social structures, and their participation in the recursive recreation of these structures through action. These are much the same assumptions about the behaviour of actors that I draw upon here.

Even this actor oriented approach version has limitations that make working on supplementary theories worthwhile. It is remains essentially a theory of behaviour rather than a theory of institutions as such. Institutions are treated as derivative of actors' behaviour without properties of their own and explanations are sought only at the level of actors. Hence, Long and Long (1992) argue "the main task of analysis.. is to identify and characterized differing actor strategies and rationales, the conditions under which they arise, their viability or effectiveness for solving specific problems, and their structural outcomes" (28).

Drawing on the Scott (1995) definition, a theory of institutions as such can begin where Long and Long (1992) leave off. Institutions are created and modified through more or less conflictual interactions among various groups, which happen at both formal and informal levels. For example, governments, faculties, students and other groups interact in patterns ways that we call a university. The institution is maintained and modified through both formal processes in which groups seek to articulate rules, ie describe the patterned behaviour they would like to see within the university, and informal processes in which rules, both formal and informal, are iteratively interpreted and expressed as behaviour and then reinterpreted in light of that behaviour (Wilson and Jentoft, 1998). These groups have different amounts of power, meaning that their abilities to effectively pattern behaviour differ.

How do institutions actually pattern individual behaviour? The answer from actor-based approaches is that institutions provide actors with incentives, options and constraints that channel how they choose to pursue their goals. The key supplement that I would like to suggest is based on the fact that actors not only must make choices to pursue their goals, they also must coordinate their actions with others. A different set of patterning mechanisms arises from this coordination problem that complements the incentives, options and constraints paradigm. These other mechanisms, however, attach to the institution rather than to the actor. The remainder of this section outlines the reasoning behind this complementary approach.

Bringing to the forefront the question of how actions are coordinated is an important step in understanding institutions and Habermas (1987a, 1987b) has been a seminal influence in moving social science in this direction. The basic idea is that an understanding of society, and of institutions, can be built around the functional requirements of coordinating action. Coordinating action requires some degree of 'mutual understanding'. This is perhaps the most misunderstood term in Habermasian scholarship because the way it sounds makes people construct it as a utopian ideal of cooperation. The reality it points to is much more pedestrian. The coordination of action requires that actors must to some degree share a construction of reality that allows them to know how to proceed. At a very mundane level they must have an understanding that is, in fact, *mutual*, meaning they must think the same thing about some aspects of their situation. This does not mean that there are not many things that actors disagree about, nor does it mean that it is not possible to 'pull the wool over someone's eyes' about aspects of a situation, although it explains the motivation for doing so. It does mean that if people are to achieve anything through their interaction there is a basic set of things that they need understand the same way.

So far these observations are commonplace. However, they provide the basis for an understanding of the properties of institutions as such. This is because if an institution is to pattern interactions it must have communicative mechanisms to address this question of creating mutual understandings so that the individuals can coordinate their actions[1].

Drawing on Habermas' Theory of Communicative Action (1984, 1987a), with

[1] Throughout the paper sentences appear in which institutions are the subject, as if institutions were actors in their own right. This is shorthand needed for clarity. It should be interpreted as meaning 'individuals as they behave in patterns that produce and reproduce an institution must draw on shared understandings that ...'.

revisions[2], I suggest that there are six such communicative mechanisms: money, authority, influence, prestige, rational communications and right-of-way[3]. All institutions use some mixture of these six mechanisms to solve the problem of creating a mutual understanding of how to proceed that allows the coordination of action. These mechanisms are not themselves institutions, but rather the techniques of communication that institutions structure. They are the media through which institutions pattern behaviour and they operate more or less instantaneously. They are the raw materials out of which institutions are constructed. People trying to 'design' institutions are describing patterns of behaviour that they hope will guide this institutional construction. Hence, while the communication mechanisms are the raw materials of institutions they are not synonymous with the design elements of institutions. While money, authority and right-of-way are often used intentionally in institutional design, prestige and influence rarely are, and rational communications should be more than it is.

The point of identifying these communication mechanisms is that the particular combination of mechanisms used by any given institution is a very important property of that institution and has profound implications for what the institution can and cannot accomplish. The key variable that determines the effectiveness of a particular mixture is the scale on which the institution operates. This scale should be understood primarily as social scale, ie the number of people involved. Geographical scale has its own interesting properties from the perspective of understanding the resource, but as they are often strongly related, the discussion can proceed without repeated reference to the two different types of scale.

The key difference among communication mechanisms is the degree to which they are embedded in what Habermas (1984) calls the 'lifeworld'. The lifeworld is the vast set of background assumptions that make communications possible. The lifeworld allows a statement to make sense. A shared lifeworld makes it possible for a speaker to say 'Maria cut the grass' and 'Lee cut the cake' and be sure the hearer knows which of the two used a lawnmower. The communication mechanisms fall on an 'embeddedness' continuum based on how much they rely on a shared lifeworld to establish mutual understandings of situations. All communications require some shared code, but an interaction involving money does not require very much. You can buy a candy from someone without uttering a word. You both understand how money works and that suffices. The interactions requires very little shared lifeworld. Markets are institutions characterised by a heavy reliance on money in their mixture of communicative mechanisms.

As we move through the mechanisms from the least embedded to the most embedded, we move from money to authority. Authority here means an understanding that A requires X of B and that B will suffer some formal sanction if she fails to do X. An example of an institution making use of authority is the Department of Fisheries requiring a landings report. This understanding requires more shared background knowledge than money, but is still relatively disembedded from shared meanings. For authority to function, B must simply understand the content of A's requirement and the structure of the authority that will invoke the sanction. A, unfortunately, does not have to understand anything about B.

---

[2] The key concepts are entirely a product of Habermas' creativity, but I have put them together in a very different way. As this is not primarily a chapter on social theory a discussion of the differences between what is described here and what Habermas says is beyond its scope. These differences are discussed in Wilson, 1996.

[3] In earlier work (eg Wilson and McCay, 1998) we have referred to these mechanisms as governance mechanisms, the change to communication mechanism is more specific and hopefully clearer.

Bureaucracies are the type of institution characterised by a very heavy reliance on authority.

Money and authority, which Habermas (1987a) refers to as 'steering media', have several things in common besides being relatively disembedded from the lifeworld[4]. These disembedded mechanisms operate by reducing the question of the institutionally compliant behaviour to simple 'take it or leave it' or 'do as you are told or suffer the consequences' decisions. They rely on rewards and punishments to constrain individual behaviour and this makes behaviour more predictable than it otherwise would be. All institutions require some predictability to pattern behaviour, so all institutions involve at least some steering medium in their mix of communicative mechansism. Furthermore, greater predictablity makes it possible for institutions to be more effective over larger scales. The behaviour of actors operating within institutions substantially governed by these media is much more predictable than most social behaviour. Behaviour within markets is so predictable, in fact, that it be quantitatively modelled with a good deal of precision, which is a very unusual institutional property. Steering media have to have some grounding in formal law, the relationship between property rights and money is a good illustration, and this is not true of the other mechanisms. To work effectively, these kinds of communications mechanisms rely on the possibility of verifying material conditions. Money relies on the ability to verify an exchange, authority relies on the ability to verify that what is being required has been fulfilled. Because they are relatively free from the encumbrances of the lifeworld, for economic institutions we describe this as having 'low transaction costs', money and authority based institutions are able to operate effectively across large scales. The general relationship between the communication mechanisms and scale is illustrated in Table One.

Influence and prestige operate are similar to authority, but are less formal and require richer shared meanings. Influence is an understanding that someone has greater command over how a situation will be understood and acted upon based on greater control of exchange values, while prestige is based on factors of personal attractiveness. Relatively informal, but often very important institutions, such as interest groups, professional associations and social networks rely heavily on influence and prestige to coordinate behaviour. Influence is less embedded, ie control over exchange values is a less culturally nuanced characteristic than attractiveness, and influence involves constraining behaviour more than prestige does.

The more embedded a communication mechanisms is in the lifeworld the more it relies on convincing people that 'X is the case and requires Y response' and the less it relies on constraining behaviour. Hence, predictability of behaviour decreases as embeddedness increases.

---

[4] Right-of-way, the mechanism that governs interactions in traffic institutions, is also a relatively disembedded steering medium that shares many characteristics with money and authority. It is of limited importance in fisheries co-management so I will not discuss it further.

Table 1. Four Continua Characterizing the Communications Mechanisms

Rational Communications	More embedded in the lifeworld of shared communicative resources	Relies more on convincing	Less predictable outcomes	Operates more effectively on smaller scales
Prestige Influence				
Authority Money	Less embedded in the lifeworld of shared communicative resources	Relies more on constraining	More predictable outcomes	Operates more effectively on higher scales

On the extreme embedded end of this continuum is the mechanism of rational communications. Habermas' work is best known for his description of an ideal model of rational communications, which he calls non-distorted communications. Non-distorted communications meet two conditions: there is no manipulation involved in the communication and that everything communicated is open to any question, from any participant, about its validity (White, 1988). But this model should not be thought of as an attempt to describe empirical conditions. His argument is more subtle than that. In communications, this ideal of non-distorted communications acts as a 'partly counterfactual' (Habermas and Nielsen, 1990:105) critical device. This means that people use it as a yardstick to evaluate the kind of communicative situation they find themselves in. While no one expects the conditions to be fully met, they have to be met to some degree if a convincing shared reality is to be reproduced (Habermas and Nielsen, 1990). No institution, even markets, can function without at least some rational communications in its mixture of communication mechanisms.

Rational communications are completely embedded in the lifeworld. Any person can draw on any meaning giving resource to put forward a claim about what is true or right, and therefore the strength of rational communications as a device for coordinating actions is that it is the most sensitive to both factual truth and ethical values. The best example of an institution that relies heavily on rational communications is science. Rejection of dogma, distrust of orthodoxies, scepticism about arguments and the creation of rational criteria for their evaluation are all part of the scientific ideal. This ideal acts exactly as a 'partly counterfactual critical device' providing the basis by which scientists evaluate scientific claims. Saying this in no way implies that the institution of science as it is actually carried out always meets these ideals, even a light perusal of the sociology of science shows that this is not the case. These ideals still act as criteria in discussions where people try to determine what is and is not science, because they are precisely what allow the institution of science to be sensitive to factual truth.

This point about science and rational communications should be understood carefully in the context of fisheries co-management. The point here is not about the differences of perceptions of natural realities that are generated by observations made at different scales and by natural processes operating at different scales. That is a separate and also critically important idea, which is discussed in Chapter 2. The point here is about the ability of institutions to understand and respond to scientific claims, regardless of the scale of the observations which generate these claims. An institution's ability to make use of any facts about nature depends on the degree to which claims about nature can be effectively challenged and evaluated. Without this facility, institutions would not have a prayer about being able to handle coordinating observations of natural processes across scales.

This sensitivity to factual truth and ethical values is the main benefit of rational communications as a communication mechanism. It also has some important weaknesses. It is a very, very slow way to coordinate actions and it takes a great deal of effort. As any group seeking to use rational communications becomes larger, prestige and influence will play a proportionally greater role in coordinating activities. This will be true as much to save time and get the job at hand completed as because of the exercise of power.

The final point in this basic outline of the coordination-problem approach, ties the idea of the communication mechanisms back to the 'more or less conflictual interactions' between groups that create and maintain institutions. These conflictual interactions between groups are contests of power, understood as coordinated social action (Arendt, 1958, Habermas, 1977)[5]. Power is a product of groups such as interest groups, corporations, chieftaincies, governments, trade unions, etc. Group members, and others who identify with the group, behave in certain ways that reproduce the group and pursue its goals. Groups are the source of the power that creates and maintains other institutions.

But groups are a kind of institution - they use governance mechanisms to coordinate action and hence to generate power and this coordinated action can be solidified into other institutions. Many groups generate power with a mix of mechanisms that is mainly coercive; many other groups generate power with a mix of mechanisms that is mainly based on convincing. This generation of power using the more embedded mix of mechanisms is what is meant by the often used but little understood term 'solidarity'. Nor, as a glance at the power that committed religious groups exercise, is solidarity necessarily a less effective way to generate power than coercive mechanisms are.

Like any institution, groups depend on different combinations of communication mechanisms to maintain themselves in being. Corporations and governments rely heavily on money and authority to maintain their ability to coordinate behaviour. A good hypothesis is that they, in turn, are interested in creating and maintaining other institutions that rely heavily on money and authority because such institutions are more tractable for them. Fishers' organizations and other groups that depend on individual commitment to maintain the solidarity of their group are dependent on more embedded mechanisms. The implications of these differences have been explored very throughly by Offe (1985) in his comparison of the ways that companies and unions hold and wield power. The strengths and weakness of the various groups involved in, for example, co-management reflect the strengths and weaknesses of the mixture communication mechanisms they use to maintain themselves.

These uses of communicative mechanisms to create and maintain power distort the contents of communications. The use of authority and money squeezes rational communications out of decision making processes. North's (1990) theory of marginal institutional changed based on the efforts of entrepreneurs to find ways to reduce transaction can be understood as a description of powerful actors seeking to drive the need for costly referrals to the lifeworld out of the market place. Wilson and Degnbol (2002) describe how legal requirements for particular types of indicators forced scientists to make

---

[5] From this perspective to speak of an individual having power is to speak of that person's ability to coordinate the actions of others, ie, to steer a formal or informal group.

public claims about a fishery resource they did not fully believe in, or at least would have liked to strongly qualify. Such distortions are common in the 'mandated' science (Salter, 1988), which is science used to support regulatory systems.

The imperatives of maintaining solidarity have their own more subtle ways of squeezing rational communications. Solidarity relies on unifying symbols and such symbols take many forms, including particular constructions of nature. Certain story lines, eg 'static gear is sustainable gear, mobile gear is not', become local ideologies, used by groups to maintain and test loyalty. Such symbolic claims may be more or less valid in particular situations, but their status as symbols shields them from criticism. An interesting hypothesis is that generating sociological research and theory only makes a difference when some group uses it to play such a role.

In summary, what is being suggested is a narrative about how institutions operate that allows testable claims[6] to be made about the properties of institutions themselves, properties that are not derived from claims about the individual options and behaviour of actors involved in the institutions. Different mixtures of small number of mechanisms of communications are used by all institutions to coordinate people's behaviour. The main difference between these mechanisms is the degree to which they rely on shared meanings to coordinate behaviour. These mechanisms have functional strengths and weaknesses that cannot be avoided. The groups involved in conflicts over institutions are also dependent on these mechanisms to mobilize the power they need to compete with other groups. The strengths and weaknesses of these groups reflect the strengths and weaknesses of the communicative mechanisms on which they depend.

The most important variable for determining the effectiveness of different mixtures of communication mechanisms is scale. As a general rule, the larger the scale of the institution the more it must rely on the more disembedded, formalized and predicable mechanisms to communicate and hence coordinate action. What decreases at larger scales is the ability to make effective use of the richer, more complex and nuanced mechanisms. The more embedded mechanisms make use of a much wider variety of shared understandings. These mechanisms rely on convincing people to cooperate with the institution rather than offering rewards and punishments. Effectiveness over larger scales is therefore reduced because reliance on convincing people leads to more unpredictable results than reliance on rewards and punishments. At higher scales, this rich and nuanced information becomes too great to process. More fluid and participatory approaches for identifying options and making decisions become too cumbersome to respond to needs.

3. SCALE AND MOTIVATIONS FOR PARTICIPATING IN CO-MANAGEMENT

Understanding this relationship between scale and institutions can help us understand both the state's and the community's motivations for being involved in co-management. We should begin by remembering that both the 'state' and 'community' are really short-hand words that actually cover many different groups who are the real forces in creating co-management institutions. This is a point that seems to be missed by people who argue that those of us that advocate increased community participation do so out of a belief that communities are homogeneous groups operating out of common interests. When closely

---

[6] Discussing the testability of the theory is beyond the scope of the present chapter, but has been addressed in unpublished work.

enough examined, no social formation, no state, no government agency, no NGO, no family is a homogeneous group operating out of common interests. Communities are nothing special in this regard, that term, like all names for social formations, is simply helpful shorthand. Any of these formations can analysed as an institution when the formation involves patterned social behaviour.

3.1. The State's Motivations for Participating in Co-management

The problem the state, and other large-scale institutions such as NGOs, faces when it confronts local communities is *accountability*. How can the state know and influence the kinds of decisions and behaviours that are going on at the local level so that they can insure that they reflect the states priorities (such as the need to ensure that the local community use more than their share of a shared resource). In creating co-management institutions, the state (or often the donor that is influencing the state) desires to make use of the richer institutional mechanisms embedded in local cultures to facilitate this. This access to richer communicative mechanisms is how co-management aids the state in facilitating access to information (Pinkerton, 1989) and increasing the legitimacy of management (Jentoft, 1989; Jentoft and Kristoffersen, 1989). Richer communications are critical for good fisheries science (Neis and Felt, 2000; Wilson, 1999) and better information about how management mechanisms work at local levels, and less reliance on expensive surveillance and enforcement devices. The state's motivation for participating in co-management begins with its need for the communities' help in dealing with aspects of fisheries management that require richer, more sensitive and subtle tools than authority.

3.2. Community Motivations for Participating in Co-management

The motivations of groups at the community level to participate in co-management are a mirror image of the state's. Communities engage in co-management efforts because they desire to tap into the state's disembedded bureaucratic authority and financial resources. Two reasons why communities want to involve the state in fisheries management are common. The first is simply a desire for the money and other resources that co-management programmes frequently make available, either directly through the state or through donors and/or NGOs working with the state. In such cases, the community will remain motivated to participate in co-management so long as the money is flowing. The second reason is that the community is already facing a situation in which a) the fisheries resource has declined and conflicts have arisen as a result or b) new fishers or old fishers using more intensive fishing techniques threaten to create a decline in the resource and conflicts have arisen as a result or c) new fishers or old fishers using gear taking up more space creates a competition for space and conflicts have arisen as a result. Holm *et al.* (2000) argue that this later conflict was the driving force the paradigmatic co-management institutions on the Lofoten Islands (Jentoft and Kristoffersen, 1989). In these common cases, community groups are interested in establishing a co-management programme to help to resolve an already conflictual situation. In most cases these conflicts are the product of forces and trends operating at a much higher than local scale. For these communities, state involvement holds two attractions: enabling legitimacy and managing conflicts.

3.2.1 Legitimacy

The first attraction of state involvement is that the state can help establish legitimacy within the community for a local group to be involved in management by establishing them as a legitimate expression of bureaucratic authority. The title of 'The River Would Run Red with Blood' (Taylor, 1987) describes the prediction of the local priest of an Irish village discussing what would happen if someone local tried to implement fisheries management. The blood would be the result if one member of the community tried to enforce a fisheries management measure that other community members were violating. In discussions of fisheries management we commonly treat the word 'legitimacy' as meaning a community's willingness to go along with particular fisheries management measures. But the legitimacy of the process by which a measure is created is also critical (Jentoft, 1993) and for many local people, process legitimacy begins with state involvement. This is an ubiquitous problem: how can the legitimacy of the local management be established to the point where it can be enforced by community members? Fisheries co-management efforts in Zambia and Malawi rely on beach committees to enforce management efforts, and these committees are constantly lobbying the Departments of Fisheries to give them increased authority. This raises the question of if a local co-management group can be given so much authority that they begin to rely too heavily on that authority and lose the benefits of more embedded mechanisms.

Some processes at the community level, however, generate legitimacy for newly created groups or, more often, existing groups taking on wider fisheries management roles. Such processes are difficult, rely heavily on generating local solidarity, and require a visible problem around which the community can rally, so they are seen in situations where conflict is intense. Donda (2000) compares co-management efforts on Lakes Malombi and Chiuta in Malawi. Where Lake Malombi co-management began with a donor and government driven effort, Lake Chiuta's began with local fishers organizing themselves into 'pressure groups' with the purpose of driving out an influx of fishers from other lakes whose use of seine nets was causing a number of problems. The pressure groups turned to the local government for help in dealing with the and then became involved the fisheries management and the Department of Fisheries. Donda (2000) argues that the Chiuta co-management has been more successful than the Malombi co-management partly because it began through this local initiatives. A number of other conditions contribute to the possibility of effective local organization. Local organization is facilitated by pre-existing communication networks (Pinkerton, 1989; Schlager, 1994; Wilson, 1982). Pinkerton (1989) suggests that a crisis of stock depletion is necessary that is sufficiently dire that the fishers are willing to contribute financially to the rehabilitation of the resource, though such crises may also arise overcrowding, gear entanglement or other direct sources of conflict (Holm *et al.*, 2000). The important issue for local organization, however, is the existing or potential conflict, not the condition of the resource itself.

Beyond this local organization, a common non-state source of process legitimacy in Asia and Africa are various forms of 'traditional authorities' a term which means essentially authorities drawing on pre-colonial traditions. In many cases resource management has been an historical function of these traditional authorities. Traditional organizations with roots in medieval fisheries guilds have been used as a legitimating base for co-management in Europe (Alegret, 1999; Marciniak and Jentoft, 1997). In many countries the state is sufficiently differentiated that more locally situated agencies can also provide a source of legitimacy for community action that is an 'alternative' to more centralized state fisheries management functions (Amos, 1999). Problems have arisen, however, with many of these various traditional and local authorities because they are 'authorities' in the sense that they are local and regional level institutions that rely very heavily on the communication mechanism of authority. This means that they can easily lose the benefits that the more embedded mechanisms offer at the local level. They are often undemocratic and unaccountable (Ribot, 1996) and, in spite of their 'cultural' status, fail contribute to making richer and more flexible communications available to fisheries management.

3.2.2 Conflict Processes Operating on Large Scales

The second attraction is that almost all conflicts operate at a wider than simply local scale. Local community groups may not be able to have any affect on destructive fishing or habitat-related behaviour simply because it is generated elsewhere. Even if the behaviour does happen locally, if it is part of a wider pattern the communities face a commons dilemma if they seek to address the problem locally without broader cooperation; the behaviour will continue in other communities and only the benefits will likely be lost to the community that took the initiative. Furthermore, groups that have particular fishing rights and use particular fishing techniques rarely base their identity only on their relationship to the fishery (Magrath, 1989). Even social categories that seem to derive purely from the fishery, such as the users of particular gears, are often linked to other categories. Social categories that have been linked to fisheries conflicts include ethnicity (Bailey, 1986), gender (Medard and Wilson, 1998), colonial domination (Johannes, 1978), nationality (Peterson, 1993) and class (Davis and Bailey, 1996). These issues may pull fisheries management issues into broader conflicts.

In many situations, such as in the events in Malawi described above, fisheries conflicts are rooted in migration. This takes the form of both incoming fishers seeking to settle and begin fishing in an already populated area or traditional, seasonal fisheries migrations becoming a source of conflict when they are confronted with growing populations and or changes in access rights. On Lake Kariba fisheries management measures have been caught up in migration-driven struggles over land, including the forced removal of fishers from islands and other illegal fishing camps into the official ones. The local chiefs supported the new plan so they could control the movement of immigrants (Jul-Larsen and Malasha, 1999). Neiland *et al.* (1996) describe a situation in Nigeria, similar to that seen in Malawi, where co-management systems were created by the locals in response to in-migration with migrants using different gears. These systems have remained essentially allocative but have managed to avoided serious conflict. Another common type of conflict is that between artisanal fishers using smaller scale and even traditional gears and often more oriented toward local markets, and industrial fishers using more modern gears who are of often more oriented toward distant markets. These conflicts are found throughout the world in both developed (Davis, 1999; Sandburg, 1996) and developing (Kurien, 1993; Plateau, 1989) countries.

Begossi's (1998) comparison of six local management regimes in Brazil demonstrates the usefulness of approaches from human ecology in understanding scale. In her model a critical factor for determining the scale over which community management is effective is the relative cost of defending the boundaries. Co-management systems, by introducing state authority, reduce these costs. Hence, communities that cooperate in co-management regimes can effectively control larger areas.

The stakeholder conflicts in fisheries are situations of both problem and possibility. It is communities searching for a response to such problems that are the most ripe for the creation of effective co-management institutions. They can be of use to the state as it tries to address fisheries problems that operate on a higher scale than the community, and the state can be of use to the community as they try to deal with the local effects of broader problems. Situations that are ripe for co-management will often be situations of intensifying conflict.

3.3. From Motivation to Action

Once the necessary motivations for co-management are in place, the question becomes how to build co-management institutions. State power made available through co-management can easily be appropriated by one or a few local groups and used against other groups. Rather than channelling conflict creatively, co-management become a weapon. Pressures exist at both the local and state levels to transform local groups cooperating in the co-management arrangement into extensions of state bureaucracy. Avoiding these problems means remaining aware of the issue of scale and its relationship to how institutions operate. What is possible at smaller scales is giving fisheries management access to institutional mechanisms that allow richer and more nuanced communications based on convincing people to cooperate. The state can use its authority to provide space for this, as has happened in the Maine lobster fishery for example (Acheson and Brewer, 2001). But the simple injection of bureaucratic authority (or market mechanisms for the same reason) into the local situation will not accomplish this. Fisheries co-management will simply be an attempt to extend disembedded institutional mechanisms (McCay and Jentoft, 1998) into smaller scale situations with little benefit.

4. CO-MANAGEMENT AND THE SOCIAL CONSTRUCTION OF THE RESOURCE

Conflicts in fisheries involve competing constructions of the condition of the fisheries resource. Several reasons for this have been discussed. Holm *et al.* (2000) argue that there are always resource conservation based discourses in fisheries debates, and believe that this probably reflects a 'generalized feeling that fish stocks are vulnerable' (358). Once a government management regime is in place or even being discussed, however, much more is at stake than generalized feelings about the actual condition of the resource. It will be indicators of the condition of the resource that will be used to justify and, in many cases, trigger management actions. Others, including this author (Wilson, 2000; Chapter 15) have argued that good science is critical part of the legitimacy of fisheries management institutions. The basic issue, however, is the reliance that authority has on verifying material conditions[7]. The ability to verify facts about the fishery is the fulcrum on which the functioning of state management institutions pivots. Even simple technical measures can be difficult to monitor in practice, and the relative simplicity of monitoring them is one of their great political and administrative attractions. More complex systems such as quotas, however, can end up be entirely driven by the problems of defining and verifying the condition that requires the authority-backed response. At larger scales the imperative of authority for clearly verifiable material conditions can overwhelm the ability of the more embedded mechanisms, science in particular, to provide truthful constructions of nature. At larger scales nature is constructed, indeed distorted, into a form 'naturally' amenable to bureaucratic management (Wilson and Degnbol, 2002).

Beyond the state's need for verifiable conditions, in fisheries conflicts people select different facts from fisheries science to put together an overall picture of the resource that fits their rhetorical needs. As discussed above in reference to distorted communications,

[7] Money, as a steering medium, shares this same characteristic. So 'market-based' solutions cannot avoid this problem of verifying material conditions, they simply structure it differently.

facts and beliefs about the fishery become symbols used to rally groups for whom solidarity is the source of their social power. At the local level the resource is constructed as, indeed distorted into, a small set of salient facts that demonstrate how 'our group believes in using real science'.

These pressures toward the socially distorted construction of nature have profound implications for co-management. For one thing co-management, if sufficiently empowered, can mitigate the higher scale pressures to distort the resource to make it easier to manage. For another, co-management efforts have a track record of forcing competing local groups into institutional contexts that require them to try to meet the conditions of non-distorted communications (Wilson, 1999) leading to a less distorted construction of the resource. Fisheries co-management efforts being built in Canada and Alaska are based on similarly protected processes of open communications, many of which importantly include scientists within these local institutions (Wilson, 1999). Positive reports are coming from these efforts that indicate a process of mutual learning leading to better fisheries management within a framework legitimated and protected by state management of higher level conflicts. This sort of approach recognizes the critical importance of scale and its implications for institutions and makes this knowledge work for better fisheries management.

5. INTERACTIONS BETWEEN SCALES

Because scale is the key variable for understanding the conditions under which various mixtures of communication mechanisms are effective, the communication mechanism model provides a useful language for discussing the problems of moving between scales. Many attempts at institutional reform can be understood as attempts to find ways to extend the effectiveness of the communication mechanisms into scales for which they are not well suited. Money is rarely an issue on small scales because the issues small-scale institutions address hardly ever involve things that valuation and exchange would help with. If resources become 'scarce' at small scales they are either simply shared, or the problem stems from a higher scale condition in the first place. Authority, however, is often at issue on small scales. People whose social power depend on authority have searched unceasingly for ways to make authority more effective at smaller scales. The problem stems from authority's reliance on verifiable material conditions (eg a behaviour, the state of a resource, etc.) and the need to mobilizing the information necessary to make the threat of sanctions credible. Both the processing and interpretation of information, and moral and ethical factors, seem to place very stringent limits on the ability of authority to penetrate to the smallest scales. Authority plays an important role in family life, particularly where children are involved, but this role a) is significantly limited in the ways in can be mobilized and b) exercised within an institution that's mechanism mix contains a large amount of rational communications. The military's requirement for immediate and accurate prediction of behaviour means that they have created an entire unique culture and legal system around the need to inject authority into the smallest possible scales. Outside the military, attempts to inject authority through large scale institutions into the smallest scales have been carried the furthest by totalitarian fascism and communism, with disastrous results.

Indeed, democracy can be understood as precisely the opposite attempt. Democracy is an effort to make rational communication effective at larger scales by forcing those exercising authority to account to the general community for their use of that authority. The difficulties of increasing the amount of rational communications in the mix of large scale institutions are as well known as those with bringing authority to smaller scales. Large scale discussions in democracies are famously stilted. The accounting of those exercising authority to the general community can be just as accurately, if not more accurately, described as marketing than as rational discussion. The problem stems from rational communication's reliance on the rules of non-distorted communications. All participants cannot (effectively) raise any question about validity claims. People quickly conclude that the situation does not approach that of non-distorted communications and exit the attempt to create mutual understandings. The 'mutual understandings' that remain are ideological positions that hyper-simplify reality.

Representation, as discussed in Chapter 16, is the best way thus far developed to extend the effective scale of rational communications. The basic idea is to set up a simple authority structure that identifies some people with an enforceable right to speak for others, and give them a chance to interact in as open and embedded a way as possible. It is critical to make sure that these people have to account to the people they supposedly represent, Ribot (1996) among others has pointed out the difficulties of co-management systems built on 'traditional' systems in which non-elected 'traditional leaders' speak for people who have no control over those who speak for them. These systems are argued to, and often demonstrably do, have a great deal of legitimacy, but legitimacy is not the same thing as open communications. Chapter 9 describes the unsatisfactory results of another co-management system, the fisheries management council system in the United States, that is based on a form of representation, in this case 'representatives' appointed by states governments to represent the fishers from that state, in which there is no real accountability of the representative and the represented. Finally, democracy is the best way we know to ensure the accountability (literally to force to explain) that injects some rational communications into larger scale institutional processes.

Nevertheless, even democratic representation is an institutional design that still requires a good deal of authority to set up, and in which influence and prestige inevitably will play important roles. It cannot carry all of the rich communicative resources available at smaller scales. This is the reality behind the value of 'subsidiarity' which recommends that decisions be made at the lowest scale, and hence most richly informed, level. A related institutional structure are 'nested' systems (Ostrom, 1990) in which representation and subsidiarity are combined. A nested system, with rigorous subsidiarity and democratic representation is the most functional approach to mobilizing rational communications for large scale management institutions. Nested systems, however, cannot function unless conflicts are managed so that they work through the system rather than constantly seek to manipulate the system.

This, then, is the critical role of the state in co-management, to provide the authority and legitimacy that support rational communications. State legal authority can be used to do this by recognizing and supporting local, open and democratic decision making process that arise in response to conflicts. This is what it means in a practical and specific sense for the state and the community to have a co-management 'partnership.' The existence of this state-conferred legitimacy is one of Ostrom's (1990) design principles and existing examples have been described (Ruddle, 1987). In this kind of partnership, the strengths and weaknesses of institutions that stem from the relationship between the mixture of communication mechanisms they use and the scale of their operations compliment each other in a way that benefits the management of the resource.

6. CONCLUSION

The current and classical critiques of community approaches to natural resource management can play an important role in helping us to remain focussed on what we mean by 'community' and what it is that we think 'community' can contribute to resource management. This chapter agrees that what it has to contribute stems neither from homogeneity nor shared norms, that indeed both of these ideas stem from a sociologically uninformed view of communities. This chapter does argue, however, that scale is central to understanding the critical contribution that those of us who advocate community involvement are pointing to. At smaller scales greater communicative resources are available that, in turn, make the steering of institutions more sensitive to the nuances of the many relevant social and ethical values and, just as importantly, better able to identify and respond to good science and other kinds of factual truth.

It is shared understandings that enable rich and varied communications that are the key to the contribution of community to management. These rich communications, in which people are free to make statements that can in turn be questioned and examined, are what make responsiveness to ethical values and factual truth possible. This is a very different perspective on the community contribution to management than that which is common in the literature. The potential contribution of community is not through higher institutional legitimacy because local institutions can have high legitimacy while still blocking open communications. Nor does this community contribution have a logical or necessary tie to geography. It applies just as much to the epistemic communities operating in international negotiations (Haas, 1989) as it does to village communities. Although it is true, as emphasized in Chapter 4, that geographical communities have some advantages in the shared meanings and common goals that they can draw upon to empower effective management.

Far from relying on the homogeneity of communities, this chapter has argued that conflict, as it intersects with scale, is the life force of co-management. Indeed, in situations where it does play a role, homogeneity provides the basis of a solidarity-generating identity directed toward some conflict. In almost every case of co-management, it is the desire of various groups in the community for the power they need to deal with conflict that provides the motivation to participate. Such conflicts may arise from resource depletion or other sources of increased competition for the resource. Absent this motivation, communities will participate in co-management only as long as someone pays them.

The main conclusion of this analysis is that co-management's chief function should be as a way for the state to use its authority to contain and channel fisheries conflicts in creative ways. This means using authority to make it possible for more open and culturally embedded communications to play an effective role in institutional decision making processes. The role that communicative mechanisms play in the generation of power is the critical source of communicative distortion, particularly that distortion that arises from different institutional perspectives and not simply from opportunistic behaviour. State authority, as described above, has been used effectively to limit and channel the exercise of power by competing groups and hence reduce communicative distortions. The creative channelling of conflict does require, however, that the state use its authority for this purpose rather than attempting to manage the fishery directly. Reductions in communicative distortions through co-management can only be achieved when the state is authentically willing to surrender real decision making power to local institutions, even while holding them accountable for their responses to the needs of the broader society. Accountability can, in fact, itself be empowering when it takes the form of outside participation in goal clarification and the evaluation of achievements. Top-down management, whether or not the word co-management appears in the title of the programme, does not release the contribution that community can make to management, ie, to make management institutions sensitive to ethical values and factual truth. Through co-management, however, the state can turn the energy of conflict into a resource for responsive management.

REFERENCES

Acheson, J.M. and Brewer, J.F. (2001) 'Changes in the Territorial System of the Maine Lobster Industry' in Dolsak, Nives and Elinor Ostrom (Eds) The Commons in the New Millennium: Challenges and Adaptation Boston: MIT Press.

Agrawal, Arun and Gibson, Clark C. (2001) 'The Role of Community in Natural Resource Conservation' pp 1-31 in Agrawal, Arun and Clark C. Gibson (Eds) Communities and the Environment New Brunswick NJ: Rutgers University Press.

Alegret, Juan-Luis (1996) 'Ancient Institutions Confronting Change: the Catalan Fishermen's Cofradias' pp 92-98 in Crean, K and D. Symes (Eds) Fisheries Management in Crisis. Oxford: Fishing News Books.

Allison, Edward H. and Ellis, Frank (2001) 'The livelihoods approach and management of small-scale fisheries' *Marine Policy* **25**, 377-388.

Amos, Chris 'Sea fisheries committees as coastal fisheries and environmental managers' in Proceedings of the Conference on Holistic Management and the Role of Fisheries and Mariculture in the Coastal Community Tjärno, Sweden: The North Sea Commission.

Arendt, H. (1958) The Human Condition Chicago: The University of Chicago Press.

Bailey, C. (1986) 'Government Protection of Traditional Resource Use Rights--The Case of Indonesian Fisheries' Chapter 21, Pp. 292-308 in Korten, D.C. (ed.) Community Management: Asian Experience and Perspectives West Hartford CT: Kumarian Press.

Begossi, Alpina (1998) Property rights for fisheries at different scales: applications for conservation in Brazil' *Fisheries Research* **34**, 269-278.

Davis, A. and Bailey, C. (1996) 'Common in Custom, Uncommon in Advantage: Common Property, Local Elites, and Alternative Approaches to Fisheries Management'. *Society and Natural Resources* **9**, 251-265.

Davis, A. (1999) 'The fishing way of life' The Sunday Herald (Halifax NS) 2/21/99.

Donda, S. J. (2000) Theoretical Advancement and Institutional Analysis of Fisheries Co-management in Malawi. Experiences from lakes Malombe and Chiuta Institute for Fisheries Management and Coastal Community IFM Research Publication No. 56.

Haas, P.M. (1989) 'Do regimes matter? Epistemic communities and Mediterranean pollution control' *International Organization* **43**, 379-403.

Habermas, J. (1977) 'Hannah Arendt's Communications Concept of Power' *Social Research* **44**, 3-24.

Habermas, J. (1984) The Theory of Communicative Action Volume I Reason and the Rationalization of Society Boston: Beacon Press.

Habermas, J. (1987a) The Theory of Communicative Action: Volume II Lifeworld and System: A Critique of Functionalist Reasoning Boston: Beacon Press.

Habermas, J. (1987b) The Philosophical Discourse of Modernity Cambridge: MIT Press.

Habermas, J. and Nielsen, T.H. (1990) 'Jurgen Habermas: Morality, Society, and Ethics: an Interview with Torben Hvid Nielsen' *Acta Sociolgia* **33**, 93-114.

Holm, Petter; Hersoug, Bjoern and Raanes, Seine Arne (2000) 'Revisiting Lofoten: Co-Managing Fish Stocks or Fishing Space' *Human Organization* **59**(3), 353-364.

Jentoft, S. and Kristoffersen, T. (1989) 'Fishermen's Co-management: The Case of the Lofoten Fishery' *Human Organization* **48**(4), 355-365.

Jentoft, Svein (1993) Dangling Lines: The Fisheries Crisis and the Future of Coastal Communities - the Norwegian Experience. Institute of Social and Economic Research, Memorial University of Newfoundland, Social and Economic Studies No. 50.

Jentoft, S. (1989) 'Fisheries Co-management' *Marine Policy* **13**, 137-154.

Jentoft, S., McCay, B.J. and Wilson, D.C. (1998) 'Social theory and fisheries co-management' *Marine Policy* **22** (2), in press.

Johannes, R. (1978) 'Traditional Marine Conservation Methods in Oceania and Their Demise'. *Annual Review of Ecology* Sept: 349-354.

Jul-Larsen, Eyof and Malasha, Isaac (1999) 'Zambian Case Study' paper presented at the International Workshop on Fisheries Co- Management, Penang, Malaysia 23-28 August.

Kurien, J. (1993) 'Ruining the Commons: Coastal Overfishing and Fishworkers' Action in South India' *The*

Ecologist **23**(1), 5-11.
Leach, M., Mearns, R. and Scoones, I. (1997) 'Environmental Entitlements: A Framework for Understanding the Institutional Dynamics of Environmental Change' Institute for Development Studies Discussion Paper #359.
Long, N., and Long, A. (Eds) (1992) Battlefields of Knowledge: The Interlocking of Theory and Practice in Social Research and Development, London: Routledge.
Magrath, W. (1989) 'The Challenge of the Commons: the Allocation of Nonexclusive Resources' World Bank: Environment Department Working Paper #14.
Marciniak, B. and Jentoft, S. (1997) 'Fisheries Management in Post-Communist Poland: Prospects of User-Participation' *Ocean and Coastal Management* **34** (1), 73-89.
McCay, B.J. and Jentoft, S. (1998) 'Market or Community Failure? Critical Perspectives on Common Property Research' *Human Organization* **57**(1), 21-29 CC.
Medard, M. and Wilson, D.C. (1996) 'Changing Economic Problems for Women in the Nile Perch Fishing Communities on Lake Victoria' Anthropologica XXXVIII (2):149-172.
Neiland, A.E., Sarch, M.T., Madakan, S.P., Ladu, B. and Hassett, D. (1996) 'The community-based approach to fisheries management in north-east Nigeria: a socio-economic analysis' University of Portsmouth, Centre for the Economics and Management of Aquatic Resources: Research Paper 99.
Neis, Barbara and Felt Lawrence (Eds) (2000) Finding our Sea Legs: Linking Fishery People and their Knowledge with Science and Management St John's Newfoundland: ISER Books.
North, D.C. (1990) Institutions, Institutional Change, and Economic Performance Cambridge: Cambridge University Press.
Offe, C, (1985) Disorganized Capitalism Cambridge: The MIT Press.
Ostrom, E. (1990) Governing the Commons New York: Cambridge University Press.
Peterson, M.J. (1993) 'International Fisheries Management' pp 249-305 in Haas, Peter M., Robert O. Keohane and Marc A. Levy (Eds) Institutions for the Earth: Sources of Effective International Environmental Protection. Cambridge: MIT Press.
Platteau, J.P. (1989) 'The Dynamics of Fisheries Development in Developing Countries: A General Overview' *Development and Change* **20**, 565-597.
Pinkerton, E. (1989) 'Introduction: Attaining Better Fisheries Management through Co-Management - Prospects, Problems and Propositions' Pp. 3-36 in *Co-Operative Management of Local Fisheries*. Edited by E. Pinkerton. Vancouver: University of British Columbia Press.
Ribot, J.C. (1996) 'Participation Without Representation: Chiefs, Councils and Forestry Law in the West African Sahel' Cultural Survival Quarterly, September.
Ruddle, Kenneth (1987) Administration and conflict management in Japanese coastal fisheries FAO Fisheries Technical Paper 273.Y1.
Salter, L. (1988) Mandated Science: Science and Scientists in the Making of Standards Dordrecht, Holland: Kluwer Academic Publisher.
Sandburg, Audun (1996) Community Fishing or Fishing Communities pp 34-44 in Crean, K and D. Symes (Eds) Fisheries Management in Crisis. Oxford: Fishing News Books.
Schlager, Edella (1994) 'Fishers' Institutional Responses to Common-Pool Resource Dilemmas' pp 247-265 in Ostrom, Elinor, Roy Gardner and James Walker (Eds) Rules, Games and Common-Pool Resources Ann Arbor MI: University of Michigan Press.
Scott, R. (1995) Institutions and Organizations Thousand Oaks CA: Sage Publications.
Searle, J. (1995) The Construction of Social Reality New York: The Free Press.
Taylor, L. (1987) 'The River Would Run Red With Blood: Community and Common Property in an Irish Fishing Settlement' in McCay, B.J. and J.M. Acheson (Eds) The Question of the Commons: The Culture and Ecology of Communal Resources Tucson: The University of Arizona Press.
White, S. (1988) The Recent Work of Jurgen Habermas: Reason, Justice and Modernity Cambridge: The Cambridge University Press.
Wilson, D.C. (1996) 'A Respecification of the Concept of Power in Habermas' Communicative Systems Theory' paper presented at the Annual Meeting of the Association for Humanist Sociology, Hartford CT, 31 October-3 November.
Wilson, D.C. and McCay, B.J. (1998) 'A Governance Model for Environmental Policy' Crossing Boundaries, The Seventh Conference of the International Association for the Study of Common Property, Vancouver, B.C. 10-14 June.
Wilson, D. C. (1999) 'Fisheries Science Collaborations: The Critical Role of the Community' Institute for Fisheries Management Research Publication No. 45.
Wilson, D.C. and Jentoft, S. (1999) 'Structure, Agency and Embeddedness: Sociological Approaches to Fisheries Management Institutions' pp 63-72 in Symes, D. (Ed.) Alternative Management Systems for Fisheries Oxford: Blackwell Science.
Wilson, D.C. (2000) 'Fisheries Management as a Social Problem' pp 153-164 in Sheppard, Charles (Ed.) Seas

at the Millennium Vol III London: Elsevier.
Wilson, D. C. and Degnbol, P. (2002) 'The Effects of Legal Mandates on Fisheries Science Deliberations: The Case of Atlantic Bluefish in the United States' Fisheries Research in press.
Wilson, J.A. (1982) The Economical Management of Multispecies Fisheries Land Economics **58**, 417-434.

Chapter 12

CO-MANAGEMENT AND MARINE RESERVES IN FISHERY MANAGEMENT

CAROLINE POMEROY
Institute of Marine Sciences, University of California, Santa Cruz, USA

1. INTRODUCTION

Marine protected areas (MPAs), especially marine reserves, where some or all extractive resource uses (eg fishing) are prohibited, have gained currency as a fishery management tool. The growing interest in marine reserves follows from the perception that conventional fishery management has failed to sustain some fisheries, in both ecological and socio-economic terms. In addition, marine reserves are consistent with the concept of ecosystem management, in contrast to other measures which focus on single species or species complexes. As a fishery management tool, marine reserves are of interest because they are expected to protect habitat and marine life within their boundaries, enhance spawning potential, and contribute to fisheries through larval transport and spillover of juveniles and adults. Moreover, reserves fit with changing views that fisheries represent a limited set of values and uses of the ocean, and that management should be more encompassing.

Concomitant with the growing interest in marine reserves has been increasing attention to co-management, whereby government and resource users - and a widening scope of other actors - share authority and responsibility for fishery management. As with marine reserves, the interest in co-management has followed from dissatisfaction with the perceived failings of conventional fishery management. Co-management is valued because it is often more effective in achieving management goals, more acceptable to stakeholders, and less costly than government-centred management (Jentoft, 1989; McCay and Jentoft, 1996; Pinkerton, 1989).

Co-management and marine reserves increasingly intersect with one another, and have several features in common. Each is a process as well as a tool for fishery management. As such, both require consideration of the human and biophysical environments and how these interact, diverse sources and types of information, the support of those affected, and financial, human and social resources for monitoring and adjustment as ecological, social and economic conditions change.

This chapter discusses the developing importance of marine reserves in fishery management, the role that co-management has played, and the prospects for co-management's future role in the design, implementation, management and evaluation of marine reserves. The focus here is on marine reserves as specialized MPAs, because they bear most directly on fisheries. The next section provides further background on marine reserves as an alternative to conventional fishery management. The following section highlights the intersection between co-management and marine reserves. A series of case studies follows, each illustrating the role of co-management in the marine reserve process, associated accomplishments and problems. The final section considers prospects for the future role of co-management in marine reserve processes, especially in light of the increased scope and scale envisioned by many marine reserve proponents.

2. MARINE RESERVES AS AN ALTERNATIVE TO TRADITIONAL FISHERY MANAGEMENT

Marine protected areas (MPAs) are 'discrete geographic areas that have been designated to conserve and enhance marine resources by an integrated plan that includes restrictions on some activities' (NRC, 2001). MPAs range from small, highly protected areas such as the 6 km^2 Big Creek Marine Ecological Reserve in California to larger multiple use areas such as Australia's 350,000 km^2 Great Barrier Reef Marine Park, where a system of marine zones is used to conserve resources, promote diverse uses of the marine environment, and limit conflict among users. Marine reserves, or 'no-take zones', are defined as MPAs 'in which some or all biological resources are protected from removal or disturbance' (NRC, 2001). Marine reserves therefore are MPAs that pertain most directly to fishery management.

'Traditional fishery management' takes on different meanings in different contexts. From the perspective of large-scale commercial fisheries, it pertains to top-down government regulation of fisheries using input controls such as gear and effort limitations, and output controls such as minimum sizes and quotas. In non-commercial and smaller-scale settings, traditional fishery management often consists of local institutions generated and maintained by fishery participants and their communities, although government also may be involved. Marine reserves are more commonly found in these latter contexts.

Interest in marine reserves as an alternative to traditional fishery management has followed from the latter's apparent failure to sustain fisheries, and its limitations in addressing the broadened scope of fishery management goals and participants. Marine reserves can serve a variety of purposes including: conserving marine biodiversity; protecting important ecosystems, habitats, species and cultural resources; supporting scientific research; enhancing fish stocks; and providing places and opportunities for tourism and education (Alcala and Russ, 1990; Bennett and Attwood, 1991; Farrow, 1996; NRC, 2001). For fishery management specifically, marine reserves are valued because of their potential to protect habitat, enhance spawning potential, contribute to larger populations through larval transport and adult spillover, and provide insurance against traditional fishery management failure (Dayton, 1998; Dugan and Davis, 1993; Roberts, 1997). In addition, marine reserves provide reference sites for evaluating the effects of fishing on the marine environment through comparison with fished sites (NRC, 2001). Although marine reserves cannot guard against other natural and anthropogenic environmental perturbations such as El Niño events and land-based and marine pollution, protected habitat and enhanced stocks within marine reserves may contribute to system-wide resilience (NRC, 2001).

Whereas some of the expected benefits of marine reserves have been observed broadly, others have not been demonstrated consistently, and appear to be contingent on multiple natural and anthropogenic factors. Ecological benefits such as increased abundance and size of species, improved habitat, and more complete community structure within, especially compared to outside, marine reserves have been well documented (Roberts *et al.*, 1995; McClanahan and Kaunda-Arara, 1996; Russ and Alcala, 1999; Paddack and Estes, 2000). Ecological benefits theorized to follow from these effects such as larval transport (or dispersal) and spillover of juveniles and adults from marine reserves to areas that remain open (DeMartini, 1993), however, have not been consistently observed. Moreover, it is difficult to determine the extent to which these processes actually enhance fisheries (McClanahan and Kaunda-Arara, 1996, but see Russ and Alcala, 1999), or to evaluate the role of marine reserves as a 'hedge' against inadequate or failed fishery management (Murray *et al.*, 1999), given the complex interactions within and among variable and uncertain biophysical and human systems.

The use of marine reserves has generated concern among fishermen and others directly affected by marine reserves about anticipated shifts in fishing effort, the timing and distribution of social and economic costs, and other practical implications. Effort displaced from closed areas may become concentrated at the edge of the closed area, benefitting those who 'fish the line', but precluding broader fishery enhancement (Dobryznski and Nicholson, 2001; McClanahan and Kaunda-Arara, 1996; McClanahan and Mangi, 2000). The concentration of effort at the edge of a reserve or in areas that remain open may cause negative ecological impacts such as depletion of relatively sedentary species and increased habitat damage (Walters, 2000) that exceed the benefits of protection. Other 'on the water' issues associated with marine reserves include the potential for crowding in areas that remain open, and safety issues for fishermen who have to transit further to reach fishing grounds. Closed areas can result in lower catches, especially in the short term (NRC, 2001), with attendant social and economic impacts. Shoreside impacts may follow if closures near existing receiving, processing and other support infrastructure prompt fishermen to seek alternative bases of operations or landing sites. Shifts away from developed sites can negatively impact localities that lose economic activity associated with fisheries; shifts to new sites can strain existing infrastructure. Alternatively, changes in use patterns can result in greater use of existing infrastructure and associated benefits. Temporally-oriented concerns follow from the likelihood that the beneficial effects within and particularly outside marine reserves will be realized in the longer term, rather than immediately (Yoklavich, 1998; NRC, 2001), and are most likely to result where stocks are substantially depleted (Sanchirico and Wilen, 1999; Hilborn, 2001). These differences in the temporal and spatial distribution of costs and benefits of marine reserves pose critical challenges to their use.

Marine reserves are as much a process as a tool for fishery management, the stages of which include conceptualization, design, implementation and management, and evaluation. Conceptualization pertains to the identification of marine reserves as an appropriate fishery management tool to address a particular problem (eg a declining fishery), a solution to a problem that has not been fully specified, or as a means to achieve alternative (or multiple) ocean management objectives such as separating conflicting or competing uses. Conceptualization also entails the articulation of general marine reserve goals and objectives. Design addresses the number, size and placement of marine reserves, and the specific goals, objectives and rules for individual sites, with attention to the linkages among sites and how they are intended and expected to function as a network. Design also entails the development of policies and strategies for administration and supporting activities.

Implementation involves putting the products of reserve design - the management plan and its constituent elements - into practice. This requires the assembly and activation of human, financial and social resources to develop or adapt institutions and procedures to administer the marine reserve and facilitate monitoring, education and enforcement. Evaluation involves determining whether marine reserves have met their goals and objectives, identifying problems, and assessing unexpected positive and negative consequences of implementation. It also provides an opportunity to learn from successes and failures, and modify marine reserves to address problems and build upon identified strengths.

Like other fishery management measures, marine reserves require substantial resources for planning, implementation, management and evaluation, adapted to particular fishery management contexts (NRC, 2001). Their success is contingent on timely, comprehensive and cost-effective ecological, social and economic information, ongoing support by those directly affected, voluntary compliance complemented by enforcement as necessary, flexibility, resilience and responsiveness to changing conditions in the biophysical and human environments.

3. THE INTERSECTION BETWEEN CO-MANAGEMENT AND MARINE RESERVES

Co-management is an institutional arrangement between one or multiple levels of government, resource users, and the larger community for sharing authority and responsibility for governance. Co-management appeals to resource managers because it is often more effective in achieving management goals, more acceptable to fishermen, and less costly than government-centred management (Jentoft, 1989; Pinkerton, 1989). It appeals to resource users and other stakeholders because it allows them to participate meaningfully in resource management by contributing their knowledge, experience, ideas and concerns directly to the resource management process.

There are several parallels between co-management and marine reserves. Although both are treated as novel management approaches, they have demonstrated historic roots (Johannes, 1984, 1998b). As marine reserves vary in their objectives, permissions and prohibitions, co-management arrangements vary in the distribution of authority and responsibility among government and other stakeholders. Where marine reserves range from least to most protective, co-management arrangements fall along a continuum of greater to lesser government authority, and lesser to greater authority of other stakeholders. As with marine reserves, interest in co-management has followed the recognition that traditional fishery management has failed to sustain fisheries in ecological, social and economic terms. Yet neither co-management nor marine reserves are a panacea for all that ails fishery management. The best fitting form of co-management for a particular situation, and its outcome, will depend on the extent to which government and stakeholders are interested in and capable of sharing authority and responsibility, and on other factors (Pinkerton, 1989; Hanna, 1994). Similarly, marine reserves are of limited utility in addressing some fishery management problems such as excessive fishing effort, and afford only limited protection to more mobile species. Finally, both co-management and marine reserves are processes that require substantial forethought, time, human capital, individual and collective knowledge, and social resources such as trust and mutual respect. Active engagement in a co-management process can reinforce and generate new resources including knowledge, mutual understanding and support for management actions that reduce the ongoing and long-term costs of management.

There are also notable interactions between co-management and marine reserves.

Long-standing community-based and co-management institutions have been documented widely, many of them with clear, if not always intended, conservation benefits (Johannes, 1984, 1999; Pinkerton, 1989; White *et al.*, 1994). These local institutions often include marine reserves in the form of temporary or permanent spatial closures of areas to fishing for particular or all species. The current interest in co-management is directed toward recognizing and taking advantage of local institutional arrangements that have persisted, and facilitating their (re)-emergence where they have been diminished or eliminated by centralized, top-down fishery management.

The move toward more cooperative fishery management entails not only the more meaningful involvement of fishermen and their communities, but also a broadening of the scope of fishery management, both ecologically and socially. In ecological terms, this is evident in movement away from single species management toward ecosystem management with greater emphasis on habitats and non-targeted species. As both cause and effect of the broader ecological scope of fishery management, relevant stakeholders now also reflect interests other than those directly involved in fishing, including scientists, environmental groups and other non-consumptive resource users. In many cases, these newcomers have introduced and promoted the establishment of marine reserves to support a variety of fishery and other marine management objectives.

In addition to being a source of marine reserve proposals, co-management offers resources to inform and facilitate the marine reserve process outlined above. For example, co-management's diverse participants can bring their particular knowledge and concerns to the process, and assist in the identification and definition of problems. Resource users and managers, scientists and other stakeholders can provide essential scientific and local knowledge on the biophysical and human dimensions of marine resource use to aid the design of marine reserves to achieve their goals and avoid unintended negative outcomes. While government, resource users and other stakeholders may be equipped to handle selected aspects of implementation and monitoring alone, they depend upon one another for information, legitimacy, support and other resources (Laffoley and Baxter, 1997; Pollnac *et al.*, 2001). Engagement in monitoring provides participants with direct and credible evidence of the strengths and weaknesses, successes and failures of marine reserves. These, in turn, contribute to evaluation, not only of maine reserves, but also of co-management.

4. CASE STUDIES OF CO-MANAGEMENT AND MARINE RESERVES

A small but growing case study literature illustrates the intersection between co-management and marine reserves. In some cases, it has given participants a stake in the outcome, thereby promoting support and 'buy in', which are critical to the success of marine reserves in achieving their goals. In some cases, marine reserve processes that have involved co-management have not resulted in such outcomes; in others, the lack of co-management, or commitment thereto, has resulted in failure to achieve marine reserve goals, or failure to establish reserves in the first place. It is important to note, however, that analyses to date have tended to focus on a limited scope of outcomes such as participants' satisfaction with the process, and whether or not marine reserves were adopted. The ultimate test of the relationship between co-management and marine reserves for fishery management, however, is whether the use of co-management in the marine reserve process results in ecologically, socially and economically sustainable fisheries, and moreso than through alternative management approaches.

4.1. Soufrière Marine Management Area, St. Lucia

Sandersen and Koester (2000) describe efforts to introduce marine managed areas (MMAs), including marine reserves, in St. Lucia in the Caribbean. The St. Lucia Department of Fisheries first conceptualized MMAs in 1986, largely as a strategy to diminish user group conflicts. The Department designed and sought to implement a set of Fishing Priority Areas and Marine Reserves. This instructive approach (Sen and Nielsen, 1996), however, was viewed as arbitrary and top-down, and was strongly resisted by local fishermen.

Two years later, the Department reconceived the process using a more cooperative strategy, coordinated with the Soufrière Regional Development Foundation (SRDF), a local NGO, and the Caribbean Natural Resources Institute (CANARI), a regional organization (Sandersen and Koester, 2000). The Department and its collaborators worked with resource users in St. Lucia's coastal communities to draft maps of marine resource use and potential marine reserves. Although these maps were similar to the ones the Department had drafted in its prior MPA effort, they were viewed as 'everybody's maps', and therefore deemed legitimate. The Department then held open meetings to begin to address conflicts through a new coastal zoning effort. In 1995, the SRDF began to promote the establishment of a MMA at Soufrière. A Technical Advisory Committee (TAC) of community representatives was assembled to advise the Department of Fisheries' planning group on the MMA's design. The St. Lucia Department of Fisheries was to retain overarching control of the MMA, but was to consult with and delegate selected management responsibilities to the TAC. The resulting MMA included four zones designed to manage fishery conflicts.

Despite the use of more meaningful co-management during the design process, however, serious problems arose during the MMA's implementation (Sandersen and Koester, 2000). Although a local fishing representative had been included on the TAC, he was from one of several gear groups, each of which had distinct interests and concerns. The broad range of resource users was not adequately represented, and the design of the MMA imposed inequitable burdens upon some of them. The Department's planning group sought to remedy the situation, but did so without consulting the TAC. This lack of transparency led to the MMA process's loss of legitimacy in the eyes of fishermen. The situation was exacerbated by inconsistent and insufficient enforcement, attributed to a lack of interagency cooperation.

Koester and Sandersen cite four lessons regarding co-management with particular relevance to marine reserves. They note that co-management requires 1) a genuine devolution of power, 2) careful consideration of differences within and among user groups, 3) effective enforcement, complementary to voluntary compliance, and 4) focussed effort to develop and maintain mutual understanding among participants and mitigate the destructive effects of power imbalances, distrust and polarization.

4.2. Channel Islands, California

Through the 1990s, increases in commercial and recreational activity, growing tensions between the two sectors, and broader concerns about rockfish declines and other marine wildlife led to proposals by private parties and government entities to establish marine reserves around the southern California Channel Islands. Several state and federal agencies have overlapping jurisdiction in the area. In late 1999, the California Fish and Game Commission (the Commission) directed the Department of Fish and Game (CDFG) and the Channel Islands National Marine Sanctuary (CINMS) to develop a consensus-based, cooperative state-federal process to consider establishing marine reserves at the Channel

Islands. A Marine Reserves Working Group (MRWG) composed of representatives of diverse interests, was created to engage in a cooperative form of co-management (following Sen and Nielsen, 1996) for this process.

Although the group was established to consider marine reserves for the Channel Islands, government agency representatives who chaired the working group and environmental NGO (ENGO) representatives focussed directly on designing marine reserves. To these stakeholders, marine reserves constituted a solution to a problem that the group had not yet agreed upon. In mid 2000, as the group was struggling to specify marine reserve goals and objectives, a subset of stakeholders, including but not limited to consumptive users, forced the issue of problem identification. After much discussion, a consensus problem statement was crafted:

> 'To protect, maintain, restore, and enhance living marine resources, it is necessary to develop new management strategies that encompass an ecosystem perspective and promote collaboration between competing interests. One strategy is to develop reserves where all harvest is prohibited. Reserves provide a precautionary measure against the possible impacts of an expanding human population and management uncertainties, offer education and research opportunities, and provide reference areas to measure non-harvesting impacts ...' (CINMS, 2002).

Having achieved consensus on the problem and marine reserves as a solution, the MRWG moved forward to design a system of marine reserves for the Channel Islands. In December 2000, the group arrived at consensus on a set of goals and objectives regarding ecosystem biodiversity, socio-economics, sustainable fisheries, natural and cultural heritage, and education (CINMS, 2002). These goals and objectives were used to guide the design of marine reserves and plans for their implementation, management and evaluation.

To assist the process, a Science Panel and a Socio-Economic Panel were established. Although the two panels were created to advise and provide relevant information to the group (following the Florida Keys National Marine Sanctuary's Tortugas 2000 process), they functioned quite differently from one another. A lack of socio-economic data necessitated field research, which consumed much of the Socio-Economic Panel's time and resources. A lack of communication and coordination between CINMS staff in California and the chair of the Socio-Economic Panel in Washington DC, and among panel members, together with some MRWG participants' reluctance to use the socio-economic information, contributed to its marginalization. In contrast, considerable biophysical scientific research and expertise were readily available to the Science Panel, enabling it to conduct its activities in a timely manner. This culminated in its recommendation that 30 to 50% of the Channel Islands' marine area be closed to all consumptive activities.

The Science Panel's recommendation was interpreted variously by different MRWG members, and revived polarization among them. The substantial mutual understanding, respect and trust built by the MRWG, however, afforded some resilience, and the group continued its collaborative effort. Design work continued through the winter of 2000-2001. Several alternative configurations of marine reserves were drafted by subsets of the MRWG and the group as a whole, and were submitted to the Science and Socio-Economic Panels to ascertain their 'ecological value' and potential socio-economic impacts, respectively. However, the MRWG was unable to reach consensus on any of these maps.

At the April 2001 meeting, facilitators pressed the MRWG to collectively design a network of marine reserves for the Channel Islands. Starting at San Miguel Island, the most remote and least contentious of the five islands, the group negotiated a series of closures.

By the end of the day-long meeting, the MRWG had designed a network of marine reserves around the three largest islands (San Miguel, Santa Rosa, Santa Cruz), which encompassed over 80% of the study area.

Several MRWG members pressed for one more meeting to complete the design, and a subsequent meeting to specify overarching rules and policies for the system. The facilitators and agency co-chairs, however, rejected this idea and instead encouraged MRWG members to meet with one another before the next meeting to resolve their differences. The group would finish the map in the first couple of hours of the May meeting, then address the last 'details' of the process the same day. The rationale for this push was that the process had already been extended from 9 to 22 months, and at considerable cost. Moreover, state and federal officials became concerned that the state's recently legislated process to establish MPAs would overtake the MRWG process, which they viewed as a precedent for the state and other MPA efforts.

During the intervening weeks, some MRWG members met with agency personnel and each other to negotiate among themselves. The push to finish the process and the use of side agreements to resolve issues related to the most contentious areas, however, contributed to the disintegration of the agreement and much of the mutual respect and trust generated through the process. At the May 2001 meeting, the MRWG was disbanded. The CINMS Advisory Council subsequently authorized the CDFG and CINMS staff to devise a preferred alternative based on the MRWG's efforts, which they presented at the Commission's August 2001 meeting. This action signalled a reversion from a cooperative to a more instructive approach which many MRWG participants and an increasingly interested public were unwilling to accept. For all its shortcomings, the MRWG process had empowered participants and observers and provided them with new social resources, tools and knowledge to more fully engage in the design process. At the August 2001 meeting, the Commission received extensive public testimony and five citizen-generated alternative proposals in addition to the agencies' preferred alternative. The Commission formally considered all of these proposals. In October 2002, it approved the preferred alternative by a vote of two to one, with two Commissioners absent.

The Channel Islands case highlights the need for investments in time to develop and sustain shared understanding, common goals and objectives to support the marine reserve process. It also demonstrates the relevance of local knowledge as a complement to scientific knowledge to inform marine reserve design, and the critical importance of engaging this information and those who provide it throughout the process. In addition, it points to practical and ethical issues associated with collecting and using local knowledge to design marine reserves. A co-management process that recognized the value of this information and accorded it a respected role might have resulted in a different outcome.

4.3. Big Creek, California

The establishment of Big Creek Marine Ecological Reserve in central California provides an example of community-based management that was subsequently institutionalized by the state (Pomeroy and Beck, 1999). In 1988, local skiff fishermen asked the manager of the terrestrial Landels-Hill-Big Creek (LHBC) reserve, managed by the University of California, Santa Cruz (UCSC), for permission to launch from the reserve beach to better access fishing grounds along the Big Sur coast. The reserve manager agreed, provided that the fishermen would voluntarily abide by a marine reserve adjacent to the terrestrial reserve. The fishermen agreed and worked with the reserve manager to design, implement and help enforce the reserve. The arrangement continued through the early 1990s, when the fishermen agreed to the manager's request that they collect data on samples of their catch when fishing from Big Creek.

The informal marine reserve was given legal status in 1994 pursuant to the 1990 Marine Resources Protection Act (MRPA). The MRPA mandated the establishment of four marine reserves to aid in the study and replenishment of nearshore rockfish. The CDFG engaged in a consultative process to identify and evaluate a dozen proposed marine reserves. Big Creek stood out among these because it had the support of local fishermen, voiced in a letter to the CDFG (Pomeroy and Beck, 1999). Recognizing the value of the collaborative arrangement for the monitoring and enforcement resources gained, the reserve manager negotiated a memorandum of understanding (MOU) between UCSC and CDFG for co-management of the reserve. The MOU gives the LHBC reserve manager responsibility for the day-to-day operations of the marine reserve, and recognizes the priority of the pre-existing arrangement that allows the fishermen to transit through the reserve in exchange for their data collection efforts. The arrangement has withstood stresses from uncertain and changing environmental, economic and regulatory conditions (Pomeroy, 2002). Its resilience is due largely to the long-standing working relationship between the reserve manager and local fishermen that is built upon mutual understanding, respect and trust. Moreover, the reserve has begun to show signs of achieving some of its goals, including increased abundance and sizes of species within its boundaries (Ven Tresca *et al.*, 1997).

The Big Creek reserve is the product of a two-tiered co-management system. It was initiated locally through cooperative management led by the reserve manager, closely coordinated with local fishermen. By-products of this process included trust between the reserve manager and local fishermen, and a commitment to work together to meet their mutual goals in this highly localized context. To insure and legitimize the local arrangement, the reserve manager directly engaged CDFG in the consultative MRPA process. The implementation, management and evaluation of the Big Creek marine reserve have followed a more cooperative model of co-management. The reserve manager's central role in daily operations follows an informative model vis à vis the state, and a cooperative model involving the local fishermen.

4.4. The California MLPA Process

California's Marine Life Protection Act (MLPA) process presents a striking contrast to the Channel Islands and Big Creek cases. The MLPA, drafted by ENGO staff and passed by the state Legislature in 1999, mandated the establishment of a network of MPAs, including marine reserves, in state waters. The legislation dictated a top-down process, informed by 'scientific experts' and public input. Despite some operational latitude, the CDFG instituted a consultative process. It established a 'master plan team' (MPT) composed of biophysical scientists and CDFG personnel to develop a master plan to 'guide the adoption and implementation of the Marine Life Protection Programme established by the bill.... and to advise and assist in the preparation of a master plan' [Fish and Game Code (FGC) Div. 3. SEC 1. Ch. 10.5 Sec. 2855 (a) and (b)(1)]. The MLPA required that the MPT include 'Five to seven members who shall be scientists, one of whom may have expertise in the economics and culture of California coastal communities' (FGC Div. 3. SEC 1. Ch. 10.5 Sec. 2855 3(b)3(B)). It also required that 'The master plan shall be prepared with the advice, assistance, and involvement of participants in the various fisheries and their representatives, marine conservationists, marine scientists, and other interested persons' (FGC Div. 3. SEC 1. Ch. 10.5 Sec. 2855 3(b)(4)).

Lacking financial resources, time and experience to support a more cooperative process, CDFG adopted a consultative approach, interpreted by many as instructive, that stood in sharp contrast to the ongoing Channel Islands process. The MPT consisted of 15 members from the resource management and biophysical science communities. Despite provisions for a social scientist on the MPT, none was appointed. Involvement of other stakeholders was sharply limited, and did not include consumptive resource users. CDFG solicited input from the broader public through a mailing to fishing license holders and others on its marine-related mailing lists. The mailing requested information on particular areas of interest or concern, known fishing grounds, and other sites that the MPT might consider in drafting MMA network maps.

In June 2001, CDFG released the MPT's 'initial draft concepts' for public review. In July, it held workshops at 10 sites along the California coast. Despite the assertion that these were *initial draft concepts* to be further shaped by public input, many fishermen and other observers viewed them as the product of a top-down (or an ENGO) effort and a *fait accompli*. They protested the 'behind closed doors' approach of the MPT, and the failure to actively engage those who could offer insights into practical marine reserve design, and those would be most affected by the proposed reserves.

Realizing that public support for a preferred alternative would not be forthcoming, the Legislature passed a bill extending the MLPA deadline, and CDFG revisited its plan and timeline. It postponed the second set of siting workshops that had been scheduled to finalize MMA boundaries, and engaged in consultative small group meetings with stakeholders throughout the state. CDFG staff then developed a revised process to more fully engage stakeholders. Seven regional working groups are participating in facilitated regional workshops to review the MLPA's goals and develop alternatives for MPA sites from the ground up (CDFG, 2002). While the workshops are open to the public, comments are being received only through locally nominated, but agency-approved, constituent representatives (CDFG, 2002). Other steps in the process include: 1) review of MLPA guidelines and establishment of a process schedule; 2) discussion of spatial alternatives for MMA networks provided by panel representatives; 3) determination of an initial range of alternatives; 4) socioeconomic and scientific review; 5) discussion of reviews and alternatives; and, 6) final draft presentation and review. These steps are expected to result in a draft marine reserve master plan to be presented to the Commission for public review and subsequent adoption.

Although California's MLPA process is not yet complete, several lessons may be derived from the experience to date. The need for some form of meaningful co-management from the start is clear, as is the potential for failure without it. CDFG's original, largely instructive approach was problematic not only in its failure to engender trust, mutual understanding and support for the process, but also in its neglect of local knowledge which could have complemented the growing, yet incomplete, scientific information available. The revised process's more cooperative approach holds promise, although two issues persist. First, the ill will and mistrust generated by the initial MLPA effort are substantial, and have reinforced many stakeholders' distrust of resource managers, scientists and ENGOs involved in the process. Second, although the revised process clearly seeks to enable meaningful co-management in the early stages of the process, it does not address implementation, management or evaluation, suggesting that once marine reserves are designed, managers will revert to a centralized, top-down management model. While this may suit the large and complex California situation, the public may be as dissatisfied with this approach as it was with the initial MLPA effort.

4.5. Vanuatu

Johannes' (1998a) examination of the shift toward government-supported, village-based management of marine resources in Vanuatu provides another example of the intersection between co-management and marine reserves. In 1990, Vanuatu Fisheries Department staff offered to provide advice on the management of trochus (*Trochus niloticus*) to communities with spatially-based fishing rights. The Department's interest was in promoting better resource management by working with villages 'to help combine local knowledge with modern research-based knowledge' (Johannes, 1998a: 166). The response to the offer was enthusiastic. Fisheries Department staff responded by conducting local trochus surveys, and engaging in dialogue with resource owners and fishermen about the conservation and fishery value of harvesting legal size animals. These discussions also sought villagers' knowledge of temporal and spatial trends in trochus populations and local nearshore currents.

A combination of advisory and informative co-management led to the conceptualization and design of marine reserves for some villages. Marine reserves were not the object or goal of the effort, but constituted one tool that Fisheries Department staff suggested (in the form of trochus breeding preserves) if and where conditions, based on combined local and scientific knowledge, warranted. Although Fisheries Department staff provided advice on these matters, decisions rested with the villagers, in recognition of their need to 'balance biological considerations with social and economic needs' (Johannes, 1998a: 167). By the mid 1990s, 23 of 27 villages Johannes surveyed had established area closures ranging from one month to 'indefinitely' for other species as well as trochus. Villages exercise primary authority for implementation and management, with the Fisheries Department continuing to provide scientific knowledge and advice and facilitate the sharing of local and scientific knowledge.

Johannes concludes that 'the benefits of closures are now widely appreciated by villagers,' although they have yet to be systematically evaluated for their ecological, social and economic benefits (1998: 183). He also offers insights into strategies and conditions that favour government-supported, village-based management of tropical small-scale fisheries elsewhere. Although he does not link these to marine reserves per se, the outcomes he describes suggest they are relevant to marine reserve processes. He recommends that government resource managers: 1) publicize their willingness to collaborate with and assist villagers on marine resource management issues; 2) focus first on villages that have strong marine tenure, local authority, community cohesion and fishing territories that are amenable to village surveillance; and, 3) concentrate on a limited set of commercially important fisheries that are relatively simple to census and monitor. He also recommends that managers: 1) support national law that legitimizes local authorities' resource management efforts; 2) provide formal legal assistance only where local dispute resolution has failed; 3) train government extension personnel to enable them to facilitate village efforts to combine local and scientific knowledge for resource management; and, 4) leave final management decisions and enforcement to village authorities.

4.6. Samoa

King and Faasili (1999) describe a similar case in Samoa, where efforts to support and encourage community-based fishery management were more formalized than those in Vanuatu. The Samoan Government's Fisheries Division developed an extension programme to encourage village communities to manage their own marine resources through the development and implementation of their own fisheries management plans (FMPs). Extension staff were trained to facilitate a process that included the following steps. First, village extension facilitators (VEFs) contact village councils to notify them of the extension programme offerings and encourage participation in the programme. If a village expresses interest, the VEFs then coordinate group meetings to facilitate the identification of problems and potential solutions, in part through a participatory survey of local marine resources. The village is then encouraged to nominate a Fisheries Management Advisory Committee (FMAC) to further consider those problems and solutions. The VEFs then assist the FMAC in drafting a FMP for discussion and approval by the village council. FMAC members work with Fisheries Division staff to finalize the plan at Fisheries Division offices, where staff can provide timely additional technical information as needed. The final product is a 'community-owned village FMP' that constitutes an agreement with the Samoan Government and specifies the respective roles and tasks of the village and the Fisheries Division. The village council then reviews and approves the FMP, and appoints a Fisheries Management Committee to oversee plan implementation. The Fisheries Division maintains contact with the Committee, and provides technical support as specified in the FMP.

King and Faasili (1999) report that within two years of the programme's initiation, 44 of 65 villages had produced their own FMPs. Among these, 35 designed and established marine reserves in part of their fishing areas. Marine reserves were not the goal of the extension process nor of the villages that adopted them, but rather a management tool identified as appropriate for addressing particular problems. These marine reserves were a product of an advisory form of co-management whereby villages retain authority for the design, implementation, management and evaluation of their FMPs, including marine reserves, while Fisheries Division staff provide support.

4.7. Philippines: Apo and Sumilon Islands, San Salvador

The relationship between co-management and marine reserves has been documented at Sumilon and Apo Islands in the Philippines by Russ and Alcala (1999) and White *et al.* (2002). The two reserves were established in the mid 1970s following efforts by researchers at nearby Silliman University. Although the two cases have much in common, differences in the co-management and marine reserve processes used to establish, implement and maintain them help explain their divergent outcomes.

In both cases, the conceptualization of marine reserves was initiated by scientists at Silliman University, located near but external to the communities associated with the two reserves. In 1974, Silliman staff proposed the Sumilon Island Reserve to residents following a marine conservation and education programme (Russ and Alcala, 1999). In their scientific argument for establishing the marine reserve, the scientists noted the potential for benefits of increased fish yields in areas adjacent to the reserve. The Santander and Oslob municipal councils authorized Silliman University to establish the reserve for marine biological research and to regulate consumptive uses within the reserve (White, 1984). A university caretaker was placed on the island to monitor and enforce the reserve, and university staff undertook research at the site. Although some fishermen reported increased yields following reserve implementation, a change of local political authority in 1980 led to local challenges to the university's authority, and the subsequent breakdown of the reserve. The university turned to the national government for support of its authority, which declared the site a National Fish Sanctuary under the Bureau of Fisheries and Aquatic Resources (BFAR). BFAR assumed legal responsibility for the reserve while the

University continued to administer protection, management and research. Local resentment for the imposition of this top-down, instructive management approach grew, and fishing in the reserve continued through 1987. In 1988, municipal ordinances recognizing the reserve were passed, but the reserve was heavily fished in 1992. In 1994, a foreign NGO initiated a new reserve management effort through a Memorandum of Agreement with the Municipality of Oslob and a regional NGO, which resulted in partial protection of the reserve. Russ and Alcala note, however, a critical lack of 'genuine community-level involvement and support for the reserve, 24 years after it was initiated' (1999:311).

Although the Apo Island reserve also was initiated by Silliman University scientists following a similar marine conservation and education programme (begun in 1976), it stands in sharp contrast to the Sumilon Island case in both process and outcome. Russ and Alcala (1999) note that the idea of a marine reserve evolved during this programme. In 1982, Dauin Municipality and the university endorsed an agreement to implement the reserve. The community was involved in co-management of the reserve from conception through management for four years before it was legally protected in 1986. In 1985, the community endorsed a comprehensive marine reserve plan that included the small reserve established in 1982 and the entire reef to 500 meters offshore. The community formed a Marine Management Committee (MMC), which was given primary responsibility for maintaining the reserve, with the Philippine Constabulary providing enforcement support and Silliman University providing scientific and management advice. Evidence cited by Russ and Alcala (1999) and White and Calumpong (1992) suggests that the reserve has strong local support and substantial compliance with regulations. They add that fishermen have a positive attitude toward the reserve, and report increases in catch since it was established. Scientific monitoring of the reserve has likewise shown ecological benefits within and outside the reserve (Russ and Alcala, 1999).

In assessing the Sumilon and Apo Island cases, Russ and Alcala (1999) note key differences between them in the nature and extent of community involvement. The Sumilon Island reserve suffered from inadequate involvement of the community, and is 'a classic example of the users of a resource feeling that a remote, bureaucratic management agency ... was unjustifiably imposing restrictions on their rights to a resource' (1999:316). In contrast, they attribute the success of the Apo Island reserve to the fact that 'community support for the reserve concept was actively maintained, largely because the original ideas and concepts evolved from, and the implementation and maintenance were achieved by, the local community itself' (Russ and Alcala, 1999:317).

5. POSSIBILITIES FOR THE FUTURE ROLE OF CO-MANAGEMENT IN MARINE RESERVE PROCESSES

Both co-management and marine reserves have gained recognition as potential alternatives or complements to top-down, state-centred fishery management in response to discontent with the limitations of traditional fishery management. Co-management is focussed primarily on addressing the human dimensions of fishery management, whereas marine reserves are directed primarily toward addressing concerns about the integrity of marine ecosystems. These distinctions are somewhat oversimplified, given the interrelatedness between human and biophysical systems, especially in the context of fisheries.

Co-management serves two primary functions relative to marine reserves. In fostering discussions among resource managers and a broad range of other stakeholders, it can lead to the identification of problems or interests, and the specification of marine reserves as a potential solution, as occurred in the examples from Vanuatu, Samoa and Apo Island. The marine reserve proposals generated through these processes were directed toward conserving marine resources in communities that are primarily dependent upon fishing. The Big Creek case presents a slightly different case, where the marine reserve was first proposed by a local actor who engaged fishermen in a collaborative process.

Co-management also constitutes an institutional arrangement to facilitate the marine reserve process, whether it is initiated internally or externally. It can provide human, social and financial resources essential to effective marine reserve processes and outcomes. Specifically, co-management can bring together diverse and complementary types of information (eg local and scientific knowledge) about fishery management problems and potential solutions, historic and prospective use patterns and outcomes, and existing institutions that are supportive or potentially problematic for the establishment of marine reserves. Co-management also can enhance support for the marine reserve process by acknowledging and responding to diverse perspectives. Support includes both buy in to the marine reserve process, resulting in greater compliance with associated rules, but also financial and in-kind contributions to marine reserve implementation, management and evaluation.

The record to date suggests numerous accomplishments, but also points to limitations, potential pitfalls and problems associated with both co-management and marine reserve processes. As Harris (1998) has noted of fishery management, neither co-management nor a marine reserve is an event; rather each is a process. Both require substantial commitments of financial, human and social resources for their initiation and maintenance. In addition, both require sufficient time to allow for the process to run its course. Co-management and marine reserves also require flexibility and adaptiveness to adjust to changes in the associated biophysical and human environments. As with top-down management, mismatches between situations and approaches used can diminish the effectiveness of co-management and its application in marine reserve processes, in both human and biophysical terms. Finally, the use of co-management in marine reserve processes is also vulnerable to internal and external pressures, such as increasing demand or competing claims, and side agreements that give some participants an unfair advantage in the process.

Nonetheless, co-management is likely to play an increasingly important role in marine reserve processes, given the sustained and growing interest in marine reserves as a fishery management tool. The recognition of critical needs for information and support to enable the effective design, implementation, monitoring, evaluation and performance of marine reserves points clearly to the need for diverse sources of information and support. Co-management has the particular, and perhaps unique, ability to provide these resources.

Critical challenges for the effective use of co-management for marine reserves lie in the diversification of management goals, both within and beyond fisheries per se, the increasing role of the non-fishing public in the management process, and management's attempts to more fully encompass ecosystems and relationships among them. This increased scope and complexity of marine reserve efforts may make co-management less tractable. Increased numbers and diversity of actors make communication, information sharing and agreement more difficult. Whereas consensus may become less likely, however, the greater diversity of perspectives and ideas encompassed by a widened scope of actors may increase the possibilities for productive negotiation. Tisdell and Broadus (1995) and others have suggested that marine reserve proponents may do better to work toward a broadly supported marine reserve design that is not ecologically perfect, rather than an ecologically perfect design that is likely to fail for lack of support.

Most critical, perhaps, is the increased geographic scope of MPA networks envisioned, for example, in the Baja California to Bering Sea Initiative (CPWS 2002). The goal of linking marine reserves within and across ecosystems and broader bioregions reflects growing recognition of the interconnectedness of such systems. Co-management has a role to play in bringing stakeholders at analogous levels of governance together to serve the same functions as in smaller scale efforts. Yet efforts to take this larger system view may disconnect co-management from its foundations in smaller scale, locally grounded institutions and local scale information which are critical to the effective interface between co-management and marine reserves. To guard against this, it will be important to look to subsidiary (Sandberg, 1999) and nested arrangements (Ostrom, 1990) to make and maintain the linkages among co-management arrangements at different scales.

ACKNOWLEDGEMENTS

I am grateful to MPA process participants, agency personnel who provided data and other support, and research collaborators for their invaluable contributions to the Channel Islands and Big Creek research, and to Monica Hunter, for her thoughtful review of a draft of this paper. Funding for the Channel Islands work reported here was provided, in part, under contract to the National Ocean Service Special Projects Office, and through grant No. NA66RG0477, project R/MA-39 National Sea Grant College Program, NOAA, U.S. Department of Commerce through the California Sea Grant College System, California State Resources Agency, with matching funds provided by the Institute of Marine Sciences, University of California, Santa Cruz (UCSC). Funding for the Big Creek research was provided by UCSC's Office of the Vice Chancellor for Research and the Monterey Bay Regional Studies Program, and the California Marine Ecological Reserves Research Program (project R/BC-2).

REFERENCES

Alcala, A. C. and Russ, G.R. (1990) A direct test of the effects of protective management on abundance and yield of tropical marine resources. *Journal du Conseil International pour l'Exploration de la Mer* **46**, 40-47.

Badalamenti, F., Ramos, A.A., Voultsiadou, E., Sanchez-Lizaso, J.L., D'Anna, G., Pipitone, C., Mas, J., Ruiz Fernandez, J.A., Whitmarsh, D. and Riggio, S. (2000) Cultural and socio-economic impacts of Mediterranean marine protected areas. *Environmental Conservation* **27**(2), 110-125.

Bennett, B.A., and Attwood, C.G. (1991) Evidence for recovery of a surf-zone fish assemblage following the establishment of a marine reserve on the southern coast of South Africa. *Marine Ecology Progress Series* **75**(2,3), 173-181.

California Department of Fish and Game (CDFG) (2002) www.dfg.ca.gov/mrd/mlpa/working_group.html. accessed April 2002.

Canadian Parks and wilderness Society (CPWS) (2002) www.cpawsbc.org/b2b/index.php accessed April 2002.

Channel Islands National Marine Sanctuary (CINMS) (2002) www.cinms.nos.noaa.gov/MRWGgoals.html. accessed April 2002.

Dayton, P.K. (1998) Reversal of the burden of proof in fisheries management. *Science* **279**(5352), 821.

DeMartini, E. E. (1993) Modeling the potential of fisheries reserves for managing Pacific coral reef fishes. *Fishery Bulletin* **91**, 414-427.

Dobryznski, T.J. and Nicholson, E.E. (2001) An evaluation of the short-term social and economic impacts of marine reserves on user groups in Key West. Master's Thesis, Nicholas School of the Environment, Duke University.

Dugan, J.E. and Davis, G.E. (1993) Applications of marine refugia to coastal fisheries management. *Canadian Journal of Fisheries and Aquatic Sciences* **50**, 2029-2042.

Farrow, S. (1996) Marine protected areas emerging economics. *Marine Policy* **(20)**6, 439-446.

Hanna, S. (1994) Co-management. Pp.233-242 in Limiting access to marine fisheries: Keeping the focus on conservation. K.L. Gimbel, ed. Washington, DC: Center for Marine Conservation, World Wildlife Fund.

Harris, C.K. (1998) Social Regime formation and participation in fisheries management. Pp. 261-276 in Reinventing Fisheries Management, T.J. Pitcher, P.J.B. Hart and D. Pauly, eds Boston: Kluwer Academic Publishers.

Jentoft, S. (1989) Fishermen's co-management: Delegating government responsibility to fishermen's organizations. *Marine Policy* **13**(2), 137-154.

Johannes, R.E. (1998a) Government-supported, village-based management of marine resources in Vanuatu. *Ocean and Coastal Management* **40**(2/3), 165-186.

Johannes, R.E. (1998b) The case for data-less marine resource management: Examples from tropical nearshore fisheries. *Trends in Ecology and Evolution* **13**(6), 243-246.

Johannes, R.E. (1984) Marine conservation in relation to traditional life-styles of tropical artisanal fishermen. *The Environmentalist* **4**, 30-35.

McCay, B.J. (1989) Co-management of a clam revitalization project: The New Jersey 'spawner sanctuary' programme. Pp. 103-124 in Co-operative management of local fisheries, E. Pinkerton, ed. Vancouver: University of British Columbia Press.

McCay, B.J., and Jentoft, S. (1996) From the bottom up: Participatory issues in fisheries management. *Society and Natural Resources* **9**, 237-250.

McClanahan, T.R., and Kaunda-Arara, B. (1996) Fishery recovery in a coral-reef marine park and its effect on the fishery. *Conservation Biology* **10**(4), 1187-1199.

McClanahan, T.R., and Mangi, S. (2000) Spillover of exploitable fishes from a marine park and its effect on the adjacent fishery. *Ecological Applications* **10**(6), 1792-1805.

Murray, S.N., Ambrose, R.F., Bohnsack, J.A:, Botsford, L.W., Carr, M.H., Davis, G.E., Dayton, P.K., Gotshall, D., Gunderson, D.R., Hixon, M.A., Lubchenco, J., Mangel, M., MacCall, A., McArdle, D.A., Ogden, J.C., Roughgarden, J., Starr, R.M., Tegner, M.J., Yoklavich, M.M. (1999) No-take reserve networks: Sustaining fishery populations and marine ecosystems. *Fisheries* **24**(11), 11-25.

National Research Council (NRC) (2001) Marine protected areas: Tools for sustaining ocean ecosystems. Washington, DC: National Academy Press.

Ostrom, E. (1990) Governing the commons: The evolution of institutions for collective action. New York: Cambridge University Press.

Paddack, M.J., and Estes, J.A. (2000) Kelp forest fish populations in marine reserves and
adjacent exploited areas of central California. *Ecological Applications* **10**(3), 855 870.

Pinkerton, E., ed. (1989) Co-operative management of local fisheries. Vancouver: University of British Columbia Press.

Pomeroy, C. (2002) Marine reserves as a resource management tool: An evaluative study of cooperative data collection at Big Creek. Marine Ecological Reserves Research Programme Research Results-1996-2001, Project No. R/BC-2. CD-ROM. La Jolla, CA: California Sea Grant College Programme.

Pomeroy, C. and Beck, J. (1999) An experiment in fishery comanagement: Evidence from Big Creek. *Society and Natural Resources* **12**, 719-739.

Roberts, C.M. (1997) Ecological advice for the global fisheries crisis. *Trends in Ecology and Evolution* **12**, 35-38.

Roberts, C.M., Ballantine, W.J., Buxton, C.D., Dayton, P., Crowder, L.B., Milon, W., Orbach, M.K., Pauly, D. and Trexler, J. (1995) Review of the use of marine fishery reserves in the US southeastern Atlantic. NOAA Tech. Memo NMFS-SEFSC 376. Miami: NOAA NMFS.

Russ, G.R., and Alcala, A.C. (1999) Management histories of Sumilon and Apo Marine Reserves, Philippines, and their influence on national marine resource policy. *Coral Reefs* **18**, 307-319.

Sanchirico, J.N., and Wilen, J.E. (1999) Bioeconomics of spatial exploitation in a patchy environment. *Journal of Environmental Economics and Management* **37**, 129-150.

Sandberg, A. (1999) State commons and local democracy in northern peripheries. Paper presented at the Second Workshop on the Workshop, 9-13 June, Indiana University, Bloomington.

Sandersen, H.T. and Koester, S. (2000) Co-management of tropical coastal zones: The case of Soufrière marine management area, St. Lucia, WI. *Coastal Management* **28**, 87-97.

Sen, S., and Raakjaer Nielsen, J. (1996) Fisheries Co-management: A comparative analysis. *Marine Policy* **20**(5), 405-418.

Tisdell, C. and Broadus, J.M. (1989) Policy issues related to the establishment and management of marine reserves. *Coastal Management* **17**, 37-53.

Walters, C. (2000) Impacts of dispersal, ecological interactions, and fishing effort dynamics on efficacy of marine protected areas: How large should protected areas be? *Bulletin of Marine Science* **66**, 745-757.

White, A.T. and Calumpong, H. (1992) Summary field report. Saving Tubbataha Reef. Monitoring marine reserves in the central Visayas, Earthwatch Expedition, Philippines. Unpublished paper, 24 pp.

White, A.T., Courtney, C.A. and Salamanca, A. (2002) Experience with marine protected area planning and management in the Philippines. *Coastal Management* **30**, 1-26.

White, A.T., Hale, L.Z., Renard, Y. and Cortesi, L. (1994) Collaborative and Community-based Management of Coral Reefs: Lessons from Experience. West Hartford, CT: Kumarian Press.
Yoklavich, M., ed. (1998) Marine Harvest Refugia for West Coast Rockfish: A Workshop. NOAA-TM-NMFS-SWFSC-255.

Chapter 13

CO-MANAGEMENT AND RECREATIONAL FISHING

RIKU VARJOPURO
Finnish Environment Institute, Research Programme for Environmental Policy, Helsinki, Finland

PEKKA SALMI
Finnish Game and Fisheries Research Institute, Enonkoski, Finland

1. INTRODUCTION

Recreational fishing and its governance has rarely been studied, partly due to the short history of the leisure hobbies in the modern society compared with the long traditions of fishing as a livelihood. However, there has been academic interest in the social dimensions of recreational fisheries, especially in The USA and Canada (Viard and Brenner, 1998). In many countries the recreational fishermen have become powerful stakeholders in the utilization of fishing waters. Yet they often form heterogeneous groups and are not self-evidently included in a participatory management system.

This paper discusses the issues and complexities that arise when recreational fishing in marine areas is added to the management mix and the implications for co-management in these cases. We will approach this question by analysing an empirical case of recreational fishing in the Archipelago Sea area (Southwest Finland) in the Baltic Sea. The analysis is based on an analytical framework that emphasizes the importance of system dynamics and the role of contextual factors in the natural resource use. In the case of recreational fishing it is especially important to take into account both local level as well as larger social processes and interactions between actors on different levels. These interactions between different levels set conditions for management of recreational fishing.

We will relate our findings to a discussion about fisheries co-management. Many of the case studies presented in the literature deal with commercial fisheries or settings where commercial and subsistence fisheries co-exist, but they seldom refer to recreational fishing. We will discuss interactions between different types of recreational fishing in coastal areas as well as recreational fishing in multiple-use contexts, which are typical in many coastal areas around the world. We also compare the findings from the Finnish case study with circumstances in other countries.

2. RECREATIONAL FISHING

Recreational fisheries can be described as those fisheries where the stock is exploited either for an individual's personal consumption or for leisure (Hickley and Tompkins, 1998). We find this kind of definition useful, because it allows our discussion to cover many types of fisheries. This definition groups leisure and sport fishers together with subsistence fishers.

Recreational fishing has taken its place as an important actor in coastal and freshwater fisheries in many Western countries. In many fisheries, recreational fishers outnumber the commercial fishers. Even the recreational catch may be higher than the commercial catch. However, it is difficult to address the importance of recreational fishing, because social and economic data are poor or absent (Hickley and Tompkins, 1998).

There is considerable variation in the popularity of recreational fishing between European countries. In western European countries, the fisheries are generally exploited for leisure and sport. In 20 years the number of sport fishers has declined, but the economic importance of recreational fishing is still very high. Recreational fishery is growing in the region. In the Nordic countries the fisheries are exploited to a large extent also on a subsistence basis and fishermen tend to use passive gear (eg gill-nets, traps and baited lines). In eastern Europe, the recreational fishery is mostly for subsistence. In southern Europe, recreational fishery is declining, except in Italy (Cowx, 1998). The relative amount of recreational fishers of the whole population is highest in the north (Table 1).

Table 1. Summary of the status of recreational fishing in European countries (Cowx, 1998)

Country	Estimated number of recreational fishermen	Recreational fishermen as % of total population	Purpose of recreational fishing	
			Subsistence	Sport
Austria	220 000	3.1	No	Yes
Belgium	290 000	2.9	No (?)	Yes
Bulgaria	180 000	2.0	Yes	Yes
Cyprus	3 000	0.5	No	Yes
Czech Rep.	281 000	2.7	Yes	Yes
Denmark	100 000	1.9	No	Yes
Finland	2 100 000	42.0	Yes	Yes
France	5 000 000	8.9	No	Yes
Germany	2 350 000	3.0	Yes	Yes
Hungary	328 000	3.2	Yes	Yes
Ireland	144 000	3.7	No	Yes
Netherlands	1 300 000	9.0	No	Yes
Norway	900 000	21.4	Yes	Yes
Poland	2 000 000	5.1	Yes	Yes
Portugal	80 000	0.8	No	Yes
Romania	200 000	0.9	Yes	Yes
Slovakia	89 000	1.7	Yes	Yes
Sweden	1 200 000	27.0	No	Yes
Switzerland	200 000	3.1	Yes (?)	Yes
UK	2 000 000	3.5	No	Yes

Recreational fishing is a popular hobby and creates economic activity in industrialised countries on other continents as well. Recreational fishing in Australia and New Zealand seems to be almost as popular as in the Nordic countries. In some estimates the proportion of recreational fishers in Australia is 33 % of the population and 20% in New Zealand. In these countries, part of the recreational fishing is subsistence fishing (Kearney, 2001; Fisheries Western Australia, 2000; MacMurran, 1999). In the United States the proportion of recreational fishers of the total population in 1996 was slightly over 13 % (National survey, 1997).

Aas and Bogelius (1998) summarize categories of interactions between recreational fisheries and other interest groups as follows: 1) between groups of recreational fishers, eg resident or non-resident fishers; 2) between fishers and the management system, eg disagreements about regulations; 3) between recreational and commercial fishing, eg disagreement about allocation of quota; and, 4) between recreational fishing and other recreational activities or user groups, eg conflicts between bird watchers and anglers.

Conflicts between recreational and commercial fishers have arisen either due to competition for space or direct competition for the catch. In the political forum, recreational fishers have taken an influential role in the constitution of fisheries policy (eg Regier *et al.*, 1999; Borg, 1999; Olburs, 2000; Pirhonen and Salmi, 1998). The interaction between commercial fishing and recreational fishing is only a part of the picture. The recreational sector consists of many types of fishing, which may compete with each other as well as with commercial fishing. An increased specialization among sport fishers has put pressure on attractive waters and caused problems between groups of recreational fishermen (Aas and Bogelius, 1998).

3. FISHERIES CO-MANAGEMENT

A lively discussion of fisheries co-management has been going on for well over a decade now. Co-management has several definitions, many of which point at the sharing of the power and responsibility in fisheries decision making between the state authorities and the user group organizations. This includes all kinds of arrangements where sole decision-making power does not lie in either end of continuum from central decision making to self-management. Some definitions take a more restricted approach that does not see simple user-participation as a sufficient condition and argue that real collaboration in decision making is needed before a management regime can be called 'a co-management'. It is not enough that the central administration creates an opportunity for a user-group consultation, because consultation does not guarantee real influence in the decision-making process (eg Sen and Raakjær Nielsen, 1996; Jentoft, 2000; McCay and Jentoft, 1996).

We will accept the 'liberal' definition of co-management in our paper. The reason for this is that our intention is to discuss the prospects and problems of the co-management of recreational fishing rather than to analyse existing co-management arrangements. This wider perspective allows us to take more broadly into account the simultaneous decision-making processes and driving forces that affect recreational fisheries.

Several comparative papers on conditions affecting success of co-management have been published (eg Pinkerton, 1994; Sen and Raakjær Nielsen, 1996; Noble, 2000; Borrini-Feyerabend *et al.*, 2000; Pomeroy *et al.*, 2001). These papers have introduced many important issues regarding co-management. The following six issues are referred to on many occasions and are general enough to allow discussion on the principles rather than the details of fisheries management regime. We will use this list in our analysis to discuss the issues and complexities that arise when recreational fishing in marine areas is added to the management mix and the roles played and implications for co-management in these cases.

User groups. User groups and the relations between them are the essence of co-management. It is necessary to identify all user groups that have a stake at the resource system, including stakeholders that are not direct users of the fish resources. Important groups are usually decision-makers, researchers and eg the nature conservationists. In many countries fishing tourist entrepreneurs form a core interest group for the development of recreational fisheries and co-management. In a detailed analysis a large number of stakeholders concerning the use of coastal resources can be identified (eg Ellegård, 1998).

Organization and representation. Co-management aims at allowing participation for the relevant stakeholders. Participation is easy to arrange for well-organized user groups. Unfortunately, many of the stakeholder groups are not organized. This is especially the case in the multiple-use system that our paper deals with. The capability of user groups to participate determines how representative the management body will be. In addition, it is important to pay attention to how the co-management system itself is organized. The way the decision making and participation is arranged may support or hamper participation of stakeholders irrespective of their capabilities and degree of stakeholder organization. Not all forms of representation are legitimate. It is important to differentiate between 'legitimate participation' and 'participatory legitimacy' where the latter refers to 'how the management process itself gains greater legitimacy and effectiveness from being participatory' (Wilson and McCay, 1998).

Existing regime and institutions. Existing management regimes should be taken into account in developing new co-management arrangements. In addition to management itself, property rights regimes exist also in most fisheries. There may be formal or informal property rights entitled to local fishers or the resource may be owned by the state. Property rights must be defined in the new management regime. It is also important that the management regime itself is supported by the state authorities.

Interests. Different stakeholders have different interests regarding the resource as well as the co-management arrangement. Some users may extract the resource in material terms, eg fish biomass, while other uses may be non-consumptive. There are also occasional users for whom the resource and its exploitation are not actually very important. Strong or weak interests in the resource or its exploitation affect stakeholders' willingness to participate in management. Interests in participating also differ among stakeholders because of expectations or fears regarding the co-management. Some stakeholders see that co-management gives them an opportunity to enhance their interests in the resource. Others may see co-management as an opportunity to protect their interests against outside pressures. In cases of already well-established local management regimes, the groups whose interests the existing regime protects may feel threatened by the co-management. However, for successful co-management arrangements all stakeholder groups should be able to feel benefits from participation.

Management tasks and goals. A management regime must be able to handle the most important issues related to the resource use system. There may be conflict over allocation of the resource between the user groups or the resource is depleted and conservation measures are needed, which easily generate allocation conflict. The resource may also be a key to local economic or social development. Management tasks and goals must be clear.

Scale and boundaries. It is important to pay attention to the ecological boundaries, such as marine areas and fish stocks, as well as social boundaries, such as ethnic groups, administrative jurisdictions, or property lines, involved in the resource use system. It may be impossible to find a perfect match for different kinds of scales, but it is important to be aware of different boundaries and the complexities this brings along. Scale affects the way stakeholders are represented in the co-management arrangement. In small-scale settings the local users may speak for themselves, but when the scale is large users must elect representatives from among their peers to speak for them (Jentoft, 2000).

4. RECREATIONAL FISHERIES MANAGEMENT IN THE FINNISH ARCHIPELAGO SEA

First we will describe the context in which recreational fishing in the Archipelago Sea area takes place to understand the dynamics of the system. Then we will compare our case to characteristics of recreational fisheries in some other industrialized countries.

4.1. Characteristics of the Recreational Fishing System in the Archipelago Sea

During recent decades the land and seascape of the Archipelago Sea Region has been transformed from a production landscape of fishing peasants to one of leisure, recreation and consumption (Andersson and Eklund, 1999). In the early 1970s half of the population got its livelihood from the primary sector, but today the majority of the population is engaged in the public welfare sector, tourism, recreational services and transport. During this change, the number of permanent inhabitants has declined, but pastime activities have become prominent in the area. Number of summer houses increased rapidly in two decades, which reflects important remote contextual factors affecting the local resource use system. There are ca 24 000 summer cottages in the Archipelago Sea area (Salmi, 2001).

In the following we will concentrate on some of the central characteristics of the recreational fishing in the Archipelago Sea area that determine recreational fishing as a resource use system. We use the three main categories introduced by Edwards and Steins (1999) that are: 1) Physical-technological characteristics of the resource system, 2) Institutional characteristics and 3) Characteristics of the user community. These findings are then set alongside with the general conditions of successful co-management presented above to discuss the potentials of including recreational fisheries in the fisheries co-management regime.

4.1.1 Physical and Technological Characteristics

Salinity in the Baltic Sea is very low, especially in the Northern part. There are both salt water and fresh water fish species in the Archipelago Sea area. The typical target species in recreational fisheries are perch, pike-perch and pike. Sport fishers catch also salmon and trout in the outer parts of the archipelago. The water quality is deteriorated in the Archipelago Sea due to eutrophication (eg Kauppila and Bäck, 2001). The eutrophication has some effects on fish species composition and, consequently, to the abundance of target species in recreational fishing. However, these changes do not seem to substantially determine capacity of the resource system. Shallow waters inside the large archipelago area provide relatively high production of several fish species. Capacity of the resource can be partly controlled by stock enhancement.

Technical characteristics of the recreational fishing have also changed during the last few decades. Most of the recreational fishermen use rods and lines, but commonly a few gill nets, in the waters close to their summer cottages. Leisure and subsistence fishers use their gill nets mostly in the same water areas and target the same fish species as commercial gill net fishers. The modern trolling boats have made larger areas of deeper waters more accessible for the sport fishers. Trolling boats made sport fishing also more visible in the archipelago.

4.1.2 Institutional Characteristics

What seems to be more important than physical or technological characteristics, however, is allocation of catches and especially who has a right to use the area. All this comes to the very central issue: distribution of decision-making power. Hence, institutional characteristics of the system have become very central in recent years, especially after the changes in legislation that altered the decision-making regarding use rights in recreational fishing. Institutional characteristics are important also because of the already existing collaborative decision-making bodies, statutory fishery associations and the Fisheries Regions (Salmi and Varjopuro, 2001).

Responsibilities concerning fisheries management in Finnish coastal waters are shared among public fisheries authorities, Fisheries Regions at the intermediate level, and private owners at the local level of the decision-making regime (Table 2). The coastal waters up to 500 metres from the shoreline are in private ownership, but are mostly managed by collectives of individual shareholders, namely 'statutory fishery associations' (SFA). In an archipelago, where there are a lot of small islands and islets, a zone that extends 500 metres from the shorelines covers nearly the whole area. Waters outside 500 metres line are state property.

Table 2. Ownership and management of coastal waters

<table>
<tr><th>Ownership</th><th>Owner</th><th>Management unit</th></tr>
<tr><td>Private</td><td>Individual</td><td>Individual</td></tr>
<tr><td rowspan="2">Private</td><td rowspan="2">Individuals
(who have formed a collective of shareholders, namely SFA)</td><td>Statutory Fishery Association (SFA) (collective of owners)</td></tr>
<tr><td rowspan="2">Fisheries Region
(SFAs, individual owners and representatives of fisher groups)</td></tr>
<tr><td rowspan="2">Public</td><td rowspan="2">Citizen</td></tr>
<tr><td>Regional and central fisheries authorities</td></tr>
</table>

Owners of the water areas are mostly either local land-owners or people who have summer cottages in the area. Usually the local owners have the largest properties, which give them a majority vote in SFAs. Depending on the management unit, owners can privately or collectively sell fishing permits to their water areas. Although the economic importance of commercial fishing in the Archipelago Sea area has decreased, the local water owners support traditional commercial water use, rather than that of the leisure people coming from urban areas (Salmi, 2001). Not surprisingly, the non-local owners often give high priority to recreational fishing interests.

In the intermediate level, Fisheries Regions cover the area of several statutory fishery associations, private waters and sometimes also small areas of public waters. The Fisheries

Region's decision-making body consists of representatives of the SFAs, private owners, local associations of commercial fishers and local associations of recreational fishers. Regional fisheries authorities have a supervisory role in the management system. Fisheries Regions have been established to guarantee better opportunities for different stakeholders to take part in fisheries decision making, although the owners were guaranteed the majority of the votes irrespective of number of other representatives. There have been problems in establishing and running of Fisheries Regions, especially at coastal areas (Sipponen, 1998). Fisheries Regions' role in the management of recreational fisheries has not been as important as the role of SFAs or private water owners.

The latest changes in the Finnish fisheries legislation have taken a substantial amount of the decision-making power away from the owners. Opportunities for leisure and sport fishing have been improved by introduction of a province-wide lure fishing permit, which allows fishing with one lure in an area of one province. These permits are sold by the state authorities and the money collected is allocated to Fisheries Regions and further to SFAs and the individual owners. The local owners protested the system because it would reduce their decision-making power about the use of their own property and they could not any more control the access to their water areas. In addition to the province-wide lure fishing fee system, also other use rights of non-owners have been protected by law: eg ice fishing with rod and angling with rod and natural bait are allowed without any fee and irrespective of the ownership of the water (so-called every man's rights).

Decisions affecting recreational fishing are made on several levels: there are local and regional bylaws and national policy decisions that determine fishing rights in recreational fishing. A new kind of demand concerning coastal resources brought along with urban way of living (and new policies that support urban recreational uses of coastal resources) have considerably changed existing resource use patterns, decision-making arrangements and power relations at regional and local levels. The process of improving leisure and sport fishers' use rights through national legislation and local owners' protests against them is a good example of an important contextual factor that has affected management of this CPR-system.

4.1.3 Characteristics of the user community

The structure of the 'user community' (Edwards and Steins, 1999) in the Archipelago Sea recreational fishery is quite complex and not very coherent. If measured by the number of fishermen, recreational fishing is now the most important fisheries activity in the Archipelago Sea area. The large number of summer cottages has provided fishing opportunities for non-local recreational fishermen.

Three major groups can be identified in the Archipelago Sea recreational fisheries:

- 'Sport fishers', who are active fishers with hand gear or trolling boats. They fish in large areas along the coast.
- 'Leisure fishers', who are most often summer dwellers in the area and fish with hand gear or gill-nets. They usually fish in waters near their summer cottages and form the largest group.
- 'Subsistence fishers', who are mostly local people who fish with passive fishing methods, gill-nets and fish traps. Their fishing have a strong recreational function, but fishing for own consumption has a long tradition in the area and is here called subsistence fishing to emphasize its cultural background.

Table 3. User group characteristics

	Sport fishers	Leisure fishers	Subsistence fishers
Locality	Non-local	Non-local	Local
Owner of water area	Seldom an owner	Often an owner	Often an owner
Importance of fishing	High	Variable	High
Position in the decision making regarding the permits	None	Weak. Shareholder in SFA, but not usually actively involved.	Strong. Individual owner of water area or a shareholder in SFA. Usually actively involved.
Degree of organization	Variable. Only minority is organized.	None or low	High
Organization's power	Nationally high. Regionally and locally low.	None or low	Nationally, regionally and locally high.
Other observations	Sport fishing represents urban way of living	Good fishing opportunities provided by legislation ('every man's rights')	Control over local resource use patterns is socially and culturally valued

There are several aspects that separate or unite these three groups (Table 3). The ownership of coastal waters is one of them. Subsistence fishers and majority of leisure fishers are often owners of water areas, which gives them a right to take part in decision making in SFAs. Sport fishers are seldom owners.

For active sport fishers fishing is an important hobby, in which they have invested a lot of money for fishing tackle, boats, etc. For the local subsistence fishers, fishing is a part of the traditional way of living, reflecting also former high economic importance of fishing for the fisher-peasants. These two groups have the highest stakes in the fishing. The leisure fishers are often satisfied with even quite moderate opportunities for fishing. They fish near their summer houses with a small number of fishing gear only a few times a year and are not passionate about the debates over fishing rights in the area. Fishing tourists, who use commercial services such as rented cottages and fishing guides, should also be included in the groups of leisure and sport fishermen. However, in the Archipelago Sea area this kind of fishing tourism is not very notable compared with recreational fishermen from the local area, those using their own summer cottages or fishers who come boating from the cities.

Third kind of demarcation line that characterises interest groups is differences between rate of organization and participation of actors. There are organizations for sport fishers, but not all sport fishers are members in the organizations. Sport fishers' organizations are influential on the national level fisheries decision making, namely in the parliament (Pirhonen and Salmi, 1998). The local subsistence fishers are well represented in the decision making as well. They are the most influential party at the local (SFAs) and regional level (Fisheries Regions) decision-making bodies. Non-local leisure fishers are often owners of fishing waters, but in practice the local owners are the most active decision makers.

There is also a fourth aspect that divides or unites the groups. Recreational fishing, especially sport fishing, is an urban hobby, taking place mostly in rural areas. The local recreational fishing - fishing for household consumption - has its roots in rural way of living. In local resource use system the different ways of living meet each other.

Using the term 'a user community' (Edwards and Steins, 1999) may be particularly awkward in this multiple-use context, because the system actually forms an interface of user communities. In the system urban sport fishers, who do not necessarily have any prior connections to the area, meet local inhabitants. Local inhabitants are influential decision-makers in the system. In addition to these two groups, also summer dwellers are involved. They bring to the system another urban way of using the coastal landscape. However, using the term 'user community' is useful in emphasizing the fact that there are several interest groups who use the same resource and who are all affected by the decisions that define user rights in recreational fishing.

5. ARCHIPELAGO SEA RECREATIONAL FISHING AND CO-MANAGEMENT

Above we listed important aspects in fisheries co-management: 1) user groups, 2) organizations and representation, 3) existing regime and institutions, 4) interests, 5) management task, and 6) scale and boundaries. If we compare the central characteristics of the Archipelago Sea recreational fisheries system with the important issues in co-management, we can spot certain problems, but also the potentials for managing this resource use system collaboratively.

The recreational fisheries system in the Archipelago Sea area consists of several **user groups**. The relevant user groups can be identified. In this respect, defining the groups to participate in a co-management regime should not be a problem. In fact the existing collaborative decision-making body - the Fisheries Region - is based on participation of the relevant user groups.

However, the uneven degree of **organization** seems to be one of the problems in the system. Sport fishers and local subsistence fishers are relatively well organized, but the leisure fishers - the largest group - are not. This hampers their possibilities to participate in the decision making. Despite of the weak organization level, the interests of the large group of leisure fishers have to a certain extent been taken into account in the political arena, which has resulted in public access rights to the private waters ('every man's rights').

In the discussion of a potential for creating or developing a co-management system **the existing regime and institutions** must be taken into account. Private ownership has been strongly protected by the Finnish constitutional law and, hence, the political actors aiming at promoting the fishing opportunities for rod fishermen have not been able to completely set this institution aside. When developing new decision-making bodies the existing decision-making bodies should be taken as a starting point.

According to the broad definition of co-management, both SFAs and Fisheries Regions are forms of co-management, with Fisheries Regions being closer to the co-management ideal. However, neither of the decision-making bodies provide a forum for full participation of recreational fishers. Owners of the water areas, especially the local ones whose interest is not to increase fishing opportunities of the non-local recreational fishers, have most of the decision-making power in both SFAs and Fisheries Regions. In SFAs the participation is based on ownership, not on belonging to the user groups.

In the recreational fisheries of the Archipelago Sea the different user groups have different **interests**, related both directly to fishing and to participation in decision making. The latter may be problematic in the Archipelago Sea context, if we aim at creating a well functioning co-management regime in the area. The local owners can already protect their own interests in the existing decision-making bodies.

On the other hand, sport fishers may find it not worth the effort to try to influence the regional or local level decision making. First of all, the present system does not give them much power at the regional and local level, so a better strategy has been to influence national level fisheries policy, which has changed local practices, eg in the case when the parliament decided about the system of province-wide lure fishing licences. Secondly, sport fishers are not dependent on any specific areas for fishing, because they can easily move to other areas where access is granted easily. When it comes to the third group of recreational fishers - leisure fishers - many of them seem to have weak incentives for closer participation in the decision making. The highly variable interests in fishing among the user groups may easily lead to a situation where common interests to pursue after are difficult to find.

In the Archipelago Sea area recreational fishing is a multi-species fishery and none of the target species seem to be seriously depleted. In addition, the state of the stocks can be partly controlled by stock enhancement activities. **Management task** for the case of recreational fishing in the Archipelago Sea area is rather finding a common ground for managing recreational fishing than the conservation of fish stocks. The existing management regime and also the recent conflicts regarding recreational fishing rights may make the management task a difficult one, especially when both the local owners and sport fishers, who are the parties in the most serious conflict, may see other strategies than co-management serving their interests the best.

In the Archipelago Sea recreational fisheries **scale and boundaries** - especially social ones - are among the key issues. The fishing waters in the area are fragmented to a complex system of relatively small water areas, which has to be taken into account when developing the resource management system in the area. Users come from different locations - some are local people, but the majority of users are non-local. In addition to these geographical boundaries, there are cultural boundaries as well. The Archipelago Sea recreational fisheries system is an interface of rural and urban identities and ways of living. In addition, a large proportion of the area is inhabited by Swedish speaking people who have a distinct culture and identity. Both geographical and cultural boundaries make participation difficult.

6. RECREATIONAL FISHING SYSTEMS IN OTHER CONTEXTS

All recreational fisheries systems have their special characteristics reflecting their physical, institutional and social environments. Unique to the recreational fishery system in Southwest Finland seems to be the coastal water tenure institutions and their close connection to local inhabitants and traditions of coastal commercial fisheries. The large number of occasional recreational fishers with an easy access to fishing waters and the growing importance of more serious hobby fishers are characteristics shared by other areas where recreational fishing is growing in importance. On the other hand, fishing tourism does not play an important role in the area. In the context of recreational fishing in the Archipelago Sea area institutional and user community characteristics are more important and at least more acute than the physical and technological aspects.

When looking at the recreational fishing systems in other countries, the picture looks different. User communities are different. In the East Coast of The USA, fishing tourism business is an important stakeholder, perhaps the most important single group within recreational fisheries. Another important group is the active sport fishers. The occasional fishing tourists form the majority of fishers. This was clearly indicated in a social and cultural impact assessment of planned policy changes regarding some of the most important target species of recreational fishing in the area (Wilson *et al.*, 1998). In many US East Coast fishing communities charter boats and fishing tournaments, as well as business affiliated with fishing tourism, are the most important sources of income.

The resource itself, the fish, has an ambiguous role. For successful fishing tournaments it is important that the participants believe that there is a chance of catching a big fish of a certain species. Some of the occasional tourists, on the other hand, are satisfied if they can go fishing and catch at least something. There is competition over the resources between fishing tourism entrepreneurs and commercial fishers over certain billfish species - the most valuable target species in both charter boat fishing and in fishing tournaments. The management issues are related to open-sea fisheries of highly migratory species, which indicates that the problem and related decision making structures have larger scale than there is, for instance, in the Finnish coastal fishery case. In the US East coast impacts of commercial fishing on the species targeted by recreational fishing is a relevant question (Wilson *et al.*, 1998).

In the USA there are also tensions between local charter boat owners and non-resident individual hobby fishers. The hobby fishers are active sport fishers who have their own boat. These tensions are related to competition over resources and fishing space. Another issue between these groups is who actually represents sport fishers on national fisheries policy arena. Recreational fishers have strong organizations, but these represent only certain type of recreational fishers - the active 'hobbyists'. The occasional fishing tourists, which is the largest group of recreational fishers, do not participate in the organizations. Also the fishing tourism entrepreneurs have found it difficult or even annoying that the voice of recreational fishing has been captured by one group of fishers (Wilson *et al.*, 1998).

In Australia we have yet another kind of recreational fishing system. For instance in the western Australia, the fishing tourism is relatively well developed and growing rapidly when compared to the Finnish case. In this sense it resembles the recreational fishery in the USA. But it also has some similarities with the Finnish case. Popularity of recreational fishing is almost as high in Australia as it is in Finland. In addition, recreational fishing with nets is common in certain areas, although the present trend is to curtail this type of fishing. One of the long-time tensions in the Australian coastal fishing has been the disputes between commercial netters, recreational netters and anglers. In the coastal areas fisheries management decisions have often supported the interests of the anglers. The new development is that fishing tourism business has become an important player, which has to be taken into account in the design of the decision making (Borg, 1999; Smith *et al.*, 1999; Fisheries Western Australia, 2000).

In the Australian Fisheries Management Authorities' (AFMA) partnership approach, recreational fishers and charter boat owners have been represented since 1994. The AFMA's partnership approach is based on regional decision-making arrangement, rather than on local arrangements. In Australia collaborative decision making has been successful in some of the resource allocation problems between commercial coastal fishers and anglers

(Kearney, 2001; Smith *et al.*, 1999).

7. CO-MANAGEMENT OF RECREATIONAL FISHING

It has been emphasized that an improvement in fisheries management can only emerge through a real partnership between recreational fishermen, the fisheries administration and other interest groups (Viard and Brenner, 1998). A multiple use situation is typical in coastal waters. Therefore, co-management approach should probably include more than one sector of state administration. This is the case, for example, when recreational fishing is managed by environmental authorities, while commercial fisheries belong to another ministry. This creates additional challenges for collaborative decision making and conflict regulation.

The existence of a multiple-use resource use system and the conflict in the Finnish case would suggest that there is a need for co-management, but there are many characteristics in the resource use system that would undermine the success of a co-management regime. This is especially the case, if we insist that only a management regime that is based on real collaboration in the decision-making can be called 'co-management'. The most critical characteristics of the system are related to low degree of organization of the largest user group and to the potential lack of interest (or even practical difficulties) to participate in a real collaborative decision-making regime. The latter applies to all user groups, but for different reasons (see above). These same problems are expected to occur in other recreational fishing systems as well. Recreational fisheries consist of heterogeneous user groups, which is evident also in the USA and Australia.

What are then the possibilities for incorporating the groups of recreational fishing into the management mix? It is important to define the sub-groups and their potential roles. The type of involvement depends on the problems and the tasks related to the management system in question and the role of the specific user group. The role of the user groups in the co-management arrangement is connected to the importance of the group in relation to the management tasks, but also to the interests and opportunities of the group members to speak for themselves. For instance, the representation of well-organized recreational fishers' groups can be arranged by their organization representatives. Well-organized groups have a possibility to participate in decision-making bodies on different scales. Usually the recreational fishers are represented rather on the regional or national decision-making bodies than in very local arrangements.

But the problem remains with the fishers who are not organized. One can ask, of course, is it really a problem if the vast number of leisure fishers, who do not actually seem to be very interested in participating decision making, are not well represented in it? It may not be a problem for, for instance, the majority of recreational fishers in the USA, the occasional fishing tourists, as well as the non-local leisure fishers in Finland, provided that they have reasonable fishing opportunities. The interests of fishing tourists can be supported by the industry that provides them fishing opportunities. If the tourism business is well represented in the decision making their interests may be guarded relatively well. This seems to be the case in Australia, where recreational fishers and also fishing tourism business are represented in the decision making.

Another possibility for facilitating the co-management body with the perspectives and information of these groups is social science research. There is a relatively long tradition of so-called human dimensions research in the management of the recreational fishing in the United States. According to Aas and Ditton (1998) the knowledge gathered with human dimension surveys is central to formulating goals and objectives, problem identification and to understanding the implications of various management alternatives. This kind of 'knowledge-based management' supported by the surveys is remote to the core idea of co-management, but could be an addition to the system especially when participation of relevant groups cannot be arranged in any other way. This kind of approach, if adopted, does not allow any straight interaction between the user group and decision makers. In addition, its influence on the actual decision making cannot be guaranteed. The role of authorities, either as one of the core groups in a co-management arrangement or as the supervisor of national fisheries policies, becomes important. It can be expected that the authorities will consider interests of all stakeholders, including the less-organized ones, since the authorities' duty is to oversee the public interests.

It is natural that one is not particularly interested in fisheries management if one fishes only once or twice a year. However, the presumable lack of interest in participation may also be a result of uneven power relations. Some of the groups of recreational fishers may not be seen as important interest groups to be included in the management regime, eg due to lack of knowledge about the sector or a hegemonic position of some of the user groups in the system. This is especially problematic in recreational fishing, because of the heterogeneous user groups. In addition, relevant decisions in recreational fishing are made on several levels. Some of the large organizations would rather concentrate their power on national level politics than wasting resources on trying to influence local level decision making. Correspondingly, powerful local level groups may not be willing to compromise their local power by getting closely involved in regional or national level decision-making bodies.

However, the situation where the puruit of common interests is rare, may even be the normal starting point in multiple-use contexts. Co-management can be built on one general idea or target - 'an integrating concept' (see Hildén, 2000) - which many of the user groups perceive as serving their own specific needs while giving a target and reason for their common effort. Truly collaborative decision making, which would comprehensively deal with all aspects of a particular recreational fishing system and involve all the major user groups, may be difficult if not impossible to achieve. Local co-management efforts in recreational fishing can be based on 'integrating concepts' where the scope of the management task is a few issues and the decision-making body does not include all user groups, especially the non-local ones. National level politics should take into account the local interests and the interests of the groups that are not organized. Authorities should create possibilities of public participation in decision making regarding recreational fishing, both on local and national levels. The democratic decision making for recreational fishing would then be based on several arrangements, which would not pass the test of truly collaborative management. However, co-management should not be a target in itself. It is a means to democratic and equal management of natural resources and reaching common goals.

REFERENCES

Aas, Ø. & Bogelius, A. (1998) Summary report of the symposium topic session on the interactions between fisheries and other user groups. In: P. Hickley & H. Tompkins (eds) *Recreational fisheries. Social, economic and management aspects*. Fishing News Books, Blackwell Science, Oxford, p. 93-96.

Aas, Ø. & Ditton R.B. (1998) Human dimensions perspective on recreational fisheries management: implications for Europe. In: P. Hickley & H. Tompkins (eds) *Recreational fisheries. Social, economic and management aspects*. Fishing News Books, Blackwell Science, Oxford, p. 153-164.

Andersson K. and Eklund, E. (1999) Tradition and innovation in coastal Finland: The transformation of the Archipelago Sea region. *Sociologia Ruralis* **39**, pp. 377-393.

Borg, J. (1999) Estuaries in Western Australia - an integrated approach to management. Paper presented in FishRights '99 conference. Australia. 8 pages.

Borrini-Feyerabend, G., Farvar, M.T., Nguiguiri, J.C. and Ndangang, V.A. (2000) Co-management of Natural Resources: Organizing, Negotiating and Learning-by-doing. GTZ and IUCN, Kasparek Verlag, Heidelberg.

Cowx, I.G. (1998) Aquatic resource planning for resolution of fisheries management issues. In: P. Hickley & H. Tompkins (eds) *Recreational fisheries. Social, economic and management aspects*. Fishing News Books, Blackwell Science, Oxford, p. 97-105.

Edwards, V. & N. Steins (1999) A framework for analysing contextual factors in common pool resource research. *Journal of Environmental Policy & Planning* **1**, pp. 205-221.

Ellegård, A. (1998) Mussel Culture at Stake: Identifying the Holders. HERS-SUCOZOMA Report 1998: 4. Department of Interdisciplinary Studies of the Human Condition, Human Ecology Section, Göteborg University.

Fisheries Western Australia (2000) Management direction for Western Australia's recreational fisheries. Fisheries Management Paper No. 136. Fisheries Western Australia, Perth, Australia. 64 pages.

Hickley, P. & Tompkins, H. (eds) (1998) Recreational fisheries. Social, economic and management aspects. Fishing News Books, Blackwell Science, Oxford.

Hilden, M. (2000) The role of integrating concepts in watershed rehabilitation. *Ecosystem Health*, vol. **6**, pp. 39-50.

Jentoft, S. (2000) Legitimacy and disappointment in fisheries management. *Marine Policy* **24**(2), pp. 141-148.

Kauppila, P. and Bäck, S. (2001) The state of Finnish coastal waters in the 1990s. The Finnish Environment 472. Finnish Environment Institute, Helsinki. 134 pages.

Kearney, R. E. (2001) Fisheries property rights and recreational/commercial conflict: implications of policy developments in Australia and New Zealand. *Marine Policy* **25**, pp. 49-59.

MacMurran, J. (1999) Property rights and recreational fishing - never the twain shall meet? Paper presented in FishRights '99 conference, Australia. 6 pages.

McCay, B. and Jentoft, S. (1996) From the bottom up: participatory issues in fisheries management. *Society and Natural Resources* **9**(3), pp. 237-250.

National survey (1997) 1996 National Survey of Fishing, Hunting, and Wildlife-Associated Recreation. US Department of the Interior, Fish and Wildlife Service and US Department of Commerce, Bureau of the Census. 176 pages.

Noble, B.F. (2000) Institutional criteria for co-management. *Marine Policy* **24**, pp. 69-77.

Olburs, C. (2000) Om uthålligt fiske och vattenbruk i skärgården. Underlagsmaterial Nr. 21. Länstyrelsen i Stockholms Län. 163 pages.

Pinkerton, E. (1994) Summary and conclusions. In C. L. Dyer and J. R. MacGoodwin (eds), Folk management in the worlds fisheries. Univ. of Colorado Press, Boulders. Pp. 317-337.

Pirhonen, M. and Salmi, P. (1998) Viehekorttikiistan argumentit. Kamppailu kalavesien käyttö- ja omistusoikeudesta. Kala- ja riistaraportteja 129. Finnish Game and Fisheries Research Institute, Enonkoski. 25 pages.

Pomeroy, R.S., Katon, B.M. and Harkes, I. (2001) Conditions affecting the success of fisheries management: lesson from Asia. *Marine Policy* **25**, pp. 197-208.

Regier, H.A., Whillans, T.H., Christie, W.J. and Bocking, S.A. (1999) Over-fishing in the Great Lakes: the context and history of the controversy. *Aquatic Ecosystem Health and Management*, **2**, pp. 239-248.

Salmi, P. & Varjopuro, R. (2001) Private water ownership and fisheries governance in Finland. Proceedings of the IIFET 2000 Conference, July 10-13, 2000, CD-ROM.

Salmi, P. (2001) Private water owners and multiple-use conflicts in the Finnish Archipelago Sea. Paper presented in the conference 'People and the sea. Maritime research in the social sciences - an agenda for the 21st century', Amsterdam, The Netherlands 30.8.-1.9.2001.

Sen, S. and J. Raakjaer Nielsen (1996) Fisheries co-management: a comparative analysis. *Marine Policy* **20**(5), pp. 405-418.

Sipponen, M. (1998) Fisheries Region - a tool for co-operation in fisheries management? In: P. Hickley & H. Tompkins (eds) *Recreational fisheries. Social, economic and management aspects*. Fishing News Books, Blackwell Science, Oxford. pp. 263-278.

Smith, A., Sainsbury, K. and Stevens (1999) Implementing effective fisheries-management systems management strategy evaluation and the Australian partnership approach. *ICES journal of Marine Science*, vol. **56**, pp. 967-979.

Viard, D. and Brenner, T. (1998) Summary report of the symposium topic session on the socio-economic and legal aspects of recreational fisheries. In: P. Hickley & H. Tompkins (eds) *Recreational fisheries. Social, economic and management aspects*. Fishing News Books, Blackwell Science, Oxford, p. 150-152.

Wilson, D. and McCay, B. (1998) How the participants talk about 'participation' in Mid-Atlantic fisheries management. *Ocean and Coastal Management*, vol. **41**, pp. 41-61.

Wilson, D., McCay, B.J., Estler, D., Perez-Lugo, M., LaMarque, J., Seminski, S. and Tomczuk, A. (1998) A Social and Cultural Impact Assessment of the Highly Migratory Species Fisheries Management Plan and the Amendment to the Atlantic Billfish Fisheries Management Plan. United States Department of Commerce, National Oceanic and Atmospheric Administration, National Marine Fisheries Service, Highly Migratory Species Office. 180 pages.

Chapter 14

THE GOVERNMENT AS A PARTNER IN CO-MANAGEMENT

ROBERT S. POMEROY
University of Connecticut, Avery Point, Groton, USA

1. INTRODUCTION

During the last decade, the approaches for management and governance of fisheries resources have undergone a significant transition. There has been a shift from traditional production and stock- and species-based management toward conservation and ecosystem-based management. Rights-based management, community-based management and co-management are in some cases replacing open-access and centralized government management systems. It is increasingly recognized that resources can be better managed when fishers and other stakeholders are directly involved in management of the resources and use rights are allocated - either individually or collectively (Berkes *et al.*, 2001). Co-management systems have emerged as a partnership arrangement using the capacities and interests of the local fishers and community, complemented by the ability of the government to provide legislation, policy, enforcement and other functions and assistance. These new approaches will require the establishment of an appropriate government administrative structure, clarification of fishery management functions, and an enabling legal and policy environment in order to promote and sustain existing local-level fisheries management systems and/or to develop new co-management systems.

This chapter will discuss the role of government, primarily national government, as a partner in fisheries co-management. The establishment of conditions for co-management by government and the role of decentralization in a strategy for co-management will be discussed. The focus of the discussion is primarily on Asia and developing countries. A case study of devolution in the Philippines will be presented to illustrate the role of government in co-management.

2. THE ESTABLISHMENT OF CONDITIONS FOR CO-MANAGEMENT

Fisheries management is the responsibility of a formal central government institution such as a ministry, department or agency. Depending upon the country, this institution may be independent or subordinated within a ministry or department of agriculture or natural resources. The task of the fisheries institution is to exercise authority and responsibility for management of the resource on behalf of the nation and its people. This central fisheries institution develops fisheries management policy and the tools for fishery management for the nation.

Centralized fisheries management has been widely criticized as a primary reason for the overexploitation of fisheries resources, although in reality the fishers should share the blame with the managers and scientists (Berkes *et al.*, 2001). Professionals have replaced the resource users as resource managers. For the most part, the fishers have done little to monitor and enforce themselves. The centralized management approach involves little effective consultation with the resource users and often leads to an antagonistic relationship between user and manager.

Increasingly, government policies and programmes stress the need for greater resource user participation and the development of local organizations to handle some aspect of resource management. In the area of fisheries, this trend is international in scope and can be observed in a number of countries in the Americas, Europe and Asia (Jentoft and McCay, 1995; Pomeroy, 1995; Sen and Raakjaer-Nielsen, 1996). A review of the international experience shows that policies favouring co-management are a necessary but not sufficient condition for successful co-management. There are only two well documented cases of long-standing marine fishery co-management arrangements that work, in Norway (Jentoft, 1985; 1989) and in Japan (Ruddle, 1987; Lim *et al.*, 1995), and both of them have a legal basis. This suggests that it may be insufficient for governments simply to call for more community involvement and fisher participation; they must also establish commensurate legal rights and authorities and devolve some of their powers. The delegation of authority and power sharing to manage the fisheries may be one of the most difficult tasks in establishing co-management. Government must not only foster conditions for fisher participation but sustain it.

A recent analysis of conditions affecting the success of fisheries co-management in Asia (Pomeroy *et al.*, 2001) concluded that if co-management initiatives are to be successful, basic issues of national government action to establish supportive legislation, policies, rights and authority structures must be addressed. If supportive legislation and policies are in place, partners tend to have less difficulty in asserting their rights and roles, particularly if the judicial system is fair and objective. The legal basis for the resource user's participation in resource management is vital and must address fundamental concerns, which include: 1) who has the right to use the resource; 2) who owns the resource; 3) what is the legal framework for implementing co-management arrangements; and, 4) who is responsible for enforcement. The co-management arrangements may be undermined in the absence of a legal basis.

The study further found that the cooperation of the local government and the local political elite is important to co-management. There must be an incentive for the local politicians to support co-management. There must be political willingness to share the benefits, costs, responsibility and authority for co-management with the community members. Co-management will not flourish if the local political ‘power structure’ is opposed in any way to the co-management arrangement. In addition to the political elite, local government staff must endorse and actively participate in the co-management process. Local government can provide a variety of technical and financial services and assistance to support local co-management arrangements, such as police for enforcement, conflict management, appeal mechanism, and approval of local ordinances for resource management. It should be noted that while the local political elite are an important actor in co-management in Asia, this may not be the case in other regions of the world, such as Europe and North America. In these 'western' countries, co-management arrangements are often between a national fisheries agency and a fisher organization, either local or national. Local governments and local political elite's play no or only a minor role.

Thus, as a first step, the national government must establish conditions for (or at least not impede) co-management systems to originate and prosper. At a minimum, government must not challenge fishers rights to hold meetings to discuss problems and solutions and to develop organizations and institutional arrangements (rights and rules) for management. Fishers must feel safe to openly meet at their own initiative and discuss problems and solutions in public forums. They must not feel threatened if they criticize existing government policies and management methods. As a second step, fishers must be given access to government and government officials to express their concerns and ideas. Fishers should feel that government officials will listen to them and take action as necessary. As a third step, fishers should be given the right to develop their own organizations and to form networks and coalitions for cooperation and coordination. Too often there has been the formation of government-sponsored organizations which are officially recognized but ineffective since they do not represent the fishers or, but these may be the only type of organization a government may allow. Fishers must be free to develop organizations on their own initiative that meet their needs and that are legitimate to them. Again, it should be noted that this discussion is related more to Asia and a developing country situation than to Europe or North America. In Europe or North America a strong organization and structure for the fishing industry usually exists and has links with government. The issue is government's willingness to share responsibility with the fisher organization and what function and form this will take.

One fundamental debate in co-management is whether resource users can be entrusted to manage their resources (Berkes, 1989). Unless governments and decision-makers who implement government policies can be convinced of the desire and the ability of users to manage themselves, not much progress can be made in co-management. It is often pointed out that government resource managers are reluctant to share authority. However, it would be a mistake to interpret this solely as a self-serving motive to hang onto political power. Many managers have well-considered reasons to be sceptical about local-level management. To convince managers that local-level management is possible, part of the responsibility falls on the resource users themselves. The ability for self-management, in turn, partially depends on the willingness of fishers to take on the new responsibilities and on the ability of the local community to control the resources in question.

Mangers' reasons for scepticism include the lack of appropriate knowledge and know-how on the part of the fishers, and the ability of fishers to organize themselves to mange for long-term sustainability. Each of these points opens up its own debate. Even in countries with high standards of education, it is true that fishers tend to have lower levels of education than the general population. But the relevant knowledge held by fishers in many areas of the world may be extremely detailed and relevant for resource management (Johannes, 1981; Freeman *et al.*, 1991; Berkes *et al.*, 1995). Indeed, it is the complementarity between such local knowledge and scientific knowledge that makes co-management stronger than either community-based management or centralized management.

Experience from the Philippines, the country with the greatest number of fisheries community-based management and co-management projects in the world, has shown that fishers have difficulty in organizing themselves for collective action. In reviewing the Philippine projects it was found that fishers often recognize that a problem exists, they will discuss the situation among themselves, and they will discuss possible solutions to the problem, but very few groups of fishers will take action to either formally organize themselves or to develop institutional arrangements (rights and rules). The review found that in less than 20 percent of the cases did the fishers take action on their own initiative to organize and develop institutional arrangements (Carlos and Pomeroy, 1996). Leadership seems to be the limiting factor for fishers to take collective action. Either no individual is willing to step forward to lead, there is no one in the community with enough credibility among the fishers to lead, or divisions within the community or group of resource users will not allow for a leader to emerge. If enough initiative exists among the fishers they may approach a supportive politician or government official and ask for assistance or they may contact an external change agent, such as an NGO, academic or research institution, to assist in community organizing and development of institutional arrangements.

In addition, not all groups of fishers have appropriate local institutions; in such cases, any co-management initiative will necessarily start with institution-building. But institution-building is a long-term and costly process. Community organizing can take from three to five years before a self-sufficient organization is in place, on the basis of cases in the Philippines (Carlos and Pomeroy, 1996), and five to ten years on the basis of a case in St. Lucia, West Indies (Smith and Berkes, 1993). In the coastal fishery of Alanya on the Mediterranean coast of Turkey, locally designed rules for resource allocation and conflict reduction, by means of rotating and taking turns at fishing sites, developed over a period of ten to fifteen years in the absence of government support or any other intervention for institution-building (Berkes, 1986). The context in which co-management exists is very important for its development. When fisheries are in a severe crisis and when a well-organized fishing industry exists, it may take less time for co-management to develop. The case of the Dutch fishing industry is a good example. The fisheries were in a severe crisis and co-management regimes were developed and implemented within two years and are still in place now.

Such experiences indicate that there often is a readiness and willingness on behalf of some groups of resource users to take responsibility for management. Thus, a key question for co-management is what management functions are best handled at the local or communal level, as opposed to the national government level. Pinkerton (1989) identified seven resource management functions that may be enhanced by the joint action of users and government resource managers at the local level: (1) data gathering, (2) logistical decisions such as who can harvest and when, (3) allocation decisions, (4) protection of resource from environmental damage, (5) enforcement of regulations, (6) enhancement of long-term planning, and (7) more inclusive decision-making. No single formula exists to implement a co-management arrangement to cover these functions. The answer depends on country-specific and site-specific conditions, and is ultimately a political decision.

The benefits sought by all actors in co-management are more appropriate, more efficient, and more equitable management. These benefits become concrete when considered in association with the following processes and goals: (1) co-management for community-based economic and social development; (2) co-management to decentralize resource management decisions; and, (3) co-management as a mechanism for reducing conflict through a process of participatory democracy. Resource users have the benefit of participating in management decisions that affect their welfare; government has the benefit of reduced challenge to its authority (Pinkerton, 1989; Jentoft, 1989).

Fishers often develop their own rules for management in addition to those created by government. For example, fishers may establish rules defining who has access to a fishing ground and what fishing gear can be used. The fishers may be able to enforce the rules as long as there is at least a minimal recognition of the legitimacy of these rules by the government. This can be formal, as through a municipal ordinance, or informal, as through police patrols to back-stop the local enforcement arrangements. If government does not recognize the legitimacy of the rules, then it will be difficult for the fishers to maintain the rules in the long run (Ostrom, 1994). Thus, one role of government in establishing conditions for co-management is the creation of legitimacy and accountability for the local organization and institutional arrangements. The government, through legislative and policy instruments, defines power sharing and decision-making arrangements. Only government can legally establish and defend user rights and security of tenure. Government is ultimately accountable for all actions undertaken through co-management. One means of establishing these conditions is through decentralization.

3. DECENTRALIZATION AND CO-MANAGEMENT

Decentralization refers to the systematic and rational dispersal of power, authority and responsibility from the central government to lower or local level institutions - to states or provinces in the case of federal countries, for example, and then further down to regional and local governments, or even to community or fisher organizations. The approach of decentralization is for the centre to delegate some measure of its power to the lower levels or smaller units in the government system or to industry organizations. Increasing local autonomy is a focal point in the decentralization process. Generally, power and authority are transferred or withdrawn by laws enacted in the centre.

Decentralization may be operationalized in varying degrees and may take a number of forms, depending upon to what extent and to whom power and authority are transferred. Decentralization can take four forms (de Guzman, 1991; Helmsing *et al.*, 1991:

1. Deconcentration is the transfer of authority and responsibility from the national government departments and agencies to regional, district and field offices of national government offices.
2. Delegation is the passing of some authority and decision-making powers to local officials, but the central government retains the right to overturn local decisions and can, at any time, take these powers back.
3. Devolution is the transfer of power and responsibility for the performance of certain functions from the national government to local governments or non-governmental organizations, voluntary organizations, industry organizations, or community associations, without reference back to the central government.
4. Privatization is the transfer of responsibility for certain governmental functions to non-governmental organizations, voluntary organizations, industry organizations, community associations, and private enterprises.

Thus, one form of decentralization is devolved management where there is a sharing of power and responsibility between government and fishers or fisher organizations. In Europe, for example, devolved management usually involves only the national government and fishers or fisher organizations, and lower levels of government are not involved. In Asia, however, devolved management usually involves national government, local government and fishers or fisher organizations.

Co-management requires a clear commitment on the part of government to the sharing of power and authority with local government and local fisher and community organizations. In many countries, government programmes and projects stress the development of local organizations and autonomy to handle some aspect of fisheries management. Seldom, however, is adequate attention given to the establishment of a administrative and policy structures that defines the legal status, rights and authorities essential for the effective performance of local organizations. Many attempts at decentralization have not delivered a real sharing of resource management power.

Initiatives in community-based resource management in Asia, for example, have been popular throughout most of this century under different names. However, as Korten (1986) explains it, 'none of these approaches to stimulating local initiatives provided a fundamental challenge to the idea that the government does development for the people, who are expected to respond with grateful acceptance of whatever guidance and assistance government chooses to offer. None challenged the nature of the government's role or the appropriateness of the structures and procedures through which government conducts its business. None confronted basic issues of local social structures and resource control.'

If new fisheries co-management initiatives are to be successful, these basic issues of government policy to establish supportive legislation, rights and authority structures must be recognized. The devolution of fishery management authority from the central government to local level governments and organizations is an issue that is not easily resolved. Legislation and policy for co-management are embedded in a broader network of laws, policies and administrative procedures, at both national and local government levels. Consequently they will be difficult to change. Government administrative and institutional structures, and fisheries laws and policies will, in most cases, require restructuring to support these initiatives.

In some cases, it may be more feasible and desirable to draw up completely new legislation, rather than to modify existing acts. A case in point is the establishment of Mafia Island Marine Park in Tanzania. The idea of a marine protected area developed in the course of an environmental assessment process regarding petroleum exploration. Local fishers were involved in the assessment process to provide information on resources. Initially, a reserve was set up to protect fishery resources; by 1991 it had evolved into a locally managed marine park. But it became obvious soon that the needs of the local people could not be met under the existing fisheries legislation. New legislation was developed with support from international conservation organizations and the FAO, and the *Tanzania Marine Parks and Reserves Act* came into being in 1994. This act provided for the formal inclusion of village council representatives on the technical committee for the Mafia Island Marine Park for co-management and for the sharing of benefits (M. Ngoile, IUCN, personal communication, July 1996; Ngoile *et al.*, 1995).

As the Tanzania case illustrates, the actual form of co-management will depend upon the type of government and the political will for decentralization. In general, co-management is consistent with the aims of democratization and empowerment. In the first place, the goals of co-management include the greater participation of fishers in the fisheries management process, more self-reliance of local level institutions, more accountability for the actions of people, and a more responsive decision-making process. The ultimate goal of decentralization is greater participation and efficiency by getting people at lower levels more involved in the decision-making processes and procedures that affect them. One assumption of decentralization is that the deployment of power and resources to the community will enhance community and economic development. Thus, the promise of decentralization is greater democratization and development of local communities. In this assumption, an important concern is the significance of intervening variables such as leadership, skills of fishers, resources, and capabilities of local institutions.

In detailing the specifics of the decentralization strategy, questions of implementation become crucial points of debate. What powers and functions, for instance, can be properly entrusted to local institutions? What are those that should be left to the central government? How is the sharing of resources to be administered? What should be the role of non-government organizations and people's organizations? What is the proper and appropriate mix of government and private sector participation? Will decentralization occur only for the fisheries bureaucracy, or will it be a government-wide initiative? This collection of issues impinges on decentralization strategies and drives the political debate associated with decentralization.

4. A CASE STUDY OF DEVOLUTION IN THE PHILIPPINES

The island settlers of what would become the Philippines had a long history of traditional fisheries rights and allocations before the archipelago was first colonized by Spain in the 17th century. The *barangay* (village) had jurisdiction over coastal resources and fishery limits were defined by them. The traditional property rights of barangays over fishing grounds were steadily eroded during the long Spanish colonial period, with community authority and rights superseded by state government control (Kalagayan, 1991). Lopez (1983) reports that under Spanish rule, the barangays were eliminated as administrative entities and with them went the territorial fishing rights claimed by each village. Under Spanish law, the fisheries and other natural resources were declared to be held by the Crown. Under both the Spanish and the Americans, traditional authority and rights were superseded by municipal government control of local fishing grounds. This administrative structure of municipal authority remains in place in the country today. Despite the historical existence of traditional fishing rights and village-based management systems in the Philippines, for the most part these systems have disappeared in the country. This is not to say that traditional community-based resource management systems, and informal fisheries rights and rules systems do not exist, for localized examples can be found throughout the country (Ferrer, 1989; Mangahas, 1994).

In the 1960s, the Philippine government, aided by Japanese advisors, undertook intensive infrastructure, technology, extension and credit programmes through the Fisheries Development Programme to 'develop' the industry (Heinan and Gonzales, 1993). In the early 1970s, the country fell under Martial Law and the centralized government control of fisheries was further reinforced through Presidential Decree (PD) 704, otherwise known as the Fisheries Act of 1975. Under PD 704, fisheries management is the responsibility of the government, both national and municipal. The management measures (mainly through regulatory instruments) undertaken by the government during this time, however, have been ineffective in promoting the sustainable development and management of the country's fisheries. In the mid-1970s, in response to decreasing unit catch of small-scale fishers, the government embarked on fishery policies and development programmes concentrated on 'use orientation', that is, increasing production and exploitation of the resource base.

In the 1980s, the government continued to support the needs of the sector through the Expanded Fish Production Programme (EFPP) from 1983-1987. In the small-scale fisheries sector, the strategy of the programme was geared towards enabling the small fishers to venture into deeper waters by equipping them with more efficient boats and fishing gears. The underlying assumption was that the fishery could support increased fishing effort, despite expert opinion as early as 1980 that it could not. Ironically, it was during this period

(1984-1988), that there was a decreased rate in coastal fish production of 1.3 percent a year, compared to the increasing rate of 6.1 percent in the preceding five years from 1979-1983 (Agbayani, 1993).

The problems in the fishery continued to worsen throughout the late 1980s and early 1990s. The management (mainly through regulatory instruments) and development (increased fishing effort) measures undertaken by the government have proven to be ineffective in promoting the sustainable management and development of the country's fisheries.

It was realized that with the increasing rate of deterioration of natural resource systems in the Philippines, there was no way the country could pursue a pathway of sustainable development. Starting in the 1960s, alternative methods of resource use and management were explored in an attempt to reverse these negative trends. Consequently, there has been a shift to forward-looking policies and strategies that advocate 'resource management' over a 'use orientation' through community-based initiatives to rehabilitate, conserve and protect the resources based on use and enhancement of local knowledge, skills, responsibility and accountability (Sajise, 1995). The irrigation sector was the first to evolve an institutional development scheme for mobilizing the active participation of water users in 1968. People-oriented programmes in the forestry sector started in the early 1970s (Serna, 1993). Community-based coastal resource management (CBCRM) started in the early 1980s. To date, well over 200 CBCRM projects have been implemented by government, NGOs, fishing communities, and academic and research institutions. No country in the world has the range of experience with CBCRM and co-management as exists in the Philippines (Carlos and Pomeroy, 1996).

The current efforts in community-based coastal resource management and co-management in the Philippines emanate from the government, NGOs and international development agencies. In 1989, President Aquino created a Presidential Commission on Anti-illegal Fishing and Marine Conservation or the Bantay Dagat Committee, which called for increased coordination among government agencies in enforcement of fisheries laws and increased participation of fishers in management (Kalagayan, 1991). In 1991, the government recognized the need to increase participation in management and to devolve control over resource access to local levels through policy and institutional reforms. Through several initiatives, the government now actively promotes devolution and community-based resource management and co-management efforts to conserve the coastal resources and diversify the income sources of the low-income small-scale fishers. These initiatives for CBCRM are embodied in the 1993-1998 Medium-Term Philippine Development Plan (MTPDP). Among its strategies are to: implement a community-based fishery management strategy; regulate fishing effort within maximum sustainable yields; promote territorial use rights for small fishers; intensify aquaculture, optimal utilization of offshore, deep sea resources; and provide diversified occupational opportunities among marginal fishers. The core programme for fisheries implemented under the plan was the Fisheries Sector Programme (FSP) from 1990 to 1995. Among the policy and institutional reforms instituted through the FSP were: (1) decentralization of authority and simplification of procedures for clearance of local fisheries management ordinances subject to national laws and/or policies; (2) strengthen the enforcement of fisheries laws through municipal-based inter-agency law enforcement teams; (3) promotion community-based initiatives to rehabilitate, conserve and protect the coastal resources and to diversify the sources of income of small-scale fishers; (4) NGOs will be engaged to assist and undertake community organizing; and (5) shift to limited access in concerned fishing areas. At the core of the resource and rehabilitation thrust of the FSP is coastal resource management. Fishers, local government units and other concerned agencies in the area were given the opportunity to determine the specific problems in their areas and to identify the management strategies to counteract these problems. These activities were continued through a follow-up project called the Fisheries Resources Management Project.

In 1991, the Philippine government enacted into law the Local Government Code (LGC) which devolved a large number of key government functions and operations to local government units (LGU) (province, municipal, barangay). The LGC marked a shift in public administration from a centrally driven system of 'top-down' management to a 'bottom-up' strategy of expanded participation and responsibility of the LGUs.

Among the functions devolved to the LGUs were resource management and environmental protection. National agencies were now required to consult with local leaders in the design of development programmes. A general operative principle is a provision that the LGUs may group themselves, consolidate or coordinate their efforts, services and resources for purposes commonly beneficial to them. The LGUs were given broad powers to generate funds through local taxes or shares in revenue from the exploitation of resources that used to be at the disposal of the national government.

The LGC granted local governments (municipalities) with a number of powers including the management of municipal or nearshore waters. Under the LGC, municipal waters were defined as all waters within 15 kilometres of the coastline. Section 35 of the LGC specifically states that LGUs may enter into joint ventures and such other cooperative arrangements with people's organizations and non-governmental organizations to engage in the delivery of certain basic services, capability building and livelihood projects, and to develop local enterprises designed to diversify fisheries, among others. The LGUs and local communities are also given certain privileges and/or preferential rights. Municipalities have the exclusive authority to grant fishery privileges in municipal waters and impose rentals, fees and charges. In terms of fishery rights, the organizations or cooperatives of marginal fishers have preferential rights to fishery privileges within the municipal waters such as the erection of fish corrals and gathering fish fry free of any rental, fee or charge (de Sagan, 1992, Tabunda and Galang, 1992).

In 1998, Republic Act No. 8550 or the Philippine Fisheries Code was signed into law. Part of the Code consolidates existing laws and guidelines previously scattered among presidential decrees, administrative orders, and local ordinances into a single consistent law regulating fishing and the protection of the aquatic environment.

Under the Fisheries Code, several sections of the LGC were clarified and supported. The Code clarified the designation of municipal waters up to 15 km for shore and the granting of preferential rights to fishing privileges in municipal waters to registered fisher organizations and cooperatives. This definition should help to resolve longstanding conflicts between small-scale and commercial fishing. In addition, the Fisheries Code called for the establishment of Fisheries and Aquatic Resources Management Councils (FARMC) at national and municipal levels. The FARMCs are formed among fisher organizations and cooperatives and NGOs with assistance from the LGU and government agencies. The FARMCs are mandated to carry out a number of management advisory functions in close collaboration with the LGU. These functions include assisting in the preparation of Municipal Fishery Development Plans, recommending the enactments of fishing ordinances, assisting in enforcement, and advising the LGU on fishery matters.

As part of the Philippines compliance with the Convention on Biological Diversity, the Congress enacted Republic Act No. 7586, An Act Providing for the Establishment and

Management of National Integrated Protected Areas System or the NIPAS Act. The Act sets aside both terrestrial and aquatic protected areas. Among other provisions of the Act is the recognition of ancestral rights to protected areas. Section 13 of the Act requires that members of concerned indigenous communities shall be consulted prior to the adoption of any regulations adopted by the Department of Environment and Natural Resources for the area, and other parts of the Act require their participation in all aspects of its management.

An example can best illustrate the importance of these decentralization activities to co-management in the Philippines. Prior to 1991 and the LGC, an ordinance to provide legally standing for a community-level marine protected area required approval by the secretary of the Department of Agriculture. A long and often unsuccessful procedure. With passage of the LGC, local municipalities had the legal right to approve an ordinance in support of a community-level marine protected area. Local government officials and fisher organizations now had the legal and administrative mandate to work cooperatively on coastal resource management. The establishment of FARMCs under the 1998 Fisheries Code, also strengthened co-management, as fishers now had authority to work cooperatively with the LGU on fisheries planning, management and enforcement.

It has not been easy for the LGUs to live up to the provisions of all this new legislation in the Philippines. While many mayors have welcomed these new rights, they also realized that they must rely less on the national government for support. Many local governments were not prepared or were unaware of their new roles, limited resources were made available to them from the national government for the transition, and it has taken them awhile to adjust to the new authority that they now have. Many have not done anything, while some have actively engaged in supporting fisher organizations and local management measures. The new administrative structure has bred success in locations that could not have been reached by national programmes. Experience has shown that the mere promulgation of legislation and polices to control resource use practices cannot in itself lead to sustainable management of fisheries resources. These efforts must be combined with capacity building and education for all stakeholders. Overall, however, those involved in fisheries conservation and management in the Philippines feel that devolution has been a positive step towards sustainable management of fisheries resources in the country (Tagarino, 1995; Fellizar *et al.*, 1997; Courtney, White and Anglo, 2000).

5. THE GOVERNMENT AND CO-MANAGEMENT

International experience suggests that fisheries co-management does not come about automatically but requires some impetus. Most commonly, it is the recognition of a resource management problem that triggers co-management. Problem recognition may be related to resource deterioration (as in the case of the Philippines and the Tanzanian marine protected area), conflicts between stakeholders (eg Norway's Lofoten cod fishery and Philippines coastal fisheries), conflicts between management agencies and local fishers (eg Canada's Atlantic coast fishery), and governance problems in general (eg Philippines, the United States *Fishery Conservation and Management Act*, and northern aboriginal land claims in Canada). In this regard, the experience with fisheries co-management is similar to the international experience with the co-management of protected areas (McNeely, 1995; Borrini-Feyerabend, 1996), forests (Lynch and Talbott, 1995), wildlife (Martin, 1986), and other resources.

In each case, governments have turned to co-management as a means of responding to a management crisis, and sometimes to a management opportunity, as in the case of resource rehabilitation projects and perhaps also in some land claims agreements. Various types of decentralization can be used by governments to establish conditions conducive to co-management. Decentralization and co-management often go together, and there are a number of similarities between their goals. The strategies of decentralization and co-management not only respond to management crises, they also offer the promise of increased democratization, and empowerment and development of regional and local communities. The goals of both co-management and decentralization are the mobilization and strengthening of people's participation in government and more equitable distribution of power and resources to local-level groups of people and communities (de Guzman, 1991).

The form and process of decentralization and co-management can be seen as a focus for user participation in management. Decentralization in a governmental context may proceed in the logical sequence of: (1) organizational and physical deconcentration; (2) administrative delegation; (3) political devolution; and, (4) popular privatization (Gasper, 1991). These modes of decentralization may occur separately or in a cumulative package. Thus, decentralization can be seen as a continuum ranging from deconcentration to privatization where more power and authority is delegated to local-level institutions as one moves along the continuum. Co-management can also be viewed as a continuum, similar to that of decentralization, based on the role(s) played by government and resource users (Berkes, 1994; Pomeroy and Williams, 1994; Sen and Raakjaer-Nielsen, 1996). In both decentralization and co-management, the central government acts to delegate power and authority to local-level institutions.

The form of decentralization will depend, like the form of co-management, on country-specific conditions. As can be seen from the examples above, there is no one 'best' form of decentralization to support co-management, as there is no one 'best' form of co-management. Decentralization can occur as a broad administrative mandate of which fisheries is included, as in the case of the Philippines, or it may occur for specific management functions, as is the case in Japan and Tanzania. Both co-management and decentralization should be viewed as an evolving process, which adjusts and matures over time. For example, the process of decentralization may proceed over time from deconcentration to devolution as more knowledge and experience is gained by the government, and as the political will for decentralization increases; this was the case in the Philippines (de Guzman, 1991). In a similar fashion, co-management systems may evolve through experience through the delegation of more and more power and authority by the government, as in the case of the Lofoten fishery in Norway (Jentoft, 1985; 1989).

The decentralization process, however, is laced with potential roadblocks and pitfalls. Politicians may be reluctant to allow greater democratization of the political system. Politicians and government agency administrators may be reluctant to relinquish their authority or portions of it in order to protect their power and positions of their own agencies. The local power and authority may fall into the hands of leaders and groups who are not committed to its basic values and goals. An important concern for the success of both decentralization and co-management are variables such as leadership, skills, resources, and capabilities of local-level organizations and institutions. In addition, fishers and/or fisher organizations must be willing to take on and to act upon their responsibility in co-management.

The process of developing a co-management system will likely involve the restructuring of national laws and policies, as well as national fisheries agencies and bureaucracies. Existing national laws and policies usually do not include specific reference to such functions of co-management as the security of local-level tenure and property rights over coastal resources, people's participation, and the recognition and incorporation of local traditional/informal/folk management systems. New laws and policies may need to be developed and/or existing laws and policies amended or reinterpreted to authorize and legitimize these functions of co-management. Both the Philippines and Thailand, for example, are undertaking such a process (Pomeroy, 1995). New laws and policies may need to be reviewed to identify compatability and inconsistency with laws and policies for resource management in other sectors and with overall administrative laws and policies. National fisheries agencies and bureaucracies may require restructuring to take on the new responsibilities and functions required of them under co-management and decentralization. Issues of coordination, communication and roles must be addressed. The government agencies must be shielded from short-term political pressures to change or dilute goals of the power-sharing arrangements under co-management.

The role of the government in co-management is to provide enabling legislation to authorize and legitimize the right to organize and to make and enforce institutional arrangements at the local level. In the case of protected marine area co-management in Tanzania, it is important to note that a series of enabling legislation was passed in the 1970s and the 1980s in support of decentralization. Although it is generally thought that the Tanzanian experiment in self-reliance and local democracy did not live up to its potential (Chambers, 1985), this legislation nevertheless enabled districts and villages to manage their own affairs, and served as the basis of new legislation for marine parks and co-management.

In addition to its role in providing enabling legislation, the government may act to address problems and issues beyond the scope of local arrangements, and to provide assistance and services (administrative, technical and financial) to support the sustainability of the local organizations and institutional arrangements. More specifically, the role of government includes ensuring accountability of co-management through overseeing local arrangements and dealing with abuses of local authority, conflict management, appeal mechanism, backstopping local monitoring and enforcement mechanisms, and applying regulatory standards. Government may also serve a coordinating role to maintain a forum or formal administrative structure for various parties in the co-management system to interact. Within a co-management system, government and fishers jointly develop an agreement on the objectives of co-management including the aims, the form, and the means. A clear understanding of the long-term goals of power-sharing is established in which the differing interests and needs of government and fishers are reconciled. Government also has a role to play as gatekeeper in case the co-management partners (fishers and government) do not act upon their responsibility.

The decision on what fisheries management functions should be handled at what level are best handled jointly by local-level organizations and national government fisheries agencies, and they will be location specific. The decision will be based on the capabilities of local-level organizations to handle certain management functions and the locus of user participation. It may be more appropriate to phase-in management functions over time as local-level organizations gain more experience and capability, rather than give them a defined set of functions. The phasing in will also depend on the form of both co-management and decentralization, but adaptive management or 'learning-by-doing' in the evolution of co-management, and feedback learning in general, is likely to be critically important (Lee, 1993).

Ultimately, whatever form of decentralization arrangement for co-management is arrived at, the process is political, involving mobilized interests and struggle for power. The government, however, holds the final authority. Co-management will not work everywhere in a country. Co-management should be viewed by government as an alternative management strategy to the centralized management system which in many cases does not work effectively anyway. Governments may not want to develop laws and policies which completely shift away from the centralized form of fisheries management, as many areas and resources within a country may still require this type of management strategy. Governments may want to consider developing a general policy within the existing legal and policy framework of the country, which allows for the existence of co-management in areas and communities, which are capable of taking on the responsibility and authority for management. Laws and policies may then be developed under a framework of decentralization, which legitimizes and authorizes co-management. There is no blueprint formula for either co-management or decentralization. Each country will need to develop a strategy based on their own needs and conditions. Several decades of international experience provides some directions for developing a co-management strategy based on decentralization.

ACKNOWLEDGEMENTS

This paper is based in part on R.S. Pomeroy and F. Berkes, 1997. Two to Tango: The Role of Government in Fisheries Co-management. Marine Policy. Vol. 21, No. 5: 465-480.

REFERENCES

Agbayani, R.F. (1993) An integrated approach to community-based fishery resource management: SEAFDEC/AQD experience in multi-disciplinary research. Out of the Shell: Coastal Resource Network newsletter. 3 June.

Berkes, F. (1986) Local-level management and the commons problem: A comparative study of Turkish coastal fisheries. *Marine Policy* **10**, 215-229.

Berkes, F. (1989) Co-management and the James Bay Agreement. In: Co-operative Management of Local Fisheries (E. Pinkerton, ed.) University of British Columbia Press, Vancouver, pp. 189-208.

Berkes, F. (1994) Co-management: Bridging the Two Solitudes. Northern Perspectives. **22**(2-3), 18-20.

Berkes, F., Folke, C. and Gadgil, M. (1995) Traditional ecological knowledge, biodiversity, resilience and sustainability. In: Perrings, C., Maler, K.G., Folke, C.S. and Jansson, B.-O. eds. Biodiversity Conservation: Policy Issues and Options. Kluwer Academic Publishers, Dordrecht.

Berkes, F., Mahon, R., McConney, P., Pollnac R. and Pomeroy, R. (2001) Managing small-scale fisheries: alternative directions and methods. International Development Research Centre, Ottawa, Ontario, Canada.

Borrini-Feyerabend, G. (1966) Collaborative management of protected areas: Tailoring the approach to the context. International Conservation Union (IUCN), Gland, Switzerland.

Carlos, M.B. and Pomeroy, R.S. (1996) A review and evaluation of community-based coastal resource management programmes and projects in the Philippines, 1989-1994. Research Report No. 6, Fisheries Co-management Project. ICLARM, Manila, Philippines.

Chambers, R. (1985) Managing Rural Development: Ideas and Experience from East Africa. Longman, London.

Courtney, C.A., White, A.T. and Anglo, E. (2000) Coastal Resource Management in the Philippines: Lessons and Directions for Sustainability. Tetra Tech EM Inc., Cebu, Philippines.

de Guzman, R.P. (1991) Decentralization for democracy and development. *Local Government Bulletin*. University of the Philippines. College of Public Administration, Diliman, Quezon City, Philippines.

de Sagan, R. (1992) The Local government code and its provisions on fisheries. Bureau of Fisheries and Aquatic Resources, Quezon City, Philippines.

Fellizar, F.P., Bernardo, R.G. and Stuart, A.P.H. (1997) Analysis of policies and policy instruments relevant to the management of fisheries/aquatic resources with emphasis on local level issues and concerns. SEAMEO Regional Center for Graduate Study and Research in Agriculture, Los Banos, College, Laguna, Philippines.

Ferrer, E. (1989) Prospects for territorial use rights in the Lingayen Gulf area. In: G. Silvestre, E. Miclat and

T.E.Chua (eds) Towards sustainable development of the coastal resources of Lingayen Gulf, Philippines. ICLARM Conf. Proc. 17. ICLARM, Manila, Philippines.

Freeman, M.M.R., Matsuda, Y. and Ruddle, K. (eds) (1991) Adaptive Marine Resource Management Systems in the Pacific. *Resource Management and Optimization*, special issue, vol. **8**, no.3-4.

Gasper, D.R. (1991) Decentralization of planning and administration in Zimbabwe: International perspectives and 1980s experiences. In A.H.J. Helmsing *et al.* Limits to Decentralization in Zimbabwe. ISS, The Hague.

Heinan, A. and Gonzales, N. (1993) Fisheries management and development in the Philippines: constraints and possibilities. Unpublished paper. Community Extension and Research Development, Quezon City, Philippines.

Helmsing, A.H.J. *et al.* (eds) (1991) Limits to decentralization in Zimbabwe. ISS. The Hague.

Jentoft, S. (1985) Models of fishery development. The cooperative approach. *Marine Policy* **9**, 322-331.

Jentoft, S. (1989) Fisheries co-management. *Marine Policy* **13**, 137-154.

Jentoft, S. and McCay, B.J. (1995) User participation in fisheries management. Lessons drawn from international experiences. *Marine Policy* **19**, 227-246.

Johannes, R.E. (1981) Words of the Lagoon. Fishing and Marine Lore in the Paulau District of Micronesia. University of California Press, Berkeley.

Kalagayan, N.V. (1991) Philippine initiatives on marine coastal conservation. Paper presented during the International Conference on Exploitation of Resources in the Pacific. Port Moresby, Papua New Guinea. Lundayan. 2(1).

Korten, D.C., editor (1986) Community Management: Asian Experience and Perspectives. Kumarian Press. West Hardford, Connecticut, USA.

Lee, K. (1993) The Compass and the Gyroscope. Island Press. Washington D.C.

Lim, C.P., Y, Matsuda and Y. Shigemi (1995) Co-management in marine fisheries: the Japanese experience. *Coastal Management* Vol. **23**, No. 3, pp. 195-222.

Lopez, M.D. (1985) Notes on traditional fisheries in the Philippines. In: K. Ruddle and R.E. Johannes (eds) The Traditional Knowledge and Management of the Coastal Systems in Asia and the Pacific. UNESCO-ROSTEA, Jakarta, Indonesia.

Lynch, O.J. and Talbott, K. (1995) Balancing Acts: Community-Based Forest Management and National Law in Asia and the Pacific. World Resources Institute, Washington, D.C.

Mangahas, M. (1994) Traditional marine tenure and management in ASEAN. In: G.R. South *et al.* (eds). Traditional Marine Tenure and Sustainable Management of Marine Resources in Asia and the Pacific: Proceedings of the International Workshop Held at the University of the South Pacific, Suva, Fiji.

Martin, R.B. (1986) Communal areas management programme for indigenous resources (CAMPFIRE). Government of Zimbabwe, Department of National Parks and Wildlife Management, Branch of Terrestrial Ecology.

Martin, R.B. (1994) The Influence of Governance on Conservation and Wildlife Utilization. Paper presented at a conference - Conservation through Sustainable Use of Wildlife. University of Queensland, Brisbane. 8-11 February 1994.

McNeely, J.A., editor (1995) Expanding Partnerships in Conservation. Island Press, Washington, D.C.

Ngoile, M.N., IUCN, Gland, Switzerland, personal communication, July (1996).

Ngoile, M.N., Linden, O. and Coughanowr, C.A. (1995) Coastal zone management in Eastern Africa including the island states: A review of issues and initiatives. Ambio **24**, 448-457.

Ostrom, E. (1994) Neither market nor state: Governance of common-pool resources in the twenty-first century. International Food Policy Research Institute, Washington, D.C.

Pinkerton, E. editor (1989) Co-operative Management of Local Fisheries: New Directions for Improved Management and Community Development. University of British Columbia Press, Vancouver.

Pomeroy, R.S., editor (1994) Community management and common property of coastal fisheries in Asia and the Pacific: concepts, methods and experiences. ICLARM Conf. Proc. 45, ICLARM, Manila, Philippines.

Pomeroy, R.S. (1995) Community-based and co-management institutions for sustainable coastal fisheries management in Southeast Asia. *Ocean and Coastal Management* Vol. **27**, No. 3, pp. 143-162.

Pomeroy, R.S., Katon, B.M. and Harkes, I. (2001) Conditions affecting the success of fisheries co-management in Asia. *Marine Policy*. Vol. **25**, No.3, 197-208.

Pomeroy, R.S. and Williams, M.J. (1994) Fisheries co-management and small-scale fisheries: a policy brief. ICLARM, Manila, Philippines.

Ruddle, K. (1987) Administration and conflict management in Japanese coastal fisheries. FAO Fisheries Technical Paper No. 273. FAO, Rome.

Sajise, P.G. (1995) Community-based resource management in the Philippines: perspectives and experiences. Paper presented during the training course in Co-management of Living Coastal Resources in ASEAN: Theory, Practice and Implications for Vietnam. Ministry of Fisheries, Hanoi, Vietnam.

Sen, S. and Raakjaer-Nielsen, J. (1996) Fisheries Co-management: a comparative analysis. *Marine Policy* Vol.

20, No. 5, pp. 405-416.

Serna, C.B. (1993) Community-based resources management: perspectives, experiences and policy issues. In: F. Fellizar (ed.) Community-based resource management: Perspectives, Experiences and Policy issues. University of the Philippines, College, Laguna, Philippines.

Smith, A.H. and Berkes, F. (1993) Community-based use of mangrove resources in St. Lucia. *International Journal of Environmental Studies* **43**, 123-132.

Tabunda, M.S. and Galang, M.M. (1992) A guide to the local government code of 1991. Mary Jo Educational Supply, Manila, Philippines.

Tagarino, R.N. (1995) Local government initiatives in fishery resources management: its implications on sustainability. Center for Policy and Development Studies, University of the Philippines, Los Banos, College, Laguna, Philippines.

Section Four

EDGE ISSUES IN FISHERIES CO-MANAGEMENT

Chapter 15

FISHERIES CO-MANAGEMENT AND THE KNOWLEDGE BASE FOR MANAGEMENT DECISIONS

DOUGLAS CLYDE WILSON
Institute for Fisheries Management and Coastal Community Development
Hirtshals, Denmark

1. INTRODUCTION

Two objectives of community-based fisheries co-management are to make management more equitable and more rational, and both of these objectives are tightly bound up with the scientific knowledge on which management is based. If the knowledge needed for management, for example about the condition of a fish stock, is more accurate and easier to get, this helps management to be more rational. If the knowledge needed for management is contributed to, shared and controlled by more stakeholders, this helps management to be more equitable. At first glance these two objectives seem to at least partially contradict one another. We cannot vote for more fish in the sea, we know this, we have tried. The reality of nature is not a function of the equity of our decision making processes, nor is knowledge of nature is equitably distributed among stakeholder groups. Meeting the knowledge needs of community-based fisheries co-management is a very complex process as it is so closely linked to these two objectives.

This tension between the need for good science and the need for participation in decision making is the subject of this chapter. It examines several related issues. The first is social construction of fisheries knowledge, ie, how working pictures of nature are assembled that can be used as the basis of fisheries management decisions. The second issue is the question of tacit versus discursive knowledge about fisheries and the institutional roles and implications of these kinds of knowledge. Then two general kinds of sources of fisheries knowledge are examined. The first is research-based knowledge that is produced by the institutions of formal science. The second is the local ecological knowledge produced by fishery workers as they go about their activities. The conclusion points to collaborative approaches to fisheries knowledge that are showing the way forward.

2. THE SOCIAL CONSTRUCTION OF FISHERIES KNOWLEDGE

Fisheries co-management is a social institution that is operates in many respects like other institutions. Knowledge plays a role in all institutions because, if an institution is to have concrete expression in people's behaviour, those people must know how to behave in accordance with an institution. In the case of fisheries co-management, this involves knowledge about nature, ie about fish, their numbers, their behaviour and their habitat. Knowledge of both society and nature must be 'socially constructed' if it is to be available to coordinate behaviour through institutions. People who study how institutions deal with[1] knowledge about nature very commonly use the phrase 'social construction of nature' to convey the important insight that institutions do not understand and respond to natural reality directly. Institutions respond to a picture of nature that someone working within institutional constraints has constructed.

Holms (2000) contrasts social construction with realism. Realism is 'based on the assumption that there is a real world out there and that it is - in principle and in practice - possible to obtain true knowledge of the world. . . . [while]. . . . social constructivism holds that nature, the real world, while it may be out there, can never be made to intervene in human affairs' (2000 pp 2). He suggests that we can look at fisheries knowledge four ways: assume that both the knowledge produced by fisheries science (ScEK) is realist and that the local ecological knowledge of fishers (LEK) is constructed; assume that ScEK is constructed and LEK is realist; assume that both ScEK and LEK are constructed; or assume that both are realist.

Understanding the relationship between knowledge and fisheries co-management requires taking Holms' (2000) third approach and assuming that both kinds of knowledge are constructed. We can take such an even handed approach, however, while still rejecting the very strong dichotomy between social construction and realism implicit in Holms' definitions. While many social scientists agree that this strong distinction is useful (Barnes *et al.*, 1996; Benton, 1994; Freudenberg *et al.*, 1995; Latour, 1987; Roepstorff, 1998) a social constructivist perspective can remain coherent, while recognizing that it is, in principle, possible to establish a fact about nature as true.

Setting lies aside, there are two basic sets of mechanisms by which the social construction of nature happens. The first is in the selection of facts that are found to be possible and useful to communicate by different individuals and social groups. These are the facts that fit institutional imperatives about how to proceed. In any encounter with nature, from a fishing boat to a laboratory, a certain set of facts is going to be selected as relevant for indicating behavioural possibilities and/or presentation to others. Even if all of these facts are established as true, their selection and the use to which they are subsequently put would still be a social construction of nature. Honest presentations of facts take the form of narratives which contains facts about which the author holds many different levels of certainty, from the author's own direct experience to something that is generally accepted as true without reflection. The second set of mechanisms of social construction are based in how we perceive facts, rather than on the facts we choose to use and communicate. Both scientists and the lay public evaluate facts in terms of their source and the social location

---

[1] To avoid awkward wording, throughout the paper I will be useing phrases such as 'institutions deal with' that sound as if instituions are actors in their own right. The reader should remember however, that institutions are shared meanings. They do not deal, think, feel, speak or otherwise act. Only people do these things and when I say 'institution A deals with' it would be more accurate, and more grotesque, for me to say 'people whose actions simultaneously respond to, create and maintain institution A deal with'.

of facts is what determines their effective validity. As Collins and Pinch (1998) point out, you and I base our personal knowledge of nature that it is not possible to fly faster than light entirely on our personal knowledge of society that facts such as these reside with physicists and not with Star Trek scriptwriters.

Neither of these mechanisms makes it logically impossible for the real world to intervene in human affairs. Weakening this dichotomy between realism and social construction allow us to introduce into the social analysis of knowledge the concepts of uncertainty and bias. As long as the conceptual basis of social constructivism is a complete rejection of realism then the research tools that are based on social construction, ie ethnographic investigations of knowledge, discourse analysis, etc. are not available to address the question of how closely a social construction of nature maps nature *in situ*.

The concepts that have been developed around research into the social construction of nature are powerful tools for understanding how uncertainty is dealt with and bias is introduced and maintained. Understanding how fisheries knowledge is socially constructed plays an important role in fisheries co-management both directly in research on the condition of the resource and indirectly by building in reflexiveness in how knowledge is used. If fisheries co-management is not make management more responsive and flexible in the face of changes in natural systems then it fails to achieve a major portion of the management benefits it promises.

3. TACIT AND DISCURSIVE KNOWLEDGE

How well fishers' knowledge can be articulated in management debates has important implications for co-management both from the perspective of mobilizing fishers' knowledge for rational management and from the perspective of equitable control over the knowledge base. The term 'discursive knowledge' means knowledge that is shared and expressed, while 'tacit knowledge' is knowledge that people have but that is not (easily) expressed. The question of tacit knowledge plays an important part in general discussions about institutions and knowledge. Some fisheries scholars, notably Gísli Pálsson (1995; 2000), have argued persuasively that much of the knowledge that fishers have is tacit knowledge.

3.1. Tacit Knowledge

Tacit knowledge plays a critical role in society. Much of our modern understanding of institutions began with the phenomenological investigations of the role that the constructions of situations play in person to person interactions (Garfinkel, 1967; Goffman, 1974). Phenomenological means looking at how people build up their conceptions of the world around them. Goffman (1974) said that we build 'frameworks' of social situations to help them us how to react and that these frameworks are built from tacit knowledge. The person 'is likely to be unaware of such organizing features as the framework has and be unable to describe the framework with any completeness if asked, yet these handicaps are no bar to his easily and fully applying it' (Goffman, 1974, pp 21). Giddens (1984) relates the distinction between tacit and discursive knowledge directly to the institutions that link people and systems. He argues that day-to-day routines that are 'grounded in practical consciousness' are the basis of the continuity of both individual personalities and long enduring social institutions (Giddens, 1984, pp 62). Habermas (1984, 1987) also awards an important place to both discursive and tacit knowledge in his communicative systems theory. Tacit knowledge is always present as a background to the communication of discursive knowledge and without this tacit background mutual understanding is not possible.

But the tacit knowledge these theorists emphasize is tacit knowledge of social roles and institutional expectations. Does tacit knowledge of nature play institutional roles as well?

Tacit knowledge of nature certainly plays a critical role in practical fishing activities. Pálsson's (1995; 2000) investigations have focussed on skippers' knowledge of the fish resource. Skippers find it hard to explain why they know things, he argues, because their knowledge comes from emersion in the everyday world. The metaphor of knowledge as a sort of mental script or 'container' is not accurate. Fishers' knowledge is part of their overall fishing skill and the 'container' approach misses what it means to engage in a skilful act. The knowledge that underlies a skill is intuitive and not easily articulated or even understood well by the possessor. New fishers learn these skills by imitating the actions of others, his research suggests that many of these practices are learned without ever involving discourse or even consciousness (Pálsson, 2000). Other researchers have observed similar processes, but also found ecological knowledge being communicated through discursive means. Ruddle (1993) describes how a Polynesian and a Latin American culture transmit ecological knowledge related to sustenance activities to their children. In the Latin American case knowledge is passed to children using much more discursive techniques than in the Polynesian one.

Roepstorff's (1998) work points to a direct link between tacit knowledge of nature and institutional maintenance. In his study of Greenland fishers, he found that the idea that knowledge is something learned by doing, and hence involving tacit as much as discursive knowledge, is an integral part of their identity. As Pálsson's research demonstrates, tacit knowledge in fisheries is directly related to the enskilling process and skill, as Tilly (1988) points out, is more than anything else a social identity with critical implications for institutional relationships. While she is not making a distinction between tacit and discursive knowledge, Maurstad (2000) points out that fishers' knowledge is passed on in one-to-one and small group situations. Thus, she argues, the distribution of fishers' knowledge is dependent on relationships between individual people. This would be particularly true of tacit knowledge, which can only be passed on this way.

So, if Pálsson, Roepstorff, Maurstad, and Tilly are correct, the knowledge that maintains and reproduces institutions is not limited to tacit social knowledge, but contains tacit knowledge of nature as well. Social roles and identities are directly linked to the skills they have, the knowledge of nature they possess, and the relationships and processes through which skills and knowledge are passed on.

This link between tacit knowledge of nature and institutions has penetrating implications for fisheries co-management. Important aspects of fishers' knowledge of nature may be both 1) part of their social identity as fishers and therefore implicated in any fishing-related institution they interact with and 2) non-discursive and hence very difficult to communicate either among themselves or with other participants in the co-management institutions.

3.2. Discursive Knowledge

A 'discourse' is a set of arguments organized to achieve some rhetorical purpose. Discourses are important because they are a means to mobilize power. For knowledge to be contributed to a discourse it must be expressed, hence 'discursive.' Two examples of how people use discourses to mobilize social power are when they try to convince people to support a change in an institution and when they use sets of statements to maintain a shared identity and, hence, the solidarity (ie capacity for concerted action) of a group. We all participate in discourses for reasons that are good or bad, honest or dishonest. Discourses about fisheries and other environmental issues almost always involve 'scientific' claims, meaning claims about nature that purport to be objectively true.

How particular discourses are 'framed,' ie how they are organized and what they choose to emphasize, is an important factor in the social construction of nature. Any attempt to involve multiple stakeholders in a co-management effort needs institutional mechanism that illuminate how the claims of all the stakeholders are framed in discourses the reflect their perspectives and economic interests. Self-consciousness about these frames is critical to meaningful cooperation.

Hajer (1995) points out that through their discursive use, certain issues and facts can become emblematic of larger things. Part of what discursive frames do is reduce the complexity of an environmental problem and suggest solutions. Hajer (1995) argues that this reduction in complexity distances the frame from the underlying facts about the environmental issue, sometimes to the point where they a recitation of the frame and its main factual arguments acquires a ritual character. The emblematic character of certain facts leads them to begin to play such an important a role in mobilizing social power (a not inconsequential example being the role of scientific facts in fund raising) that goes well beyond any part they may play in developing an accurate picture of the actual situation in nature.

A discourse analysis by Palmer and Sinclair (1995) investigates perceptions of the fish resource and management among skippers from the Newfoundland cod fishery. They find little unity among these skippers on either of these subjects and enough systematic differences to suggest that vessel size and its accompanying divergence of interests and perspectives plays an important role in how the perceptions vary. While focussed on management more than LEK, their case demonstrates their main argument that 'appeals are sometimes made to incorporate .. local knowledge into fisheries management, without addressing the problem of exactly whose knowledge is to be considered most appropriate when there is disagreement. There is simply no single local vision or knowledge that is waiting to be heard - even within a single sector of the fishing fleet' (Palmer and Sinclair, 1995, 268). As Davis (2000) emphasizes, it is as important to look for variation in knowledge within groups as between groups because of designating some group as the embodiment of 'knowledge' has strong implications for its relative power.

In fisheries, people select different facts from fisheries science to put together an overall picture of the resource that fits their needs. Facts about nature get used in many ways to attain objectives that are institutionally driven. As Hajer (1995) makes clear, facts get used as symbols; different groups use them like flags to rally around an issue or raise money. Facts also get used to trigger government decisions, providing objective criteria for the invocation of rules (Wilson and Degnbol, 2002, discussed also in Chapter 11). The ways institutions shape and distort discourses by making certain facts more valid and relevant has been has been explicated the most systematically by Habermas (1984, 1987).

The bottom line is that it really matters if the knowledge of fishery workers is more tacit than discursive. Tacit knowledge is that much more difficult to mobilize in these discursive competitions to construct pictures of nature. Fisheries co-management must involve open communications and research functions that help stakeholders articulate their tacit knowledge and provide a reflexive understanding of interests in discourses.

4. RESEARCH BASED KNOWLEDGE AND FISHERIES CO-MANAGEMENT

But isn't it the scientists' job to construct an accurate picture of nature that everyone can agree to? This is the understanding on which most of our fisheries management institutions are established. This section, therefore, examines studies in the sociology of science to understand the relationship between the institutions of formal science and the social construction of nature.

Robert Merton (1968a, 1968b) is the dominant figure in post-war sociology of science. He argues that modern science reflects four sets of institutional imperatives: universalism through the evaluation of truth claims with pre-established criteria; disinterestedness; organized scepticism; and, communism, meaning that data and findings are fully shared. Merton has been much criticized by scholars who suggest that these four imperatives are utopian ideals rather than accurate descriptions of how science is actually done. What such criticisms miss, however, is that our understanding of what is a 'scientific fact' is not based on our own experience but our understanding of the social sources of valid scientific knowledge. It is the degree to which such sources meet Merton's imperatives that undergirds our evaluation of them as a source of valid knowledge.

Merton had already made most of his contributions when Kuhn's ideas appeared. Kuhn (1970) argued that science is not a cumulative enterprise building up a picture of nature piece by piece, but rather a series of paradigms that ask fundamentally different questions about nature. The idea that sociologists picked up from Kuhn was that these intellectual and scientific paradigms also reflected concrete social communities (Mendelson, 1987), which they began to analyse in terms of power and class driven social structures that were dominant sociological paradigm of the time. Barnes and MacKenzie (1979) began to explore the sociological implications of these communities, arguing along with Habermas (1971), that concepts and paradigms are not abstract theories but resources or instruments that can be put to use to serve interests (Mendelson, 1987).

The modern intersection of science, power and interests is, perhaps ironically, a product of democracy. As Porter (1995) argues, before democracy there was no great need for objectivity in the pursuit of social legitimacy. It is the rule of law that requires something beyond noble judgement, objectivity becomes an issue in a political democracy where there are bureaucracies that are vulnerable to criticism. In management institutions that hope for the cooperative involvement of many stakeholders, the question 'how do you know that?' is a critical one. Research-based science is the institution that has been built around being able to answer that question. That institution's stamp of approval on the objective validity of a fact is a critical source of legitimacy in democratic regimes. That legitimacy stems from being able to account for the source of the knowledge. Science is about the transparency of the process of knowing. This is the social basis of the importance of quantification in science - quantification is the best tool for giving transparency to answers to the critical 'how do you know that' question.

Irwin (1995) points out, models of science as the expression of particular interests is hardly the sole domain of sociologists. The general public routinely evaluates technical statements in terms of who is making them. Because of science's critical role in the legitimation of environmental policies, for example in debates over definitions of safe pollution thresholds, politics cloaks itself in scientific jargon (Dickson, 1984). Scientists respond to the distortions arising from the legitimating role of science with what Gieryn (1995) calls 'boundary work' - defining what is really science. This takes the form of struggles over who is in and out of the relevant networks and prestige and authority. Latour (1987) demonstrates that such networks include scientific institutions, people both in and out of laboratories, as well as particular findings, technologies and standards of validity on which the science is based.

The degree to which political controversies distort science should not be overstated because 'scientific disputes presume a sense of correct outcomes, outcomes true as far as can be determined with the facts at hand' (Englehardt and Caplan, 1987 pp 22). However, scientists are also involved in the definition of controversies and these are political rather than scientific definitions. Dietz *et al.*'s (1989) survey of different types of stakeholders in environmental conflicts showed that participants in environmental policy subscribe to definitions of policy conflicts that legitimate the resources they have in relative abundance. They found four major definitions of environmental conflicts as essentially struggles between: people who understood the issues and people who did not; conflicting vested interests; conflicting value systems; and, arrogant experts versus the public. Scientists tended to subscribe the first definition as it legitimized their having greater influence on the outcomes, in the same way that the other groups defined the issues to legitimize their own influence.

In the 70s and 80s a series of studies of scientists at work began to raise questions about how objective scientific results can really be. These studies treated laboratories as places where knowledge is constructed and science as a process of progressive selection of what works based on what has worked in the past and what is likely to work (Sismondo, 1993). Woolgar, for example, argues that the closer you get to the site of production of knowledge the less possible it becomes to determine if the knowledge is objective or a judgement call (Aronowitz, 1988). So many minute decisions are involved in experiments from how to calibrate the measurement instruments to how to communicate the data that few experiments can ever be truly replicated. Indeed, Collins and Pinch (1998) emphasize the fact that many experiments involve specific and challenging skills at the heart of their operations. Some of the skills involved in cellular biological research, for example, can take years to learn. Through the back door of complex skills, tacit knowledge can enter the ideally entirely discursive and explainable world of the scientific experiment.

These studies point to the important role that cultures of the scientific communities play in determining the relevance of scientific findings (Barnes *et al.*, 1996; Collins and Pinch, 1998; Shapin, 1994). In the context of constantly having to make judgement calls and the difficulty of true experimental replication, it is the trust provided by scientific cultures that makes scientific inference possible. What this means is that scientific scepticism takes concrete social shapes. A theory that is found to be antecedently implausible will have a hard time competing with one that seems initially very plausible and this plausibility comes from the consensus of the scientific community - i. e. the local cultural tradition. Scientific cultures, for example, will treat anomalous experimental results in one of two ways based on their internal standards: as a nuisance, probably arising from how the experiment was conducted, or as a significant challenge to the relevant theory. Work on the role of trust and local scientific cultures is distinctly applicable to fisheries science; the precautionary principle, for one thing, is a cultural consensus among fisheries scientists that has been actively incorporated into scientific inference (Wilson and Degnbol, 2002).

The sociology of science helps illuminate problems related to the knowledge base of fisheries co-management. The interests in the fishing industry, bureaucrats and environmentalists, of course, play an important role in both the scientific claims made by various stakeholders and in the activities of scientists (Weeks, 1995). While our knowledge of fisheries science is growing, replication in fisheries science is, if anything, more difficult than in many other sciences. Claims by scientists to objective scientific knowledge of the condition of fish stocks are extremely hard to establish and, in fact, rarely made. A tendency in fisheries management to exaggerate the potential for science also exists within the scientific and management agencies, as the findings from Dietz *et al.* (1989), discussed above, would suggest. Most challenging from the perspective of fisheries co-management is the need of scientists and decision makers for 'boundary maintenance' that delineates what is and is not science. Fisheries co-management will fail if it is based on the idea that no such line can be drawn. Democracy requires transparency and accountability. Fisheries co-management will also fail if it is based on the idea that that line can only be drawn by certain stakeholders.

5. LOCAL ECOLOGICAL KNOWLEDGE

Local Ecological Knowledge (LEK) is a concept with a number of names. Many people use the term 'indigenous ecological knowledge' which emphasizes how this knowledge is situated in a culture. Another term ' traditional ecological knowledge' emphasizes how it is rooted in a long community memory. The term chosen here 'local ecological knowledge' emphasizes that it is place-based knowledge. Neis (2000) uses the term 'fishery workers' social-ecological knowledge' to emphasize that this knowledge is held by many members of the fishing community, not only fishers, and the knowledge has both social and ecological components. And a fifth term 'anecdotal data' emphasizes its non-systematic production and, whether intending to or not, implicitly delegitmates community knowledge in comparison with research-based knowledge. All of these terms, even the last, communicate an insight into the nature of ecological knowledge.

5.1. The Usefulness of LEK

Over the last three decades, the idea that LEK is a useful resource deserving of attention and respect has grown from a radical idea to an operating orthodoxy in international development circles. Grenier (1998) defines indigenous ecological knowledge as 'the unique, traditional, local knowledge existing within and developed around the specific conditions of women and men indigenous to a particular geographic area'. This understanding of LEK as both unique to small social groups and a useful resource for development first emerged in Third World contexts where it was attached to issues such as intellectual property rights (Juma and Ojwang, 1992), the conservation of biodiveristy (Juma, 1991), and gender (Khasiani, 1992), as well as to resource management (Wamalwa, 1989). The mainstream emphasis is on locating local information that can supplement the generalizable knowledge produced by scientists. This is justified in two ways. The first is that traditional, local knowledge reflects the thinking of local people and, therefore, is critical to producing a picture of nature that will be accepted by them. The second is that the local knowledge, while not generalizable beyond the local context, is just as valid in that context as that produced by formal science. In fact it is more useful in some ways because it is based on more detailed information and more continuous observation than is usually available from research based sources (Fischer, 2000).

Fishers can be expected to have good knowledge of those aspects of fishing on which they draw, including, as the full range of populations and precise information on where and when fish congregate (Neis *et al.*, 1994). Guest (2000) looks at demographic and experience variables to see which predict greater ecological knowledge and found that the experience variables were the stronger. Number of years living in the area and number of years fishing both predict knowledge but the number of years fishing is more important. However, the data also indicate that LEK can be acquired relatively quickly and that people acquire competence quickly without actually fishing or having lived in an area for more than a few years.

Nor is LEK limited to fishers. Power's (2000) research on women's knowledge of an Atlantic fishery found that extensive information about the fishery is available to women through both processing plant work and through tasks related to family fishing businesses. Some women have a better grasp of changes in gear and catch than fishers do because they are the ones who do the bookkeeping. They are also able to infer quite a bit about where and how a fish was caught by its condition at the processing plant. The way that social power and institutions control knowledge sharing and the construction of effective versions of nature is reflected in the fact that the women also felt that they had a lot of knowledge about what was happening during the cod fish collapse, but they were afraid to raise questions for fear of their jobs.

Many scholars have described how local ecological knowledge (LEK) is used by local people in fisheries management (Berkes, 1993; Johannes, 1978), or believe that it should be used more (Smith, 1995; Grafton and Silva-Echenique, 1997) particularly in conjunction with co-management type programmes. In a seminal article, Pinkerton (1989) argues that fishers who do not trust the data that management decisions are based do not cooperate and may even develop confrontational postures. Another team of researchers that have played a central role in enabling the use of LEK in management, found that data from fishers contributes to management by: 1) providing additional indices for use in stock assessments and scientific debates; 2) providing data on responses by fishers to management measures and on the status of poorly understood species 3) suggesting novel hypotheses and 4) enhancing long-term legitimacy of the management regime (Neis *et al.*, 1995). Pederson and Hall-Arber (1999) found that fishers in New England have and can communicate useful knowledge of the sea floor, habitat structure, and fish distribution.

Several complications exist with the use of local knowledge in management. It is difficult to evaluate LEK without a fairly broad knowledge of the local context (Felt, 1994). Pederson and Hall-Arber (1999) caution that many fishers are reluctant to share knowledge because it might be used against them and they saw some information as proprietary. People in fishing communities tend to view the resource in much smaller temporal and spatial scales than it is conceived of by managers (Smith, 1995). They often see fisheries as systems in which small perturbations may have substantial future consequences (Smith, 1990) and are likely to emphasize the importance of habitat over population dynamics (Berkes, 1993; Pinkerton, 1989). Both of these viewpoints can be incongruous with management. In order to be useful for management, ecological complexity must be simplified to a point where decisions can be made.

5.2. The Two Cultures Theory of Fisheries Knowledge

Observations about differences between 'Western', i. e. research-based and LEK (Berkes, 1993; Felt, 1994; Pinkerton, 1989; Smith, 1990,1995) have led to the emergence of a commonly heard 'two cultures' theory of LEK and research based science. The argument emphasizes how different experiences and cultural backgrounds of scientists and fishers lead them to see the resource in very different ways and then seeks to identify these ways. Smith (1995) argues that both 'sides' see the other as violating 'plain common sense'. She quotes a fisher criticising transect surveys: 'Jeesus! Don't they understand that fish swim' while, in the same study, she finds that scientist believe that the fishers do not consider increases in fishing power in their 'world view'. Pálsson (1995) description of fishing skippers' reaction to transect surveys is almost the same. Roepstorff (2000) in his study of fisheries science in Greenland suggests that fishers 'focus on fish as a living being' and think of them as 'mass nouns' while the scientist sees the fish as a 'count noun', meaning that the individual fish is a representative of the stock in the sense that the stock is the arithmetic sum of the single fish. Berkes (1993), among many other scholars, subscribes to a picture of LEK as a set of beliefs associated with indigenous and tribal societies handed down through generations and suggests that these 'systems of knowledge' share characteristics distinct from the Western system (Berkes, 1995). Others, though, have observed similar processes among people of European extraction. Finlayson (1994) in his book on fishers knowledge and the collapse of the Canadian cod reports that Department of Fisheries and Oceans scientists 'willfully dismissed' the insights of the inshore fishermen because of dissimilar cognitive cultures. They used alien rules, norms, and language in the negotiation of validity. 'Knowledge claims by members of each culture were literally heard as incoherent by the other' (103).

The problem is not that the interpretations these scholars are offering do not reflect what is going on in the situation they are studying. Problems arise with the two cultures theory of fisheries knowledge in three ways. The first problem is the idea that there are, basically, *two* cultures. As even the short review of the sociology of science offered above should make clear, science is made up of many communities with different scientific cultures and standards of validity. Any group of local people may have their own knowledge culture, a hypothesis that can be examined by investigating if people have a consensus view of the resource in question (Romney *et al.*, 1986). All of the authors mentioned above would, no doubt, concede that the idea of two cultures is an analytic simplification, but I would argue that fisheries social science has become uncritically accepting of this simplification.

The second problem is the assumption that these two cultures have essential differences and will fit into ideal types. As Agrawal (1995) argues, this will inevitably stereotype and denigrate local knowledge while idealizing scientific knowledge. 'Primitive cultures are more embedded in their environments. . . less prone to analytic reasoning . . . more closed . . . . and less subject to change in the face of contrary evidence' is how Agrawal (1995 pp 420) describes the ideal type of LEK as it has emerged in development circles. Doing so, he is quoting the early anthropologist Claude Levi-Strauss while very nearly quoting Apffel-Marglin and Marglin (1990) on the virtues of LEK, save that the latter's phraseology, while meaning the same thing, is couched in more positive language.

Finally, the 'different cultures' explanation is often incomplete and even misleading because institutional reasons why different stakeholders are saying different things are often much more important than cultural ones. The different cultures explanation assumes that the main differences are in the way that groups understand things. While this is often true, the ways interactions among and between groups are structured can also prevents the effective raising of arguments that, if and when raised, would be mutually understood (Wilson, 2003). Yet, to explain the difficulties between fishers and scientists, cultural differences are routinely offered as the common sense explanation (eg Dobbs, 2000) while institutional sources of distortions of communications are ignored.

The implication of this for co-management is that the assumption that we are dealing with a 'scientific' viewpoint and a 'local' viewpoint about fisheries science matters is often an unhelpful oversimplification. Attention must be paid to the multiple viewpoints within both research communities and user group communities. Nor can we assume that if people understand each other, or even agree, that this will translate into an effective use of the available knowledge.

6. CONCLUSION: COOPERATIVE APPROACHES TO SCIENCE FOR CO-MANAGEMENT

The discussion thus far suggests both the great complexity of the scientific basis of co-management, and hope that ways forward can be found. The complexity stems from many sources. One is the degree to which knowledge is bound together with issues of social power and relationships among stakeholders. Another, related source is the amount of knowledge in fishing communities that is tacit and difficult to put to work in the service of either the communities' interests or developing an accurate understanding of the resource. A third is how distortions of communications arise from both problems in mutual understandings among stakeholders and through institutional needs that push scientific efforts away from the ideals (Merton, 1968a, 1968b) that allow us to judge a thing as being a scientific fact.

What is hopeful is that when stakeholders do not understand each other they have a remedy: they can discuss their differences in understandings. When the knowledge of some stakeholders is tacit and difficult to mobilize such discussions are how the tacit knowledge can become discursive. And when institutional rules have the effect of distorting the communications on which science depends, stakeholders also have a remedy: they can change the rules that are distorting the communications. The greatest difficulty with these remedies is how they can be implemented in heavily conflictual situations and on larger scales, as discussed in Chapter 11.

On smaller scales there have been a number of experiences where extended discussions have taken place and rules have been changed in pioneer efforts to create institutions to provide a knowledge base for fisheries co-management.

Programmes of collaborative research where fisher and other fishing community members participate in research, usually in a research assistant type role, are now very common. Shellfish and reef scientists have been heavily involved in this kind of work and have put a lot of effort into evaluating the scientific value of amateur volunteers in areas as diverse as the United States (Mumby *et al.*, 1995), Tanzania (Darwall *et al.*, 1996) and the Philippines (White *et al.*, 1997). These studies did not find that the use of volunteer observers resulting in highly valid data. However, Ticheler *et al.* (1998) report very good success using fishers to gather catch and effort data in rural Zambia as do Poizat and Baran (1997) in Guinea and many other biological studies around the world (Neis and Felt, 2000).

Many collaborative research programmes have been longstanding successes. Tagging studies are the most common form of collaborative research. Perhaps the most venerable of these programmes is the Apex Predators Cooperative Shark Tagging Programme that has been in place since 1962 and has tagged more than 147,000 fish (Kohler, *et al.*, 1998, Shark Tagger, 1997). Scientists using fishing boats as research platforms has a very long history, with many current efforts focussed on bycatch reduction (Wilson, 1999).

Some of these efforts have moved beyond fishers acting as research assistants to truly collegial relationships. This is the point at which people have really begun to create institutions to provide a knowledge base for fisheries co-management In the Fishermen and Scientist Research Society (FSRS), 156 fishers and 42 scientists are participating in a real give and take collaboration. The FSRS sets its own research priorities, but it also supports itself through specific research projects done on contract. King (1999) reports that building trust between the fishers and the scientists has been both their greatest challenge and accomplishment. This is only one of a number of such efforts that have appeared recently in North America demonstrating the importance of trust building where a joint scientific culture is given time to emerge (Wilson, 1999).

The lesson drawn here about institutions to provide a knowledge base for fisheries co-management is that they need to be an integral part of the co-management effort. Knowledge and institutions depend on one another. Tapping into the benefits of co-management for fisheries management means tapping into the knowledge of all stakeholders about the resource. The challenges presented by tacit knowledge, the use of scientific claims in the opportunistic pursuit of interests, and preserving a transparent institutional space for objective knowledge to be protected and grow are challenges that are being met, albeit slowly and only on small scales, in fishing communities around the world. These communities are showing that, on closer examination, there is no institutional contradiction between equitable decision making processes and scientific accuracy. Both, in the final analysis, are based on exactly the same principle: allowing a full and open examination of the issues involved.

REFERENCES

Agrawal, Arun (1995) 'Dismantling the Divide Between Indigenous and Scientific Knowledge' *Development and Change* **26**, 413-439.

Apffel-Marglin F. and Marglin, S. (eds) (1990) Dominating Knowledge: Development, Culture and Resistance Oxford: Clarendon Press.

Aronowitz, S. (1988) Science as Power Minneapolis: University of Minnesota Press.

Barnes, B. and MacKenzie, D. (1979) 'On the Role of Interests in Scientific Change' pp 49-66 in Wallis, R. (Ed.) On the Margins of Science: The Social Construction of Rejected Knowledge University of Keele: Sociological Review Monograph 27.

Barnes, B., Bloor, D. and Henry, J. (1996) Scientific Knowledge: A Sociological Analysis Chicago: University of Chicago Press.

Benton, T. (1994) 'Biology and Social Theory in the Environmental Debate' Pp 28-50 in Redclift, M. and T. Benton (Eds) Social Theory and the Global Environment London: Routledge.

Berkes, F. (1993) 'Traditional Ecological Knowledge in Perspective' Pp 1 - 9 in Inglis, J. T. (Ed.) Traditional Ecological Knowledge: Concepts and Cases Ottawa: International Programme on Traditional Ecological Knowledge, International Development Research Center.

Berkes, F. (1995) 'Indigenous Knowledge and Resource Management Systems: A Native Canadian Case Study from James Bay' Pp 99-109 in Hanna, S. and M. Munasinghe (Eds) Property Rights in a Social and Ecological Context: Case Studies and Design Applications Stockholm: The Beijer Institute.

Collins, Harry and Pinch, Trevor (1998) The Golem: What you Should Know about Science Cambridge: Cambridge University Press.

Darwall, W. R. T. and Dulvy, N.K. (1996) 'An evaluation of the suitability of non-specialist volunteer researchers for coral reef fish surveys. Mafia Island, Tanzania -- a case study' *Biological Conservation.* **78**(3), 223-231.

Davis, Anthony (2000) 'Thoughts on Approaches to Designing and Conducting Ecological Knowledge Social Research' paper presented at the Ecological Knowledge Working Seminar II 'Exploring and reconciling competing systems of ecologically-framed knowledge: lessons from Maritime Canada and North Atlantic Fisheries' June 17th-21st, 2000 St. Francis Xavier University Antigonish, Nova Scotia.

Dickson, D. (1984) The New Politics of Science New York: Pantheon Books.

Dietz, T., Stern, P.C. and Rycroft, R.W. (1989) 'Definitions of Conflict and the Legitimation of Resources: The Case of Environmental Risk' *Sociological Forum* **4**(1), 47-70.

Dobbs, David (2000) The Great Gulf: Fishermen, Scientists, and the Struggle to Revive the World's Greatest Fishery Washington DC: Island Press.

Englehardt, H. T. and Caplan, A.L. (1987) 'Patterns of controversy and closure: the interplay of knowledge, values, and political forces' p 1-23 in Englehardt, H. T. and A. L. Caplan (Eds) Scientific controversies: case studies in the resolution and closure of disputes in science and technology Cambridge: Cambridge University Press.

Felt, Lawrence (1994) 'Two Tales of a Fish' in Dyer C. L. and J. R. McGoodwin (Eds) Folk Management of the World's Fisheries Boulder: University of Colorado Press.

Finlayson, A. C. (1994) Fishing for Truth: A Sociological Analysis of Northern Cod Stock Assessments from 1977 to 1990 Institute of Social and Economic Research, Memorial University of Newfoundland, Social and Economic Studies No. 52.

Fischer, J. (2000) 'Participatory Research in Ecological Fieldwork: A Nicaraguan Study' pp 41-54 in Neis, Barbara and Lawrence Felt (Eds) Finding our Sea Legs: Linking Fishery People and their Knowledge with Science and Management St. John's, Newfoundland: ISER Books.

Freudenberg, W. R., Frickel, S. and Gramling, R. (1995) 'Beyond the Nature/Society Divide: Learning to Think about a Mountain' *Sociological Forum* **10**(3), 361-392.

Garfinkel, Harold (1967) Studies in Ethnomethodology Englewood Cliffs, NJ: Prentice Hall.

Giddens, A. (1984) The Constitution of Society Berkeley: University of California Press.

Geiryn, Thomas F. (1995) 'The Boundaries of Science' pp 393-443 in in Jasanoff, S. J, G. E. Markle, J. C. Petersen and T. Pinch (Eds) Handbook of Science and Technology Studies Thousand Oaks CA: Sage.

Goffman, Erving (1974) Frame Analysis Boston: Northeastern University Press.

Grafton, R. Q. and Silva-Echenique, J. (1997) 'How to Manage Nature? Strategies, Predator-Prey Models and Chaos' *Marine Resource Economics* **12**, 127-143.

Grenier, Louise (1998) Working with Indigenous Knowledge. A guide for researchers. Ottawa: International Development Research Council (IDRC).

Guest, Greg (2000) 'Market Integration and Rapid Acquisition of Local Ecological Knowledge' unpublished manuscript.

Habermas, J. (1971) Knowledge and Human Interests Boston: Beacon Press.

Habermas, J. (1984) The Theory of Communicative Action: Volume I Reason and the Rationalization of Society Boston: Beacon Press.

Habermas, J. (1987) The Theory of Communicative Action: Volume II Lifeworld and System: A Critique of Functionalist Reasoning Boston: Beacon Press.

Hajer, M. A. (1995) The Politics of Environmental Discourse: Ecological Modernization and the Policy Process Oxford: Clarenden Press.

Holms, Petter (2000) 'Realism and Contructivism in the Science of Ecological Knowledge' paper presented at the Ecological Knowledge Working Seminar II 'Exploring and reconciling competing systems of ecologically-framed knowledge: lessons from Maritime Canada and North Atlantic Fisheries' June 17th-21st, 2000 St. Francis Xavier University Antigonish, Nova Scotia.

Holm, Petter; Hersoug, Bjoern and Raanes, Seine Arne (2000) 'Revisiting Lofoten: Co-Managing Fish Stocks or Fishing Space' *Human Organization* **59**(3), 353-364.

Irwin, A. (1995) Citizen Science: A study of people, expertise and sustainable development London: Routledge.

Johannes, R. (1978) 'Traditional Marine Conservation Methods in Oceania and Their Demise'. *Annual Review of Ecology* Sept: 349-354.

Juma, C. (1991) 'Managing biological diversity in Kenya' Pp. 125-154 in Kiriro, A. and C. Juma (Eds) Gaining Ground: Institutional Innovations in Land-use Management in Kenya Nairobi: ACTS Press.

Juma, C. and Ojwang, J.B. (1992) 'Technology Transfer and Sustainable Development' Nairobi ACTS Press and Moi University School of Environmental Studies, Ecopolicy series No. 2.

Khasiani, S. A. (Ed.) (1992) Groundwork: African Women as Environmental Managers Nairobi: ACTS Press.

King, P. (1999) 'Practitioners' Profile: the Fishermen and Scientists Research Society' *CPR Digest* **49**, 9-11.

Kohler, N. E. , Casey, J.G. and Turner, P.A. (1998) 'NMFS Cooperative Shark Tagging Programme, 1962-93: An Atlas of Shark Tag and Recapture Data *Marine Fisheries Review* **60**(2), 1-87.

Kuhn, T. S. (1970) The Structure of Scientific Revolutions (2nd ed.). Chicago: University of Chicago Press.

Latour, B. (1987) Science in Action Cambridge: Harvard University Press.

Maurstad, Anita (2000) 'Obstacles to Mapping Fisher Knowledge' paper presented at the Ecological Knowledge Working Seminar II 'Exploring and reconciling competing systems of ecologically-framed knowledge: lessons from Maritime Canada and North Atlantic Fisheries' June 17th-21st, 2000 St. Francis Xavier University Antigonish, Nova Scotia.

Mendelsohn, E. (1987) 'Political anatomy of controversy in the sciences' p 93-124 in Engelhardt, H. T. and A. L. Caplan (Eds) Scientific controversies: case studies in the resolution and closure of disputes in science and technology Cambridge: Cambridge Univesity Press.

Merton, R. K. (1968a) 'Science and the Social Order' Chapter 17, Pages 591-615 in Social Theory and Social Structure New York: The Free Press.

Merton, R. K. (1968b) 'Science and Democratic Social Structure' Chapter 18, Pages 604-615 in Social Theory and Social Structure New York: The Free Press.

Mumby, P. J. Harborne, A. R. Raines, P. S. Ridley, J. M. (1995) 'A critical assessment of data derived from Coral Cay Conservation volunteers' *Bulletin of Marine Science* **56**(3), 737-751.

Neis, B. J. Hutchings, R. Haedrich, and L. Felt (1995) 'Scientific Debates, Lumpy Lumpfish and Slubby Nets: Fishers' Vernacular Knowledge and Adaptive Management' paper presented at the annual meeting of the

American Fisheries Society, Tampa Florida.

Neis, B., Felt, L., Schneider, D.C., Haedrich, R., Hutchings, J. and Fischer, J. (1996) 'Northern Cod Stock Assessments: What Can be Learned from Interviewing Resource Users' *Canadian Journal of Fisheries and Aquatic Science* **56**, 1944-63.

Nies, Barbara (2000) 'Draft Research Design and Methodology for FEK Project on Lobster' paper presented at the Ecological Knowledge Working Seminar II 'Exploring and reconciling competing systems of ecologically-framed knowledge: lessons from Maritime Canada and North Atlantic Fisheries' June 17th-21st, 2000 St. Francis Xavier University Antigonish, Nova Scotia.

Neis, Barbara and Felt, Lawrence (Eds) (2000) Finding our Sea Legs: Linking Fishery People and their Knowledge with Science and Management St John's Newfoundland: ISER Books.

Palmer, C. T. and Sinclair, P.R. (1995) 'Perceptions of a Fishery in Crisis: Dragger Skippers on the Gulf of St. Lawrence Cod Moratorium' *Society and Natural Resources* **9**, 267-279.

Pálsson, G. (1995) 'Learning by Fishing: Practical Science and Scientific Practice' Pp 85-97 in Hanna, S. and M. Munasinghe (Eds) Property Rights in a Social and Ecological Context: Case Studies and Design Applications Stockholm: The Beijer Institute.

Pálsson, Gisli (2000) 'Finding One's Sea Legs' 26-40 in Neis, Barbara and Lawrence Felt (Eds) Finding our Sea Legs: Linking Fishery People and their Knowledge with Science and Management St. John's, Newfoundland: ISER Books.

Pederson, J. and Hall-Arber, M. (1999) 'Fish Habitat: A Focus on New England Fishermen's Perspectives' *American Fisheries Society Symposium* **22**, 188-211.

Pinkerton, E. (1989) 'Introduction: Attaining Better Fisheries Management through Co-Management - Prospects, Problems and Propositions' Pp. 3-36 in E. Pinkerton (Ed) Cooperative Management of Local Fisheries. Vancouver: University of British Columbia Press.

Poizat G. and Baran, E. (1997) 'Fishermen's knowledge as background information in tropical fish ecology: a quantitative comparisions with fish sampling results' *Environmental Biology of Fishes* **50**, 435-499.

Porter, Theodore M. (1995) Trust in Numbers: the Pursuit of Objectivity in Science and Public Life Princeton: Princeton University Press.

Power, Nicole Gerarda (2000) 'Women Processing Workers and Resource Users' pp 189-203 in Neis, Barbara and Lawrence Felt (Eds) Finding our Sea Legs St John's Newfoundland: ISER Books.

Roepstorff, Andreas (1998) 'Virtual Stocks, Experts and Knowledge Traditions' in Dorias, Jacques Louis, Murielle Nagy and Ludger Muller-Wille (Eds) Aboriginal Environmental Knowledge in the North Quebec: GETCI Universite Laval.

Roepstorf, Andreas (2000) 'The double interface of environmental knowledge' pp 165-188 in Neis, Barbara and Lawrence Felt (Eds) Finding our Sea Legs: Linking Fishery People and their Knowledge with Science and Management St. John's, Newfoundland: ISER Books.

Romney, A. Kimball; Weller, Susan C. and Batchelder, William H. (1986) 'Culture as Consensus: A Theory of Culture and Informant Accuracy' *American Anthropolgist* **88**, 313-338.

Ruddle, K. (1993) 'The transmission of traditional ecological knowledge' In Inglis, J. (Ed.) Traditional ecological knowledge: concepts and cases. International Programme on Traditional Ecological Knowledge; International Development Research Centre, Ottawa, ON, Canada. pp. 17-31.

Salter, L. (1988) Mandated Science: Science and Scientists in the Making of Standards Dordrecht, Holland: Kluwer Academic Publisher.

Shapin, Steven (1994) Social History Of Truth; Civility And Science In Seventeenth-Century England University of Chicago Press.

Shark Tagger (The) (1997) Newsletter of the Apex Predators Cooperative Shark Tagging Programme, Narragansett, RI.

Sismondo, S. (1993) 'Some Social Constructions' *Social Studies of Science* **23**, 515-53.

Smith, M. E. (1990) 'Chaos in Fisheries Management' *Marine Anthropological Studies* **3**(2), 1-13.

Smith, M. E. (1995) 'Chaos, Consensus and Common Sense' *The Ecologist* **(25)**, 80-85.

Ticheler, H. J. Kolding, J. and Chanda, B. (1998) 'Participation of local fishermen in scientific fisheries data collection: A case study from the Bangweulu Swamps, Zambia' *Fisheries Management & Ecology* **5**(1), 81-92.

Tilly, C. (1988) 'Solidary Logics' *Theory and Society* **17**, 451-458.

Wamalwa, B. N. (1989) 'Indigenous Knowledge and Natural Resources' Pp. 45-68 in Kiriro, A. and C. Juma (Eds) Gaining Ground: Institutional Innovations in Land-use Management in Kenya Nairobi: ACTS Press.

Weeks, P. (1995) 'Fisher Scientists: The Reconstruction of Scientific Discourse' *Human Organization* **54**(4), 429-436.

White, A. T. and Bay of Bengal Programme for Fisheries Management (1997) 'Collaborative and community-based management of coral reef resources: Lessons from Sri Lanka and the Philippines'. Pp 253-274 in Nickerson, D. J. and M. H. Maniku (Eds) Workshop on Integrated Reef Resources Management

in the Maldives. Male, Maldives. 16-20 March.
Wilson, D. C. (1999) 'Fisheries Science Collaborations: The Critical Role of the Community' Institute for Fisheries Mangagment Research Publication No. 45.
Wilson, D. C. and Degnbol, P. (2002) 'The Effects of Legal Mandates on Fisheries Science Deliberations: The Case of Atlantic Bluefish in the United States' *Fisheries Research* **58**, 1-14.
Wilson, D.C. (2003) 'Examining the Two Cultures Theory of Fisheries Knowledge: the Case of Bluefish Management' *Society and Natural Resources* forthcoming.

Chapter 16

REPRESENTATION IN FISHERIES CO-MANAGEMENT

SVEIN JENTOFT
Department of Planning and Community Studies, University of Tromsø, Norway

KNUT H. MIKALSEN
Department of Political Science, University of Tromsø, Norway

HANS-KRISTIAN HERNES
Faculty of Social Science, University of Tromsø, Norway

1. INTRODUCTION

Co-management builds on the active involvement in regulatory decision-making of those people whose livelihoods depend on the viability of the natural resource. In contrast to government agencies in a top-down, command-and-control role, co-management schemes divide the responsibilities of regulatory functions between government and user-groups based on the principles of collaboration and partnership. The expected outcome is a more equitable, legitimate and effective management system. The promise of co-management should, however, not lead us to underestimate the problems and risks involved that may lead to failure. Given the state of most fisheries and fish stocks, and the questionable legitimacy of current management regimes, this is something co-management systems can hardly afford.

Therefore, the problems that might occur, and the difficult choices facing users and managers, should be investigated. Both managers and users need to prepare themselves for the many challenges of co-management. For this, there are two sources of learning. First there are lessons to be drawn from previous and actual co-management experiences - in fisheries and other industries that exploit common pool natural resources. Second, managers may learn from the much broader discourse on public governance and democratic institutions where issues of representation, legitimacy and compliance loom large. Many of the questions and dilemmas facing managers and users are neither new nor unique to fisheries co-management but have a long history of contemplation and theorizing. For instance, social and political thinkers have long struggled with the idea of representative government: what does it mean, and when can we say that a government is truly representative of its constituents? Precisely the same question could be asked with respect to fisheries co-management and the relationship between government and resource users and stakeholders. What does it take to make co-management institutions fully representative?

We suggest that the representation issue should be discussed with regard to four questions: First *who* can legitimately claim to be recognized as a user or stakeholder? Fishers are not the only group affected by management decisions; the number of potential stakeholders is currently increasing as fisheries have come to be regarded less as a matter of production and more as an environmental issue. The increasing emphasis on sustainability in fisheries management widens the range of legitimate interests considerably, and represents a significant challenge for institutional design (Dobson, 1996). Second, *in what capacity* should users and stakeholders be represented? Should they be represented as members of a particular interest group, a local community or simply as concerned citizens? There is also the possibility that representatives should enjoy considerable independence and discretion. To whom the members of, say, a management council, are held accountable will certainly have an impact on how they act as representatives. Third, *how much* involvement? More involvement may not always be better. Not all issues should be decided through popular participation. Also, ideal as it may seem from a democratic perspective, direct participation in decision-making is not always possible. Scale can make representation, ie participation by proxy, imperative, but representation need not always be an inferior substitute for direct participation. Fourth, *how should representation be done*? The act of representing requires certain skills and capabilities - technical as well as rhetorical - that can be learned. The question of which factors count and what means are needed for increasing users' ability to become effective co-managers, is begging for scrutiny. In what follows, we address each of these questions in more detail.

2. WHO SHOULD BE REPRESENTED?

A basic principle in democratic theory is that those affected by a decision should have an opportunity to be heard (Dahl, 1989). If this principle were fully applied in fisheries management, there are many stakeholders besides fishers who would have a legitimate claim to representation. In the case of Norway, for instance, Mikalsen and Jentoft map a whole range of fisheries stakeholders - close to 20 in number - who would be likely candidates, although only a few of these are actually represented on the boards where management decisions are made (Mikalsen and Jentoft, 2001).

Although there are many who may argue that they should be allowed to participate more directly in the process, the strength of their stakes will vary. Some are more affected than others, and this - one may argue - should be taken into consideration when deciding who should have a seat at the table. For instance, although consumers may claim that they are affected by management decisions, they have far less at stake than, say, fishers. This is also true when comparing commercial fishers to sports fishers, or to producers of sports fishing equipment or services. Also, management systems must make decisions that differ in kind, and stakeholders are more or less affected according to subject matter. Consequently, if the principle of greater representation for those most affected should be fully applied, stakeholder representation would vary from one type of decision to another. In other words, and ideally, stakeholder representation should vary during the process. However, there is hardly any 'objective' measure of who is most affected, and stakeholders value their interests differently. Thus, one should not expect that stakeholders would necessarily agree on what their respective stakes are and how much they should amount to when it comes to influence and representation.

This demonstrates the fundamental political nature of fisheries management and the relevance of perceiving the management system as a 'political coalition', rather than a hierarchical structure working according to instrumental values and technical considerations only (Cyert and March, 1963). Although management systems aspire to rational and optimal outcomes, decisions will - in practice - reflect the choices that stakeholders can agree on. Thus, rather than maximizing some instrumental value - such as economic efficiency - the management system is 'satisfizing', looking for acceptable rather than optimal solutions (Simon, 1957). It seems reasonable to expect that the greater the number of stakeholder groups involved, the more time and resources will be spent on political 'logrolling', and the more likely is it that the outcome would be some kind of compromise. To 'get', a stakeholder must always be prepared to 'give', a demand must be balanced by a contribution as decision-making would otherwise be perceived predominantly as a zero sum game.

A fisheries co-management system that aims to approach the democratic ideal that affected interests should somehow be involved, must relate to the fact that stakes are different in kind and strength. The analysis of this issue, and the subsequent discussion of how to design a workable co-management system, may benefit from stakeholder theory as outlined in the literature on business administration (Mitchell *et al.*, 1997). Here, a 'stakeholder' has been defined as 'any group or individual who can affect or is affected by the achievement of the firm's objectives' (Freeman, 1984: 25). Stakeholders may, however, score differently on three attributes: *legitimacy, power* and *urgency*. In our context, legitimacy would refer to the fact that some groups have a legal, moral or presumed claim on the fisheries management system. The power attribute indicates that some stakeholders are in a better position than others to influence management decisions, while urgency refers to the fact that some stakeholders have demands that may require immediate attention by the management system. One may, of course, argue about the exact score a stakeholder should have on each of these attributes in order to justify some form of representation. In principle, however, it should be possible to agree that the higher the score, the stronger the claim to stakeholder status. The implication of this is that the degree of involvement in the management process would have to be decided on from case to case. One could argue that it is fair that the co-management system should differentiate between the role that different stakeholders are allowed to play.

However, one should be aware of the dangers involved. For instance, some stakeholders may have a high score on power and urgency but a low score on legitimacy. Such stakeholders may easily obtain a status within the management system that they do not deserve from an equity and fairness point of view. Also, the ranking of stakeholders is not permanent but is subject to change over time. Stakeholders will often strive to obtain a higher score than they actually have, in order to become more efficient in pursuing their interests. They may, for instance, try to increase their power by building alliances with other stakeholders, or try to create the impression that their particular demands and interests are more urgent and therefore require immediate attention. In many countries, environmental groups have succeeded in improving their stakeholder status in this way (Hernes and Mikalsen, 1999). In Norway, for example, they have obtained representation on the Regulatory Council (the national fisheries management board), where they sit at the table with representatives of various fisheries interest organizations and the government when quotas and other regulatory measures are discussed. They have certainly been vocal in stressing the urgency of the conservation issue. Interestingly, they have also sided with small-scale fishers against the bigger offshore operators, thus gaining some support in many coastal communities. Although they have not in any way been elected or appointed to represent small-scale fishers and coastal communities, they make it sound as if this is indeed the case. The effect is both a strengthening of the legitimacy of this particular constituency and a stronger position for environmentalists in fisheries management.

3. REPRESENTATION - AS WHAT?

Representation can take many forms, as can the role that representatives are expected to perform. A representative may, for instance, be expected to play the role of a delegate who votes in accordance with instructions from his constituency, or he may have freedom to negotiate and deliberate. In some cases, representatives meet with a fixed mandate, in others they are - in the spirit of Edmund Burke - allowed to make their own judgment as to how to perform, argue and vote (Arblaster, 1987). This also has a bearing on how representatives are recruited, for instance whether they are elected or appointed, and who they are, ie the particular qualities they possess with respect to their personal skills and qualifications. This again may reflect the nature of the decision-making process, whether representatives are supposed to deliberate, or simply expected to cast their vote. Representation comes in three forms or categories: *functional, territorial* and *virtual.*

We find all of them in use in fisheries co-management systems, sometimes in combination - as in the Norwegian fisheries management system where representatives of the Norwegian Fishers' Association both wear the functional and the territorial 'hat.' Functional representation means that the interest represented is a particular activity or working pattern. In fisheries, the users of a particular gear or technology will have certain interests in common and thus form a constituency along functional lines. Different functional groups such as long liners, net users or trawl fishers may thus have different, often opposing, interests. The representation of all functional groups is therefore seen as necessary in order to create a balance among competing interests in fisheries co-management. Territorial representation refers to the situation where representatives are expected to speak for a particular geographical area, usually their home district or their local community. In Norway, for instance, geography has always been an important factor in fisheries policy-making.

Management decisions are expected to be made with a view to their implications for the distribution of resources among regions, and to how they may help or hinder the preservation of a decentralized settlement pattern. Conflict often pivots around the south versus the north, and the government and the fishers' association must always try to balance competing regional interests when management decisions are made. The conflict stems from the fact that although fishers from the north and the south often harvest the same stock, the structure of the fleet is different in the two areas. The bulk of the so-called 'industrial' fleet is located in the south while most of the inshore, fixed gear fleet has its home base in the north. This is the main reason why the Norwegian Fishers' Association is careful to make sure that its representatives on management councils carry both functional and territorial interests - although formally they represent a functional group. There are bound to be conflicting expectations here, and a possible role conflict for some of the representatives. How this is played out in the management decision-making process, and how it may influence decisions, are issues that should be researched.

The role of the representative as defined by the management system or by constituents, will reflect the dominant perception of the meaning of representation. If representation is defined as 'acting in the interest of the represented', one may conclude as Pitkin does, that the representative 'must act independently; his action must involve discretion and judgment; he must be the one who acts' (1957:209). This is much in line with a classical argument held by the 18th century political philosopher Edmund Burke, who was a strong advocate for virtual representation - ie representation based on virtue and wisdom. For Burke, to be independent of a particular interest was regarded as key. To his constituency, the representative owed his judgment, wisdom and reason, not just his voice and vote. Consequently, Burke was hostile to the idea of a fixed mandate, because it would discourage deliberation among those who made the decision. He was even opposed to the principle that the representative was accountable to his constituency and could be recalled. The idea that representatives could and should be truly independent was supported by people such as William Blackstone and John Stuart Mill among others (Christopherson, 1963) but also refuted by prominent thinkers such as Rousseau who, in his book on 'the social contract', argued against representation as such and in favour of direct citizen participation.

Quaint as this debate may now seem, it nevertheless has a clear relevance for the design of co-management systems and for the role of representatives within them. Opting for some form of co-management one must decide what the basis for representation should be: function, territory or wisdom/virtue or some combination of the three. While the Norwegian co-management system, and the community quota scheme of Alaska, are examples of a functional model 'softened' by the territorial principle, the US and Canadian systems lean more towards virtual representation with independence and autonomy as important values. Members of management councils are not necessarily elected to represent a particular group and equipped with a fixed mandate from it. They represent themselves and they are therefore entitled to speak their mind - even though they, informally, may regard themselves as representatives of their peers. In contrast to the Norwegian system, council meetings in the US and Canada are open to the public. But then, the idea of popular participation in fisheries management seems to have broader support in North America than in Norway.

4. REPRESENTATION OR PARTICIPATION?

One conception of democracy - stemming from Rousseau among others - holds that representation is always inferior to direct participation. Democracy is good, and more of it is always better! The division of labour entailed by representation poses the danger that the representatives may become detached from their constituents and misrepresent their interests. They will, furthermore, acquire information and knowledge that puts them in a privileged position vis-a-vis their constituents. The latter will not always have the information necessary to grasp the complexities of the issues involved, the full implications of their own demands, and the arguments and interests of other stakeholders. Also, a sense of group solidarity - a 'we' - may develop among representatives, creating a narrow and self-serving management elite along the lines of Robert Michels' (1962) famous 'iron law of oligarchy' which may distort even the most well-intended co-management effort. In that case, the average user and other stakeholders would be excluded from participating in, and learning from, the discursive process of fisheries co-management. And legitimacy will suffer as a consequence (Jentoft, 2000). For reasons such as these, classical political thinkers (Bentham and Rousseau among others) believed that democracy must mean direct participation rather than representation. For these thinkers, then, any form of representation - virtual, functional and territorial would be inadequate.

One should, however, also consider the arguments against making direct participation an essential requirement of a genuinely democratic co-management system (Weale, 1999). First, there are opportunity costs involved in participation. Users and stakeholders have other things to do than just talk and discuss in co-management forums. They are not in the privileged position of the ancient Greeks who could deliberate all day long and have women and slaves perform the more 'menial' tasks of cooking, cleaning, child rearing and production. The time needed for participation in a direct democracy is well depicted by Robert Dahl (1999): If, say, ten people speak for 30 minutes, a meeting will last five hours - provided it runs non stop. Fifty people who speak for ten minutes would have to sit eight hours in the room. If they were given thirty minutes each, they would have to sit listening for three eight-hour days. (The reader may well continue this arithmetic exercise or consult Dahl's table on p.107).

Thus, it is easy to see that direct participation, ideal as it may be, becomes increasingly unrealistic with scale - even though in the electronic age there are ways to help lessening the problem. However, no technology can - in itself - expand the amount of time available to a group of stakeholders to discuss issues of common interest. Consequently, limitations of time and space (to assemble all stakeholders) do call for some form of representation. In most countries and fisheries, the sheer number of stakeholders would increase the numbers in Dahl's calculus substantially. In Norway, for example, it would simply be physically impossible for the fifteen thousand or so full-time fishers to meet and deliberate on management issues. In some US fisheries, attempts at accommodating demands for participation have created a costly and cumbersome process for crafting management policy that fully illustrates that there is, indeed, a price to be paid for democracy.

Moreover, there may be occasions where the opportunity costs of participation are such that all would be better off if none participated (Weale, 1999). In some cases, not altogether unknown in fisheries management, stakeholders may find themselves locked into a series of meetings simply because they know that others - with conflicting interests - will turn up. Participation then becomes necessary in order to prevent others from damaging one's own interests. Given that all stakeholders have an interest in economizing with time, and that participation in management decision-making would erode the time available for other activities, all would be better off with some form of representation.

If not only time, but also ability, skills and specialized knowledge are important factors in politics, there are obvious benefits in representation. Due to their size and complexity, most fisheries management systems require bureaucratic officials to run and implement programmes and policies, and these officials will acquire detailed knowledge of the management process. Unless checked by other stakeholders equally knowledgeable, their monopoly of information will hamper stakeholder control. One solution to this problem of accountability is the appointment of representatives who are given the opportunity to acquire the specialist knowledge needed to challenge the expertise of bureaucratic officials. Besides, providing opportunities for all stakeholders to learn the 'tricks of the (management) trade' will improve the quality of any system and strengthen the legitimacy of management decisions.

Another possible solution to the democratic problem posed by size and complexity is one often prescribed in co-management theory: decentralization and delegation. Why not leave management to the local community - or to regional organizations of users? In England and Wales for example, the management of inshore fisheries takes place through a system of regionally based Sea Fisheries Committees composed of representatives from constituent county councils, fishers' organizations and (even) from the Environment Agency

(Phillipson, 1998). In the Netherlands quotas are exchanged and catches monitored by groups of fishers, leaving the allocation of shares among vessels as well as the enforcement of quota limitations to the users themselves (van der Schans, 2000). Delegating management tasks, the reasoning goes, will not only enhance participation and democracy, but also lower the costs of enforcement by increasing the acceptance - among users - of rules and regulations.

Although there is much to be said for delegating management tasks, there are certainly limitations to be aware of. The most important lies in the fact that the boundaries of ecosystems rarely overlap completely with the borders of local communities as the former usually cover a much larger geographical area. Fish is a highly migratory resource, and fishers often operate far from their local communities. Decentralization would thus turn the fisheries management system into a patchwork of overlapping small-scale regimes, and would hardly work without some form of central coordination. At this level, however, the sheer number of stakeholders makes direct participation in decision-making impractical, at least as a principal form of involvement. That said, it is still hard to see why a management system such as the Norwegian one could not be structured so as to allow for broader participation. There are only eleven representatives on the Regulatory Council, a number that could easily be extended without undermining the need for efficient decision-making. Besides, the transparency of the management process could be increased either by allowing stakeholders currently excluded to sit in at council meetings, or - as in the Canadian system - by conducting public 'hearings' before decisions are taken. These are strategies that may increase the legitimacy of co-management without extending representation as such.

Niklas Luhmann has made an interesting point regarding the time requirements of indirect participation. Representation, especially along functional and territorial lines, leads unavoidably to an increase in the number of decisions that will have to be made since those who elect representatives must make decisions in order to provide the representative with a mandate. In other words, one must make decisions on how to make decisions (Luhmann, 1990). This is the situation in a two-tier management system. In the Norwegian Fishers' Association for example, deliberations and decisions on fisheries management are made at three levels - municipal, county and national. Whether representation in this situation really reduces the costs of decision making is an open question - given the amount of time spent on contemplation, deliberation and decision-making. But then, democracy was never meant to be time saving. Nevertheless, the dilemma is real - in fisheries co-management as well as in democratic politics in general. To quote Robert A. Dahl (1999:110):

> 'The smaller a democratic unit, the greater its potential for citizen participation and the less the need for citizens to delegate government decisions to representatives. The larger the unit, the greater its capacity for dealing with problems important to its citizens and the greater the need for citizens to delegate decisions to representatives.'

If delegation cannot be avoided because of the size of the system, attention must be paid to the relationship between 'representee' and 'representor'. The co-management process must be structured so that deliberation takes place within and among all levels in the system. Representatives should be provided with a mandate and be held accountable for their actions. Within that mandate, however, they must be allowed some discretion. If not, deliberations will be ceremonial at best. In that case, the impact of stakeholders will not be dependent on the quality of the argument but on the number of votes waiting to be cast. In this situation, the minority interest will easily be disregarded.

5. HOW TO REPRESENT?

The relationship between 'constituents' and 'representatives' may, however, vary from one group of stakeholders to another. Not all stakeholder representatives are accountable to a particular constituency, as industry representatives are. Some, like environmentalists, are primarily representing a certain cause or point of view. By definition, representation means that one thing 'stands for' another, but as Weale reminds us we must distinguish between 'standing up for', 'standing out for', and 'standing in for' (Weale, 1999). In the first instance, the representative is representing an identifiable interest - a group or a territory. In the second, the representative is voicing a certain opinion or conviction. In the third situation representatives mirror the statistical composition of the society or the relative numerical strength of all the stakeholder groups. Here, the representative is acting as a mere substitute of the population at large, and the assembly (management council or board) would simply assume the character of a microcosm. These are not distinct categories, but they do reflect the grounds on which representatives are chosen and how they are expected to act - argue and vote - when representing.

The question of how to represent (in order to be an efficient representative) does not only pertain to the relationship between representatives and their constituents. It also has to do with personal ability. Knowledge and technical expertise are, of course, important here, but in addition efficient representation often requires rhetorical skills - the command of words and arguments. Rhetoric is the art - or science even - of speaking effectively, ie the ability to persuade and sway an audience. Charismatic leaders usually have this ability. Indeed, it is perhaps their most important attribute, the very quality that makes them charismatic. As Max Weber observed, charisma is among the qualities that make people speak with authority.

Fishers can be charismatic orators who speak directly and in a non-convoluted fashion in public settings. Among them there will always be some who, despite the lack of formal education, have the rhetoric talent and skill. Their arguments often combine technical knowledge and moral categories, they frequently employ metaphors, which they draw from the fishery itself, and they know how to effectively blend technical and rational argumentation with both ethos and pathos. Compared to most other stakeholders, fishers have the advantage of a high score on stakeholder legitimacy and confidence gained from their experiences at sea. Through their work and experience they have acquired knowledge of the fishery itself that no other stakeholder - managers included - will have. Having attended several meetings of management councils and boards, we have witnessed the exercise of authority by fishers. When they speak, government officials certainly listen, sometimes almost in awe. Thus, the image of the humble, diffident and speechless fisher who allows him to be patronized by other stakeholders is false. If and when they do not understand the technical language of bureaucrats and scientists, they will certainly say so and insist that talks be conducted in a common idiom.

Rhetorical skills are certainly needed to effectively represent a constituency. It is, in fact, an important aspect of the very act of representing, but it does not guarantee a genuinely deliberative process. Deliberative decision-making requires that stakeholders not only talk, but also listen, and let themselves be persuaded by the better argument and change preferences accordingly (Knight and Johnson, 1994). Stakeholders may not necessarily agree but, in the spirit of Habermas, they should seriously try to understand the reasons why other stakeholders hold the views they do. Mutual understanding and respect, however, are impossible unless every representative can exercise a minimum of discretion. If not, attempts at deliberation will invariably turn into a series of 'collective monologues' -

as psychologists call it - where people speak to each other but do not respond to the arguments of others.

There is, of course, always the danger that the display of rhetorical skills may be counterproductive and inhibit consensus. First, there is the temptation of demagoguery - as Socrates warned of in Plato's *Gorgias* - where form takes precedence over substance. Also, rhetoric is no guarantee for rationality and justice, as Aristotle aimed to correct for in his own *Rhetoric*. Second, there is the danger - inherent in the postmodern perspective - that 'everything' is rhetoric, that truth, knowledge and facts are all socially constructed (see Chapter 15). This would, for instance, easily de-legitimize the voice of scientists, reducing the knowledge of fish stocks to a matter of the best-packaged opinion. Neither would the practical knowledge of fishers count for much. In short, if the 'reality' of fisheries management is nothing but a social construct, stakeholders will have few incentives for deliberation and interactive learning: 'My view is as good and valid as that of anyone else, so why bother listening?' Since there is no truth, and science is little more than artful rhetoric, research on the impact of management on fish stocks and local communities is a waste of time.

However, the position taken by fisheries stakeholders, and scientists in particular, is that there is a reality 'out there' which is independent of human conception and that true knowledge about this reality is possible (Finlayson, 1994). Effective presentation of this knowledge requires rhetorical skills, which then become a political asset, empowering stakeholders, providing for more effective representation and improving the deliberative process. In that case, there is an obvious incentive to improve the ability of stakeholders to present their version of reality and deliberate effectively. To our knowledge, this is something co-management systems seldom do. Representatives often owe their appointment to their formal position within their organization rather than to the rhetorical talent they may possess, or they are chosen solely for their technical expertise. This, however, does not necessarily qualify them for the challenges they will face as members of management councils or boards[1].

As co-management systems are opening up for more stakeholder groups to take part in the decision-making process, the 'rhetorical situation' will become more complex, and complicate the deliberative process. First, one should expect rhetorical skills to become a more important asset, particularly for those stakeholders whose legitimacy is not obvious. Rhetorical talent could well compensate for this deficit. Second, your role as a representative will be totally different if you represent, and are held accountable to a constituency, compared to a situation where your primary task is the promotion of a particular cause or principle. The latter type of 'representative' is usually not restrained by a fixed mandate and therefore free to deliberate and (perhaps) more susceptible to the views of other stakeholders. This is probably the major difference between the situation of a fisher representative and an environmentalist member on a management board. With new stakeholders achieving representation, there is thus reason to believe that the 'modus operandi' of (co-) management boards will change. That said, the effects of broader

[1] Bonnie McCay and Caroline Creed (1999) have written a pamphlet for member-representatives on the US fisheries management regional councils on how to participate, where rhetoric is taken seriously. The pamphlet contains detailed advice on how to prepare for, behave and speak during meetings. It also describes meeting procedures, the legislation behind the system and its purpose. In essence, the pamphlet is a handbook in constructive, civilized rhetoric that could serve as a recipe for co-management systems in other settings even if the particular circumstances would vary from case to case.

participation on management, will ultimately depend on the behaviour of fishers as they are, by far, the most important stakeholder group. The fear of being both outnumbered, outvoted and 'out voiced' may well prompt reactions by fishers, as was the case when the Norwegian government decided to grant observer status on the national management board to Friends of the Earth, Norway. The decision met with vociferous opposition from individual members as well as local associations of the Norwegian Fishermen's Association, who found the inclusion of environmentalists ludicrous and unacceptable. 'We've had enough', said one member, suggesting that the association reconsider its representation on the board while others feared that public interest groups with little knowledge of fishing - and with an agenda totally alien to sensible management - would come to dominate management policy-making (Mikalsen, 1998). Conflicts and complexity, however, are not just a consequence of the growth in the number and types of stakeholders; they also have to do with the fact that the groups involved tend to have their own 'conceptualizations' of fisheries management - bringing different rhetorical 'logics' to the table. This is one of the major challenges in designing appropriate co-management institutions.

6. CONCLUSION

Co-management in the fisheries seems to have acquired the characteristics of an institution in Philip Selznick's sense of the term: arrangements that are *'infused with values beyond the technical requirements of the task at hand'* (Selznick, 1957). As an institution, co-management has come to symbolize participation, collaboration, democracy and fairness - and for that reason it seems to prevail almost irrespective of how well it performs. Like democracy, co-management is not only justified because it serves some specific, 'technical' or 'political' purpose. It is also valued in itself, as a symbol of partnership and democratic decision-making, and as a way of acknowledging the rights and duties of stakeholders. Co-management is simply 'the right thing to do'.

Attractive as they may appear, though, co-management arrangements are certainly no panacea. As is the case in democratic politics, co-management involves selection - of participants and perspectives - raising some fundamental problems of representation and deliberation that we have tried to address in this paper. In this day and age, when fisheries management is rapidly becoming an environmental issue, the question of who qualifies as 'co-manager' is a pertinent one. The concept of 'stakeholder' has by and large replaced the term 'user-group' in debates on representation, and there are strong pressures to expand the scope of legitimate participants. How the boundaries of co-management systems are defined may well decide the overall legitimacy (and efficiency) of management policies. With regard to representation the trend certainly seems to be in the direction of more inclusive systems.

Involving new stakeholders in management decision-making may bring in participants with (widely) different views on the nature and purpose of representation. Among fishers, representation has largely been conceived as a way of looking after one's interests as members of a functional group, and representatives have been expected to act as delegates constrained by a (fixed) mandate. That said, there are certainly co-management systems where representatives of users enjoy substantial independence vis-a-vis their constituency, and where their interests as members of local communities loom larger. Other stakeholders - notably environmentalists - see participation as a way of countering the self-interest of users and the technical approach of government officials. They are committed to a cause, and to the public interest as they define it, rather than to a particular constituency. The interesting question here is whether the inclusion of new stakeholders will alter the balance between accountability and independence, redefine the purpose of representation and affect the nature of the co-management process as such.

Although a democratic ideal, direct participation is rarely feasible in (political) systems above a certain size. The conception of democracy advocated by the ancient Greeks and elaborated by 'modern' thinkers such as Rousseau, is alien to the idea of representation. To them, representation means that many will be excluded from actively participating in the political process and democracy will suffer as a consequence. With some exceptions, fisheries co-management has rarely been designed to facilitate direct participation by all stakeholders. The impracticalities of universal participation and the advantages of representation have, it seems, been all too obvious. And there are, as we have pointed out, certain advantages of representation - given the sheer number of stakeholders involved. The crucial issue, of course, is whether representation produces a co-management system that is sensitive to all the political interests and social values involved, and one that allows all conceptions of 'reality' to be played out. To accomplish this, representatives must enjoy sufficient independence to deliberate and, if necessary, yield to the power of the better argument.

Here, we believe that scale not only requires some form of representation, it also has a bearing on how representatives act and interact within a given co-management system. In small systems representatives may be unable to distinguish between their personal interests and those of the larger group or system - with impartiality suffering as a consequence. A probable asset in day-to-day problem-solving, 'myopia' is clearly dysfunctional when important values and principles are at stake. Community level management, however desirable, needs to be supported and corrected by institutions at higher levels. Representatives within regional or nation-wide systems will have to take the broader view and face general issues and principles that no fisheries co-management system can ignore.

While there may sometimes be a tension between deliberation and representation, we do not think that these 'activities' are incompatible. Fisheries co-management must allow for genuine deliberation, but, as Gargarella holds, we are entitled to know the identity and loyalties of those who deliberate (Gargarella, 1996). The question of constituency should always be addressed: whom do the representatives represent? Their functional group, their local community, the public interest or simply themselves? Ideally, the basis on which representatives are selected, and the principles that govern their work, should vary with the goals of the co-management system, with the knowledge required for efficient management, and the interests considered being at stake. In practice, the selection of representatives as well as the rules for deliberation and decision-making are highly contentious issues. It follows from this that the question of how to make a co-management system truly representative, and how to facilitate genuine deliberation, is a highly political one. There is, in other words, no fixed, neutral formula on how to design for stakeholder participation in a fisheries co-management system.

REFERENCES

Arblaster, A. (1987) Democracy, Milton Keynes: Open University Press.
Christophersen, J.A. (1963) 'Representant og velger'. *Tidsskrift for Samfunnsforskning*, nr. 4, s. 215-240.
Cyert, R.M and March, J.G. (1963) A Behavioral Theory of the Firm. Englewood Cliffs, Prentice Hall.
Dahl, R. A. (1989) Democracy and Its Critics. New Haven and London: Yale University Press.
Dahl, R.A. (1999) On Democracy, New Haven: Yale University Press.
Dobson A. (1996) 'Representative democracy and the environment' in Lafferty, W.M. and J. Meadowcroft (eds):

'Democracy and the Environment: Problems and Prospects', Cheltenham: Edward Elgar, p. 138.
Finlayson, C. (1994) Fishing for truth: a sociological analysis of the northern cod stock assessments from 1977 - 1990, St. John's: ISER-Books.
Freeman, R.E. (1984) Strategic management: A stakeholder approach' Boston: Pitman.
Gargarella, R. (1996) Full Deliberation, Representation, and Impartiality. Bergen: LOS-senteret, notat 9607.
Hernes, H.K. and Mikalsen, K.H. (1999) From Protests to Participation. Environmental Groups and the Management of Marine Fisheries. Paper presented at the 27th ECPR Joint Sessions of Workshops, Mannheim, 21-31 March.
Jentoft, S. (2000) 'Legitimacy and Disappointment in Fisheries Management'. *Marine Policy*, Vol. **24**, pp. 141-148.
Knight, J. and Johnson, J. (1994) 'Aggregation and Deliberation: On the Possibility of Democratic Legitimacy'. *Political Theory*, Vol. **22**, No.2, pp. 277-296.
Luhmann, N. (1990) Political Theory in the Welfare State. Berlin: Walter de Gruyer.
McCay and Creed (1999) Fish of Cut Bait: How to Participate in the Fisheries Management System, New Jersey Marine Sciences Consortium.
Michels, R. (1962) Political Parties: A Sociological Study of the Oligarchical Tendencies of Modern Democracies, New York: The Free Press.
Mikalsen, K.H. (1998) Regulation and Representation: Institutional Challenges in Fisheries Management, in Tim S. Gray (ed.): The Politics of Fishing: London: MacMillan Press Ltd.
Mikalsen, K.H. and Jentoft, S. (2001) 'From User-groups to Stakeholders: The Public Interest in Fisheries Management.' *Marine Policy*, **25**, pp. 281-292.
Mitchell, R.K., Agle, B.R. and Wood, D.J. (1997) 'Toward a Theory of Stakeholder Identification and Salience: Defining the Principle of Who and What Really Counts.' *The Academy of Management Review*, Vol. **22**, No. 4, 853-886.
Phillipson, J. (1998) 'The Political Culture of Fisheries Management: an Anglo-Danish Comparison of User Participation', in Tim S. Gray (ed.): The Politics of Fishing, London: MacMillan Press Ltd.
Pitkin, H.F. (1967) The Concept of Representation. Berkeley: University of California Press.
Selznick, P. (1957) Leadership in administration - a sociological interpretation, Berkeley: University of California Press 1984, p. 17.
Simon, Herbert (1957) Administrative Behavior. A Study of Decision-Making Processes in Administrative Organization. New York, The Free Press.
Van der Schans, J.W. (2001) Governance of marine resources: Conceptual clarifications and two case studies. Delft: Eburon.
Weale, A. (1999) Democracy, London: MacMillan Press Ltd.

Chapter 17

THE PLACE OF CIVIL SOCIETY IN FISHERIES MANAGEMENT:
A Research Agenda for Fisheries Co-management

SVEIN JENTOFT
Department of Planning and Community Studies, University of Tromsø, Norway

BONNIE J. MCCAY
Department of Human Ecology, Rutgers the State University, New Jersey, USA.

1. INTRODUCTION

The management of natural resources is an area attractive to all the social sciences. Although they all share the same root model concerning the role of the institution of property rights, or the lack thereof, depicted by Garrett Hardin in his seminal article on the 'Tragedy of the Commons' (Hardin, 1968), there is no consensus as to the analytic merits of this model. Neither do they agree on priorities and substantive conclusions. In this presentation we summarize the main issues and arguments in this discourse - in particular as they pertain to the question of regime design. Where are we at this point in time and what themes are begging for more social research in the years to come?

Marine fisheries resource management research has become a truly multidisciplinary area. It used to be an interest primarily of marine biologists. Although biologists still enjoy a hegemonic status, they are no longer alone in claiming to have knowledge of relevance to this subject. Fisheries management is a social issue because it unavoidably must address the behaviour of resource users, and in many contexts it must also take into account the consequences for fishery-dependent communities. Therefore, fisheries management should and increasingly does attract interest from researchers of all the social science disciplines: sociology, political science, law, social geography, philosophy, economics, and social anthropology. The research approaches taken are now as diverse and fertile as our disciplines. Explicitly or not, fisheries management would always rest on some underlying assumptions pertaining to the relationship between the individual and the society, nature and culture, agency and structure, autonomy and order, social justice and democracy, and tradition and modernity.

Fisheries management involves some of the most substantive issues in social science, such as the roles of the state, the market and the civil society and the interdependencies and interplay of those agents of governance. Fisheries management can fruitfully be regarded as comprised within the general theme of sociopolitical governance (Kooiman *et al.*, 1999). A social researcher with an interest in fisheries management will recognize the relevance of the German sociologist Claus Offe's general statement: 'Our problem is most definitely not Lenin's problem as captured in his famous question of 'What is to be done?' Instead our problem can be formulated as the logically prior question of 'who', ie what configurations of agents, might at all be capable of doing whatever 'is to be done' (Offe, 2001).

In other words, the task of defining what fisheries management must do is not the most difficult question. More difficult is the question of who is best suited for carrying out the task. Which of the three - the state, market or civil society - is best equipped for doing whatever needs to done in fisheries management? Since the 19th century, policies have favoured the state as manager, often dismantling local and other structures of governance. In the past 30 years the market emerged as manager through devices such as ITQs. The neo-liberal perspective emphasizes this approach. What about civil society?

Civil society can be defined as 'the space of uncoerced human association and also the set of relational networks formed for the sake of family, faith, interest, and ideology - that fill their space' (Waltzer, 1995:7). Civil society appears in fisheries management in many ways, and it is at the core of the 'communitarian' perspective on fisheries management, which emphasizes the roles of communities and other organizations of civil society in management. Fisheries management debates are often polarized explicitly or implicitly between neo-liberal and communitarian perspectives, as well as the more traditional emphasis on the authority and responsibility of the state. Increasingly, participants in such debates recognize the need for more 'mixed' systems of governance.

In this paper we will summarize the main perspectives and arguments in the social science discourse on fisheries management and the institutional solutions that are currently heralded within our disciplines. We distinguish between two basic paradigms, the *neo-liberal* and the *communitarian perspective*. It will be made clear that there is little agreement on basic assumptions, substantive conclusions and research priorities between the two paradigms. We also identify some themes that have hitherto received little attention among social scientists and managers alike, which we contend are begging for more social research. Our main argument is that fisheries management research needs more focus on the role of civil society and its institutions and how they restrict and enable management systems in reaching their goals.

2. THE NEO-LIBERAL PARADIGM

2.1. Individual Transferable Quotas (ITQs)

Economists perceive over-fishing as the aggregate external effect of the maximizing behaviour of individual users. Like the biologist Garret Hardin in his paper on the tragedy of commons, economists perceive the lack of clear-cut property rights as the root of the problem. Fish as a natural resource share many characteristics of other natural 'common pool' resources such as grazing land, ground water, the air, tropical forests, and wildlife when they are treated as public or open access resources. Typically, these resources are vulnerable to over-exploitation. In all these instances, resource management therefore requires some form of collective action (Ostrom, 1990). Common to these resources is a contradiction between what is rational from the perspective of the individual and what is rational for the larger group. The 'free rider' problem appears, in so far as it does not make sense for any individual to act to conserve the resource because the benefits of her action can be shared by others. A 'second-order' free rider problem is that no individual has an incentive to engage in collective action to come up with a management system because the benefits of such a system are like a public good, available to everyone (Ostrom, 1990). Consequently, collective action is often difficult to arrange, even in the most urgent of situations when the resource is on the brink of collapse.

From the neo-liberal perspective, the logical solution to this social dilemma is to replace open access or common property (which neo-liberals conflate) with private property or private-property-like arrangements. What Hardin describes with reference to a classic dramatic 'tragedy', wherein trouble inexorably follows from the lack of property rights, economists explain by 'market failure'. The market mechanism - the 'invisible hand' - is by itself incapable of providing a social optimum. For the market to work, privatization is essential. This requires the active involvement of the state, which is the only force that can legitimately install and guarantee private property. Hence, both market and state modes of governance are called upon. The management approach deriving from this analysis can be illustrated as follows.

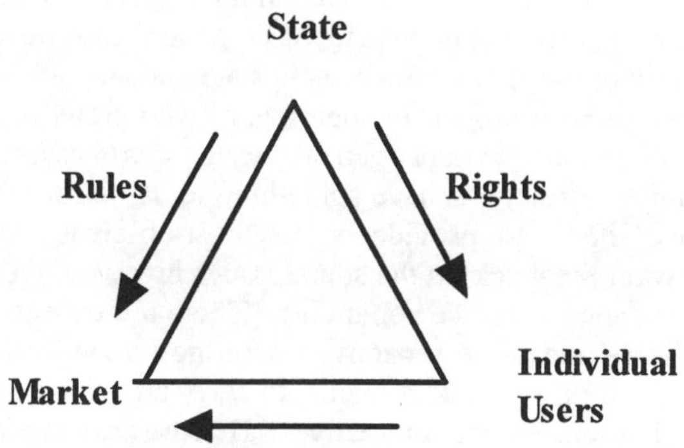

Figure 1. The neo-liberal perspective

The model has market failure as its underlying problem definition, and the state is called upon to interfere. The government then makes use of two mechanisms. First, it allocates resource rights - in fisheries these are usually in terms of quotas and licenses to individuals who can prove a history and a continued interest in the fishery. Equipped with these rights, fishers enter the market with the ambition to maximize profit. Here, however, freedom is not complete. In order to secure social order or to prevent excessive usage, the government as a second remedy sets rules pertaining to gear, space, timing. In fisheries, the rules tend to be more numerous, complex and dynamic than in most other industries, and this complicate their enforcement.

Economists argue that fish quotas should be made exclusive, divisible, and exchangeable, that fishers should be allowed to buy and sell them, - ITQs. ITQs are perceived less as a measure of resource conservation than as a way of avoiding what some economists call 'capital-stuffing' - ie over-investment in means of production and also as a method of generating the maximum resource rent (Copes, 1982). According to this theory, in fisheries that are managed by means of ITQs the less efficient resource users are driven out of the fishery, as those who are more efficient will buy up their resource rights. Hence, the total cost of fishing will be lowered.

The use of ITQs in fisheries is a good example of how management prescriptions are becoming increasingly globalized. In fisheries the system was invented by resource economists at University of British Columbia in the early 1970s, then exported to elsewhere in Canada, Australia and New Zealand by the early 1980s, and after that adopted in Iceland, the Netherlands, the United States, and other countries. There are several reasons why this model has become so popular. First, the analysis is simple and the solution concrete. The solution follows logically from the premises of the analysis: if open access is the problem, then some form of access restriction is needed. This simple answer led to limited license programmes, but by the 1980s they were found to do poorly in promoting more efficiency, hence the shift to ITQs, which allot specific (and transferable) amounts or shares of a quota to participants in a fishery. Secondly, ITQs provide an answer to a serious problem in fisheries, the problem of overcapitalization - the fact that today there is far too much fishing capacity for the resources available. As shown in the surfclam and ocean quahog ITQ system of the United States (McCay and Brandt, 2001), once ITQs are implemented, fleet tonnage, a measure of capacity, can decline markedly as vessel owners economize. These are the two most widely recognized reasons for the popularity of ITQs. Thirdly, ITQs are in perfect harmony with current neo-liberal economic policies and the belief in the supremacy of the market. Fourth, the popularity of ITQs may also be related to the much stronger involvement and prestige of economists in state bureaucracies compared to other social scientists such as anthropologists or sociologists who prefer other solutions to the commons problems. When management agencies need expert advice, they typically ask biologists and economists. Biologists have something to say on the conservation issue, while economists have inputs to provide on the cost-efficiency issue. If, however, government agencies want feedback on the social issues involved, they often consult the industry directly, not the social science community. There are exceptions. In the United States, a large set of laws and more proactive courts now require the federal fisheries management system, based on regional councils, to carry out social as well as economic impact assessments of management alternatives. Hence, growing numbers of social scientists are involved, but consultation with the industries and other interest groups remains central to actual decision-making.

2.2. Eco-labelling

The most recent proposal for combating ecologically unsound harvesting practices using market sources of governance is the so-called eco-labelling strategy. The initial move towards this solution in fisheries dates back to the early 1990s when three of the USA's largest tuna fish processors announced that they would no longer buy tuna from the eastern Pacific where the methods of capture were detrimental to the dolphin population (Hersoug *et al.*, 1999). Eco-labelling has developed further, following the model of certification for sustainable forestry. In 1996 the World-Wide Fund for Nature (WWF) collaborated with the multinational corporation Unilever - the world's largest frozen seafood buyer - to create the non-profit, non-governmental 'Marine Stewardship Council' (Long, 1999). The council has developed a number of criteria of what they perceive as sound management practices, and they issue certificates to those fisheries that show through third-party certifiers that they are managed according to these standards. Those who get through the screening process are permitted to display a logo on fish products as advice to environmentally conscious consumers. In this way, through market demand, consumers will put the necessary leverage on the industry to be more ecologically responsible and to be more receptive of government management initiatives.

The experiment is still in its initial phase, and it remains to be seen what effects it may have on fish ecosystems, fishing industries and communities. The idea has triggered much interest and support globally, not only within the academic community (Sutton, 1998) but also from rather unusual and unexpected quarters such as the British House of Lords. But it has also spurred criticism. Because certification is voluntary; its success is dependent on demand. The system is vulnerable as its reputation may easily be undermined if a certified fishery should unexpectedly collapse, which is a possibility that cannot be ruled out due to the highly unpredictable and sometimes chaotic nature of fish-stocks and ecosystems.

Certification fits within the neo-liberal perspective. Emphasis is on market-based governance. The ITQ model is targeted at the first link in the fisheries chain, the fisher; eco-labelling is addressing the last, the consumer. The perspective that lies beneath here is also one of market failure. John T. Sproul (1998:139) argues:

> 'Mostly, our current seafood markets fail to incorporate information on social and environment practice associated with processes necessary in bringing commodities from their point of extraction to consumption. In the contents of presenting defensible ecological information between harvesters, producers, and consumers, there is a clear market failure in the international trade of marine products.'

By bending consumer behaviour, to make them their purchases in an ecologically 'responsible' fashion, the idea is to influence harvesting practices. Indirectly the system also puts pressure on the government to become a more forceful manager. As one WWF spokesman said, quoted by Hernes and Mikalsen (2002): 'One thing is certain: Where industry and the market lead, governments will likely follow'. - via the management system. It involves a strong link between market and individual, on the one hand, and market and state, on the other (Figure 1).

3. INTERLUDE: MARKET FAILURE OR COMMUNITY FAILURE?

From a governance perspective, a conspicuous feature of the model depicted in figure 1 is the absence of civil society. It has only three points: state, market, and individual. The model is insensitive to the fact that individual resource users are members of social groups, human communities, families, social networks, and organizations. Sociologists and social anthropologists object to the portrayal of the people involved as isolated individuals who are out in the market place and the fishing grounds just for themselves. They note the absence of consideration for gender, age, class, ethnicity, for social identity and cultural knowledge. They are concerned that the model's solid foundation on rational choice theory and methodological individualism leaves little room for appreciation of the embeddedness of resource users in social and moral fields of meaning and action.

The absence of civil society in the traditional and neo-liberal model of fisheries governance has some profound consequences. When managers ponder about how to manage, fishers are perceived either as anonymous contributors to some level of fishing mortality or, from the economic perspective, as acting only out of the incentives and values of the market place. Fishers are not regarded as social persons carrying a baggage of cultural values, norms, knowledge and relations 'uploaded' within the civil institutions from which they depart and return continuously throughout their career. Consequently, the possibilities of individual or collective stewardship are ruled out of the equation. Moreover, management agencies are frequently caught by surprise when they find that fishers respond in unexpected ways, for instance when fishers break the rules that are imposed on them.

Unable or unwilling to explore reasons for non-compliance and resistance, the only countermeasure managers see involves stricter rules, more enforcement and harder penalties. But if we regard breaking of rules as a basically moral challenge, civil society cannot be disregarded.

One would be hard pressed to challenge the assertion that successful fisheries management requires honesty, trust, solidarity, and respect. But these are values that people acquire in civil society, not in the market nor from government. Overfishing and disregard for the rules are as likely to be due to 'community failure' or the failure of civil society as to 'market failure' or the lack of appropriate market signals due to imperfect property rights. When fishers deliberately over-fish, it is not the result of a failing market. Rather, it is a sign of a failing civil society, a civil society that is unable to make people willingly abide by norms of moderation and honesty. Emile Durkheim claimed that markets only encourage people to behave according to their individual interest and motives, not according to normative standards and collectivist orientations. As to the state, he argued: 'The state is too remote from individuals; its relation with them too external and intermittent to penetrate deeply into individual consciences and socialize them within. Where the state is the only environment in which men can live communal lives, they inevitably lose contact, become detached, and thus society disintegrates' (Durkheim, 1964:28).

Contrary to the neo-liberal paradigm, a communitarian approach to fisheries management would stress the importance of crafting management systems that support and depend on civil society institutions (McCay and Jentoft, 1998). Social scientists have invested much effort in proving that Hardin is wrong in predicting that tragedy is inevitable for natural resources held in common. They argue that resource users, when faced with the commons dilemma, do not always behave as he anticipated. Users are sometimes capable of managing their common resources without the active involvement of an external authority such as the state, even in the absence of private property. This is because there are civil institutions at community or regional levels that fill the void.

Sociologists and anthropologists criticize current management regimes for doing the exact opposite of crafting management institutions that support civil society. They destroy rather than build communities, thereby turning the tragedy of the commons into a self-fulfilling prophecy. When the state and the market are allowed to dominate, the capacity of communities and other civil institutions to manage is undermined. Cooperation is replaced by competition, and users stop regarding each other as a 'we' and start viewing each other as a distant 'they' (Etzioni, 1998). Then their social relations become subject to individual calculus where relations are maintained only as long as they serve some instrumental purpose.

Instead, management systems should be designed so that they serve as a counter measure. They should aim at building communities, rather than contributing to 'community failure' and its negative effects on any management regime. Communitarian social scientists have advocated two ways of making civil society part of the equation: - community based management and co-management.

4. THE COMMUNITARIAN PARADIGM

4.1. Community-based Management

Community-based management refers to a highly decentralized management system whereby some or all authority rests at the local community level - variously defined. The focus on community-based management arose from appreciation of the varieties and complexities of systems of 'sea tenure' throughout the world. Consequently, this approach also means greater appreciation of the socio-cultural, political, and ecological intricacies of local systems of fisheries use and management, based on empirical documentation of the many ways that local communities and traditional societies conceptualize, allocate, regulate, and defend rights to marine resources, and the widespread existence of distinctive and elaborate systems of property rights contrary to the dominant view in fisheries management that people who fish treat the sea as an open-access frontier. These community-based institutions often have salutary functions in mitigating conflict, spreading the fishing effort over wider areas, and reducing the take of some species or protection fish in certain spaces or times of the year. These findings have contributed to the policy of the Food and Agricultural Organization (FAO) of the United Nations and other development agencies, through the characterization of 'Territorial Use Rights in Fisheries' (TURFs) as important management techniques.

Sea tenure studies have contributed directly and indirectly to the recent proliferation of projects in Southern countries, and with indigenous peoples in the North, designed to protect, restore, or create anew TURF-like systems of controlled access to marine resources - now under the rubric of 'community-based fisheries management'. It is a sharp counter-point to 'the tragedy of the commons' and the neo-liberal way of understanding control over resources, focussing on the capacities of local communities under some special circumstances, to develop effective system of common resource management (McCay and Acheson, 1987). The emphasis on community-based management in fisheries also comes from wider policy shifts toward community-based approaches in rural development and resource management in the third world, particularly for forests, water, and wildlife.

The focus on sea tenure and common property rights, while productive, has tended towards 'question-begging' research programmes (Vayda and Walters, XXXX). The tendency is to claim that any and all sea tenure institutions are valuable because they promote marine conservation, despite evidence that many of them are intended to have very different functions, such as reducing conflict or maintaining the wealth and power of local elites, and that evidence for the ecological benefits of these and others is scant. Clearly, there is danger in reading a 'conservation message' into the local systems of property rights or sea tenure without more information about their views of the world, the workings of the institutions, and their effects on natural systems. But despite the fact that they where never intended to manage the resource from a conservationist perspective, they all address issues that also a conservationist oriented management system must address (McCay, 1980). They demonstrate that an alternative agent like the state is not always needed, and they suggest that civil institutions at the community level may play a greater role in fisheries management if they are allowed and equipped to do so. In a third world context communities may even be the only viable alternative due to the absence or weakness of prevailing government institutions.

4.2. Co-management

Many people use the term co-management to include everything from community-based management to consultative management, but it is useful to take a narrower definition. Co-management means that some or all management responsibilities are explicitly shared between government management agencies and user-organizations as well as other groups such as the scientific community. Contrary to the ITQ prototype that was originally constructed on paper 'from the arm chair' and then tried out in practice, the co-management alternative is grounded on, and generated from, empirical research. Co-management in one form or other exists in many countries and in many industries, and in some instances has a long history (Jentoft and Kristoffersen, 1989; McCay, 1998). Social anthropologists and sociologists have been among the most active in demonstrating the existence of co-management arrangements, explaining their origin and describing their functioning. Interestingly, the co-management alternative seems to have caught attention not only from social scientists but also from managers, NGOs, and organizations such as the World Bank and FAO. As the present volume demonstrates, there are now many experiments going on in many countries both in the North and South, and social scientists are helping to study and implement them.

Co-management is an arena of governance where there may be a genuine break-through of the communitarian perspective and argument of anthropologists and sociologists. The widespread interest in co-management may have something to do with the fact that the message is equally as simple, clear and constructive as the one promulgated by economists on ITQs. It also helps that the argument is solidly founded on empirical research. Something that already exists in reality and in many instances has proven to work well cannot be easily dismissed as utopian. But the enthusiasm may also stem from the fact that most other models have failed in sustaining the resource, and co-management may now be worth a try.

In addition, unlike community-based management, co-management does not totally challenge the authority and responsibilities of the agencies of the state. Although co-management in contrast to other management systems involves civil society, it also draws on the capacities of state and market institutions. The mixture of these forces, however, differs from case to case. Figure 2 tries to capture this.

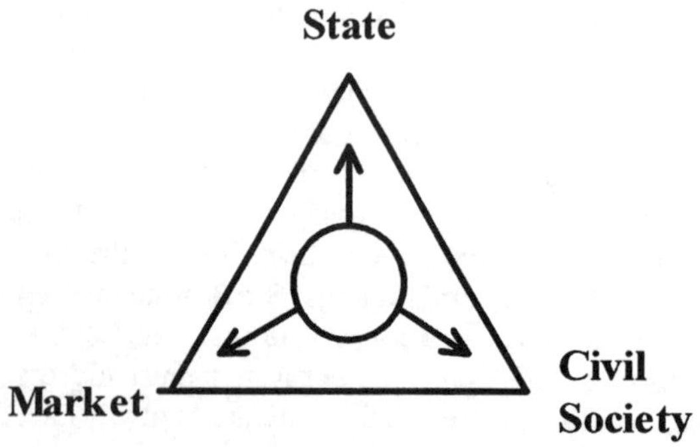

Figure 2. Co-management

In many instances, fisheries co-management systems are closely connected with civil society, like the community cooperatives in Japan (Ruddle and Johannes, 1989), the beach village committees on the lakes of Malawi (Donda, 2000), the community development quota system in Alaska (Ginter, 1995), the community management boards of Nova Scotia (Apostle *et al.*, 1998), and the 'Panchayat' system on the Coromandel Coast of India (Bavinck, 2001).

The Dutch Biesheuwel groups (van der Schans, 2001), the IQ groups in Nova Scotia (Apostle *et al.*, 2002), and the 'Clam Committee' of the US Mid-Atlantic region are examples of co-management systems intimately linked with the market through the fact that participants are holders of ITQs. In the Dutch case, fishers are obliged to pool their quotas by forming groups, within which participants are allowed to transact their quotas according to rules they have decided themselves. In Nova Scotia user-groups participated in designing the system, by providing advice to the government on quota allocations and they are themselves actively involved in enforcement and monitoring. In the US Mid-Atlantic, the ITQ owners formed a committee to cooperate in marketing but also in developing consensus for their interactions with the government management agency and, most important, in pooling funds to carry out stock assessment research in cooperation with the government scientists.

The Norwegian (Hoel *et al.*, 1996) and Danish co-management systems, on the other hand, lean towards the state pole of the triangle as they build on the involvement of fisher unions in the management decision-making process. However, the arrangement remains rather centralized and state driven, placing unions in a reactive, advisory role and individual fishers at the receiving end.

5. RESEARCH AGENDA

Social research has focussed heavily on management institutions; how they emerged, are designed and work, and the impacts they have on fish resources, habitats and human communities. In our view, this effort has produced a lot of great social science (McCay, forthcoming). Among social scientists co-management has been at the forefront. This volume demonstrates this very well. Several co-management research issues have also been developed elsewhere (Jentoft and McCay, 1995; Jentoft *et al.*, 1998). A number of questions have been discussed, and we could name a few here: One pertains to representation (cf Jentoft, Mikalsen and Hernes, chapter 16 this volume): Who are the stakeholders and how can they be effectively involved in fisheries management (Mikalsen and Jentoft, 2001)? Another issue regards scale: What is the relationship between scale and co-management effectiveness (Wilson, chapter 11, this volume)? A third question pertains to scope: Which management functions are better handled cooperatively among user-groups than by government and which functions must remain a government responsibility? (Pomeroy, chapter 14, this volume). A fourth issue is property rights: Which property rights regimes, or combinations of property rights, make co-management works better - private, collective or state? Each and every question entails hard choices. In essence these questions are more of a political than a technical nature, and they need to be analysed with due consideration of the particular context within which co-management is introduced.

We would not suggest that these areas of social research have now been fully exhausted, but we may have reached a point where further studies of these issues provide just more of the same. We know fairly well how to analyse management institution designs, what the basic alternatives are, the potentials they embody, and the risks they hold. A problem with the focus on management institutions, at least from the 'rules of the game' perspective from which it is often carried out, is also that it makes us put too much emphasis on institutional fixes. It makes us believe that if we only can find the right mix of carrots and sticks, the management system will succeed. Although incentives and rules are important, sustainable fisheries management needs a broader focus. We need a perspective on institutions in fisheries that also emphasizes their normative and cognitive elements as well as their social and cultural underpinnings (Jentoft *et al.*, 1998; Scott, 1995).

The impediments to sustainable resource management are often more of a political nature. There are many hard choices in natural resource management. Conflicting economic interests and worldviews are usually prevalent and active. The number of stakeholders who want a slice of the pie or a say in the process has increased, and the likelihood that they should all agree on how to best manage fisheries is rather slim. Thus, we need to focus more on the political ecology of fisheries management. Questions such as; who are the stakeholders and what are their political assets, how are public and private interests played out, how do unequal distribution of power among stakeholders and user-groups impact on decisions made, and who benefits and who loses from fisheries management, all beg for more social research.

The need for social impact analyses remains, and social scientists are increasingly called upon by management agencies to provide theoretical and methodological expertise for this purpose. The "colonization" thesis of Habermas (1987) should be a focus for fisheries management research. Market based management devices like ITQs or eco-labelling in fisheries affect local communities and peoples' lives in unexpected ways. So also do state based management policies and systems. As mentioned, it has been argued that state centred fisheries management systems inspired by the Hardin idea of the Tragedy of the Commons has a tendency of becoming self-fulfilling. By eroding the qualities of community essential to cooperation and solidarity, they produce the very conditions under which the tragedy occurs. Michael Taylor has argued that the state by doing so has made itself more necessary than what would otherwise have been the case if local communities were kept intact (Taylor, 1987).We should not for a moment think that community-based and co-management systems have only positive implications. For instance, Davis and Bailey have warned against the risk of co-management entrenching already existing conflicts and inequities in the local community (Davis and Bailey, 1996). To what extent are current state-based, market based and co-management systems eroding community is an import issue for social research. What is the precise nature of these impacts and the mechanisms behind them? How are they experienced at the community level? And what can be done to restore communities that have suffered from these impacts? Therefore, as social scientists we should continue to assess impacts along the state-market-civil society axis, as shown in figure 3. In addition to empirical research, a major challenge for social impact assessment in fisheries is also to develop theory and method for empirically grounded and socially sensitive projections of the future impacts of management alternatives on communities, marginal groups, and other social entities, rather than to simply fill in the blanks of legal guidelines.

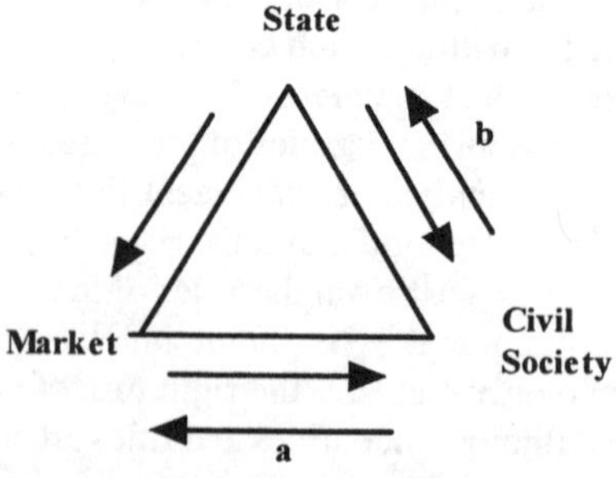

Figure 3. Research emphases

The market has traditionally been the focus area of economists, the state of political scientists and lawyers, while civil society has caught the attention of primarily sociologists, social anthropologists and social geographers. Since co-management draws on resources from all these three institutions social research pertaining to this management model must be interdisciplinary. All the social sciences mentioned here have analytical tools and research experience relevant to this discourse. We believe that the analytical framework represented by this triangle should facilitate people from different disciplines should find common ground. We need a comprehensive analysis of the interactions, exchange and impacts that occur along the three axes in all directions (figure 3). So far, however, some axes have attained much more attentions than others.

We should be equally as interested in the state-civil society- market axis as in the state-market-civil society axis. Social scientists have long argued that civil society is important for the functioning of markets. Max Weber alluded to this insight in his famous book on capitalism and religion (Weber, 1930). Emile Durkheim contended that non-contractual relationships are conducive to contractual relationships (Durkheim, 1964). The sociologists Mark Granovetter and Richard Swedberg argued the same when they claimed that economic action is socially situated, 'enmeshed in economic and non-economic institutions and networks of ongoing social relations' (Swedberg and Granovetter, 1992). The economist Kenneth Boulding noted that a market could not function in the absence of trust and honesty (Boulding, 1970). Arguments have also been made with respect to the importance of civil society for the working of democratic institutions at the state level. Most famous is Alexis Tocqueville's (1956) classic treatise on 'Democracy in America'. Civil society allows people to participate and learn how democratic institutions work. Tam's point that it is only when people have a sense of belonging that deliberative decision-making processes can work well, applies to fisheries co-management as well as to other situations where decisions are made collectively (Tam, 1997).

These arguments are well captured by the 'embeddedness' thesis of Granovetter and others. Applied to fisheries management, the embeddedness thesis emphasizes first of all the impacts that occur along the civil society-market axis (a, in figure 3), as demonstrated by the social geographer Bernard Nietschmann (1997:p 201), who for many years worked among the Miskito people in Nicaragua. He observed:

> 'In sea-fishing communities such as the Miskito's, the people's culture is used to guide, instruct, and govern, and to determine who can and cannot enter the community's sea territory or who is abusing a resource. Instead of top-down external management laws or rules to prevent or deter extreme behaviour of a few individuals (characteristic of a hierarchical land culture society based on individual ownership), egalitarian sea culture societies use bottom-up internal cultural norms and values to guide the behaviour of the entire community'.

Secondly the embeddedness thesis is also influential along the civil society - state axis (b, in figure 3), as suggested by Roger Bennett (1996: 164), who studied the implementation of costal zone planning in a number of Norwegian coastal municipalities:

> 'The Herøy plan in particular invites the reflections that cultural context can make an important contribution, consensus being facilitated by common codes of communication, tradition of cooperation in making a living and social cohesion within a small community where it is important to maintain good neighbours'.

Nietschmann's and Bennett's observations should inspire more research in other marine settings. If the social and cultural conditions that exist external and prior to the management system are crucial for its success, they should certainly attain more emphasis among social scientist. However, tight social networks, cohesiveness, common cultural values, a sense of common destiny, and long traditions of collaboration do not always characterize communities. They are, on the contrary, sometimes ridden with social fissures, inequities and conflict. If the latter is the case co-management may have small chances of becoming a widespread tool in sustainable fisheries management. Therefore, at least three questions should be addressed by social researchers: A) To what extent could co-management work under less ideal circumstances? B) If these integrative qualities of community are missing, how could they be provided? C) How can shortcomings in the social and cultural environment be compensated for in the actual management system design?

In fact, co-management is in itself a way of correcting for failings of community and small scale, for instance by bringing in the support of government (Pomeroy, chapter 14, this volume). Democratic standards and procedures, such as the one member-one vote principle, are also countermeasures to inequity and power differentials that may exist at the community or industry level. Douglas Clyde Wilson (chapter 11, this volume) contends that the complexity of fisheries, and the heterogenious nature of communities in general, makes the central question how to make use of shared understandings without distorting them, rather than a search for shared norms. These are important research issues that we believe would benefit from theories of 'discursive democracy' (Dryzek, 1990) theories of social/organizational learning (Pateman, 1970) and the recent interest in the concept of social capital among sociologists, anthropologists and social geographers (Putnam, 1993).

We have few objections to social scientists in an advocacy role in fisheries management, but only in so far as their work as social scientists (versus as citizens or resource users themselves) is firmly rooted in empirical research and is meant as providing input to a learning process that involves all user-groups and stakeholders from the bottom up. We also support a humbler social science, a science that is carefully investigating management problems in their particular ecological, social and cultural contexts before launching management solutions. Too often, the approach is the opposite. Whether they intend to or not, social scientists are prominent in the globalization of ideas. Recipes for solving the commons dilemma - whether ITQs or co-management - are no exception to this rule. Generally, however, the approach is more of a standard solution seeking a problem regardless of context, rather than a problem seeking a solution - therefore the many surprises from and backlashes to well-intended management initiatives.

6. CONCLUSION

Usually we think that initially the state has to define the over-all quota limit, and then allocate the resource shares to fishers according to some pre-defined criteria of who are the rightful beneficiaries of such privileges. Then, and only then will fisheries communities, with all their civil institutions, be kept alive and healthy. This certainly makes sense. But, following the embeddedness argument, we argue that the opposite is no less true. Civil society and indigenous culture should not only be regarded as dependent variables in the management equation. The community and the culture of fishing are not a derivative - or a 'superstructure' (in the Marxist jargon). Rather, communities and fishing cultures are a reservoir and a social capital 'basis' for sustainable fisheries resource management. Therefore, fisheries management must begin with civil society and not end with it. Civil society must be supported prior to the establishment of the management system. Government must assist building communities and civil organizations as an integrated part of their sustainable management policy. This is important regardless of paradigm. The market requires a normative foundation, and civil society is the only place where market actors will internalize the moral rules and values on which the smooth functioning of markets depends. As the sociologist Alan Wolfe (1989:233) points out: 'Civil society, if understood as the place where people pause to reflect on the moral dilemmas they face, is necessary if individuals are to possess those capacities of agency that will enable them to make rules as well as follow them'. The standard argument is that if government helps the market work, the civil society will take care of itself. Our argument turns this upside down: If the government helps civil society work well, the market will take care of itself. This is, of course, not entirely true, but it is at least equally true as the statement that claims the opposite. Therefore, our perspective is dialectic - it emphasizes the mutual relationship between market and civil society, which is contrary to the standard view of the primacy of the market (cf figure 3). The same holds true for state and civil society. By helping civil society to work better, government management institutions help themselves. If strengthened institutionally, fisheries communities and user-organizations may become important counterparts and allies of government rather than its adversaries. This is also a basic condition for effective co-management.

Here, we think, is our main challenge for social research on natural resource management in the years to come. This is where we should turn our focus. We should do all we can to examine 'the colonization thesis' - the tendency of the state and market to erode the qualities of civil society that are essential to management. We should also do our utmost to examine the embeddedness thesis, how the civil society - market axis and civil society - state axis are active in fisheries management. But we should also pay attention to the state-civil society axis, what government actually does and can do in order to solidify those civil institutions that play a crucial role in sustaining the social and cultural conditions that are fundamental to fisheries management. Co-management should certainly be analysed from this perspective also. Thus, our analytic perspective on natural resource management must be deeper and broader than it is today, and it should be turned around. Currently there is too much emphasis on the state-market-civil society axis relative to the state-civil society-market axis. We believe that social research should not limit itself to analysing the many impacts of fisheries management on civil society, but also should study how civil society impacts on, and provide a social and cultural footing for management.

REFERENCES

Apostle, Richard, Barrett, G., Holm, P., Jentoft, S., Mazany, L., McCay, B. and Mikalsen, K. (1998) Community, State, and Market on the North-Atlantic Rim. Challenges to Modernity in Fisheries. Toronto: University of Toronto Press.

Apostle, Richard, McCay, B. and Mikalsen, K. (2002) Enclosing the Commons: Individual Transferable Quotas in the Nova Scotian Fishery. St. John's: ISER Books.

Bavinck, Maarten (2001) Marine Resource Management: Conflict and Regulation in the Fisheries of the van der Schans Jan Willem Governance of Marine Resources: Conceptual Clarifications and Two Case studies. Delft: Eburon, 2001.Coromandel Coast. New Delhi: Sage Publications.

Bennett, Roger G. (1996) "Challenges in Norwegian Coastal Zone Planning". GeoJournal, 39:2.

Boulding, Kenneth E. (1970) Beyond Economics. Essays on Society, Religion, and Ethics. Ann Arbor: The University of Michigan Press.

Copes, Parzival (1982) "Implementing Canada's Marine Fisheries Policy: Objectives, Hazards and Constraints" *Marine Policy*, pp. 219-236. July.

Davis, Anthony and Conner Bailey (1996) "Common in Custom, Uncommon in Advantage: Common Property, Local Elites, and Alternative Approaches to Fisheries Management." *Society and Natural Resources*, Vol. **9**(3), pp. 251-266.

de Tocqueville, Alexis (1956) Democracy in America. New York: Mentor Books.

Donda, Steven J. (2000) Theoretical advancement and institutional analysis of fisheries co-management in Malawi. Experiences from Lakes Malumbi and Chiuta. Ålborg University and Institute for fisheries management and coastal community development (IFM), Denmark.

Dryzeck, John S. (1990) Discursive Democracy: Politics, Policy, and Political Science. Cambridge: Cambridge University Press.

Durkheim, Emile (1964) The Division of Labor in Society. London: The Free Press, p. 28.

Etzioni, Amitai (1998) The Moral Dimension: Toward a New Economics. New York: The Free Press.

Ginter, J.J.C. (1995) "The Alaska Community Development Quota Fisheries Management Program." *Ocean & Coastal Management*, **28**(1-3), pp. 147-63.

Habermas, J. (1987) The Theory of Communicative Action: Volume II Lifeworld and System: A Critique of Functionalist Reasoning Boston: Beacon Press.

Hardin, Garrett (1968) "The Tragedy of the Commons". *Science*, **162**, pp. 243-1248.

Hernes, Hans-Kristian and Mikalsen, K. (2002) "From Protests to Participation? Environment Groups and the Management of Marine Fisheries." Mobilization Special issue, 2002 forthcoming.

Hersoug ,Bjørn, Holm, P. and Rånes, S.A. (1999) "Three Challenges to the Future of Fisheries Management in Norway: ITQ's, Regional Co-management and Eco-labelling." In David Symes (ed.): Alternative Management Systems for Fisheries. Oxford: Fishing News Books.

Hoel, Alf H., Jentoft, S. and Mikalsen, K.H. (1996) "User-Group Participation in Norwegian Fisheries Management. In Robert M. Meyer *et al.* (eds): Fisheries Resource Utilization and Policy. Proceedings from the World Fisheries Congress. New Dehli: Oxford & IBH Publishing, pp. 287- 303.

Kooiman, Jan, van der Vliet, Martijn and Jentoft, Svein (eds): Creative Governance: Opportunities for Fisheries in Europe. Aldershot: Ashgate Publishing.

Jentoft, Svein and Kristoffersen, Trond I. (1989) "Fishermen's Co-management: The Case of the Lofoten Fishery." *Human Organization* Vol **48**, No. 4, pp. 355-365.

Jentoft, Svein and McCay, Bonnie (1995) "User Participation in Fisheries Management: Lessons Drawn from International Experiences." *Marine Policy* Vol. **19**, No. 3, pp. 227-246.

Jentoft, Svein, McCay, Bonnie J. and Wilson, Douglas C. (1998) "Social Theory and Fisheries Co-management. *Marine Policy*, Vol. **22**, No. 4-5, pp. 423-436.

Long, Amanda (1999) "The Marine Stewardship Council Initiative: The Development of a Market Incentive Approach to Achieving Sustainable Fisheries." In David Symes (ed.): Alternative Management Systems for Fisheries. Oxford: Fishing News Books.

McCay, Bonnie J. (1980) "A Fishermen's Cooperative Limited: Indigenous Reseource Management in a Complex Society." *Anthropological Quarterly*. Vol. **53**, No.1, pp. 29-38.

McCay, Bonnie J. and Acheson, James A. (1987) The Question of the Commons; The Culture and Ecology of Communal Resources. Tucson: University of Arizona Press.

McCay, Bonnie J. (1998) Oyster Wars and the Public Trust: Property, Law, and Ecology in New Jersey History. Tucson: University of Arizona Press.

MCay, Bonnie J. and Jentoft, Svein (1998) "Market or Community Failure? Critical Perspectives on Common Property Research." *Human Organization*, Vol. **57** (1), pp. 21-29.

McCay, Bonnie J. (forthcoming) "Emergence of Institutions for the Commons: Contexts, Situations, and Events." In E. Ostrom *et al.*, (eds); The Drama of the Commons. Washington D.C.: National Academy Press.

Mikalsen, Knut H. and Jentoft, Svein (2001) "From User-groups to Stakeholders? The Public Interest in Fisheries Management.". *Marine Policy*, Vol. **25**/4, pp. 281-292.

Nietschmann, Bernard (1997) "Protecting indigenous coral reefs and sea territories, Miskito Coast, RAAN, Nicaragua." In Stan Stevens (ed.): Conservation Through Cultural Survival: Indigenous Peoples and Protected Areas. Washington D .C.: Island Press.

Offe, Claus (2001) "Civil Society and Social Order: Demarcating and Combining Market, State and Community." *Archives Europeennes de Sociologique*. Vol. **41**, 1, pp. 78-9.

Ostrom, Elinor (1990) Governing the Commons. The Evolution of Institutions for Collective Action. Cambridge: Cambridge University Press.

Pálsson, Gísli (1991) Coastal Economies, Cultural Accounts: Human Ecology and Icelandic Discourse. Manchester: Manchester University Press.

Pateman, Carole (1970) Participation and Democratic Theory: Cambridge: Cambridge University Press.

Pitcher, Tony J., Paul J.B. Hart and Pauly, Daniel (1998) Reinventing Fisheries Management. Dordecht: Kluwer Academic Publishers.

Putnam, Robert D. (1993) Making Democracy Work: Civic Traditions in Modern Italy. Princeton: Princeton

University Press.
Ruddle, Kenneth and Johannes, Robert E. (1989) Traditional Marine Resource Management in the Pacific Basin: An Anthology. Jakarta: UNESCO/ROSTSEA, Jln. M.H. Thamrin No. 14.
Scott, W. Richard Institutions and Organizations. London: Sage Publications.
Sproul, John T. (1998) "Green Fisheries: Certification as a Management Tool." In Tony J. Pitcher et al: Reinventing Fisheries Management. Dordrecht: Fish and Fisheries Series 23, Kluwer Academic Publishers, p. 139.
Sutton, Michael (1998) "Harnessing Market Forces and Consumer Power in Favor of Sustainable Fisheries." In Tony J. Pitcher et al: Reinventing Fisheries Management. Dordrecht: Fish and Fisheries Series 23, Kluwer Academic Publishers, pp.125-135.
Swedberg, Richard and Granovetter, Mark (eds) (1992) The Sociology of Economic Life. Boulder: Westview Press, p.9.
Tam, H. (1997) Communitarianism. A New Agenda for Politics and Citizenship. London: MacMillam, Basingstoke.
Taylor, Michael (1987) The Possibility of Cooperation. Cambridge: Cambridge University Press.
Vayda, Andrew P. and Walters, Bradley "Against Political Ecology." *Human Ecology* **27**(1), pp. 167-179.
Walzer, Michael (1995) "The Idea of a Civil Society", in M. Walzer (ed.): Toward a Global Civil Society. Providence: Berhaham Books, p. 7.
Weber, Max (1930) The Protestant Ethic and the Spirit of Capitalism. London: Allen & Unwin.
Wilson, Douglas Clyde and Jentoft, Svein (1999) "Structure, Agency, and Embeddedness: Sociological Approaches to Fisheries Management Institutions". In David Symes (ed.): Alternative Fisheries Management Systems. Blackwell.
Wolfe, Alan (1989) Whose Keeper? Social Science and Moral Obligation. Berkeley: University of California Press.

Conclusion

THE FUTURE OF FISHERIES CO-MANAGEMENT

SUSAN HANNA
Department of Agricultural and Resource Economics, Oregon State University, Corvallis, Oregon, USA

1. INTRODUCTION

The early twenty-first century is a time of transition for fishery management. The transition takes place within a new public context of values and preferences expressed through more rigorous requirements at national and international scales. These requirements affect how information is generated, how fishery interests are coordinated and how management decisions are made. And in these effects, the changes engage co-management. This chapter addresses the nature and sources of change affecting fishery management worldwide. It applies emergent themes from the chapters of this book to identify the dilemmas facing co-management in the context of change, the properties of co-management that will influence the resolution of the dilemmas, and the requirements for sustaining co-management over the long term.

2. THE NATURE OF CHANGE

Fishery management today is the product of a history that has been a search for balance between fish stock protection, management effectiveness, social equity and economic efficiency. It has also been a search for balance in the distribution of authority and power.

The equilibrium of these competing interests has been a delicate, and often unstable, one. Maintaining management equilibrium is inherently difficult because of the dynamism and uncertainty of marine biophysical systems. The difficulty is further exacerbated by the uncertainties associated with the human components of fisheries, in particular the intrinsic conflict between the long-term societal objectives of fishery sustainability and the short-term objectives of individuals and governments. While individual and collective goals may coincide in the long-term, they are often at odds in the short-term.

Social equity is complicated by the diversity of objectives, interests and values among fishery stakeholders and the larger public. What is considered fair and equitable is often a matter of perspective, and perspectives among groups vary widely. Attaining economic efficiency in management is similarly elusive. Various sources of uncertainty may elevate costs of management associated with information gathering, coordination, implementation, conflict resolution, monitoring and enforcement to levels that exceed the benefits they generate.

Global interdependencies and broadened constituencies are introducing new complexities into fishery management. Changes in management philosophies at the international scale are felt at national, regional and community scales through political and market linkages between fishing economies. Changes in social values toward fisheries and marine ecosystems also extend across interest groups and national boundaries.

What is the role of fishery co-management as it enters this complicated arena at a time of change? The answer is in part determined by international trends in fishery management, and part by the particular properties of co-management. The future path for co-management lies in the interaction between the two.

3. THE SOURCES OF CHANGE

The development of 20th Century fishery management has been shaped by both national and international influences. While national paths of fishery management have varied in the particulars of sequencing and timing, they share similar general properties that have created international management trends.

In many countries, the post-war race for fish led to over-investments in fishing capacity that undermined the effectiveness of national regulations and international agreements. Investment in the knowledge base of management was concentrated in research to understand the dynamics of fish stocks. Investment in management capital - the knowledge and skills necessary for effective decisionmaking - was similarly concentrated in professional manager classes.

The result of this investment strategy across individual nations was a global condition of fishery management subjected to the pressures of too much fishing capital demanding fine-resolution estimates of allowable catch from specialized fishery managers. In the late 1990s the FAO reported that in the period between the early 1950s and the early 1990s, world fisheries had been transformed from being about 60% under-exploited to about 60% over-exploited (FAO, 1997).

Widespread recognition that such a condition was unsustainable led to new international agreements and changes in management practices. The Rome Consensus on World Fisheries, the Code of Conduct on Responsible Fishing, and the Kyoto Declaration adopted at the Conference on the Sustainable Contribution of Fisheries to Food Security contain guidelines for ending overfishing, reducing bycatch and discards, reducing fishing capacity, strengthening governance, and strengthening the scientific basis for ecosystem management (FAO, 1997). These guidelines reflect international consensus on the need for a more precautionary management approach to fisheries that focuses on conservation, accommodates variation and explicitly attempts to manage risk.

Within nations, other important trends reflect the poor state of fishery management and create conditions favouring co-management. Interest in cost-recovery strategies for management is increasing due to rising costs of management related to expanding management responsibilities, escalating numbers of stock assessments, and intensifying conflicts among user groups. Cost recovery creates an intrinsic pull toward co-management processes, as user groups funding research and management demand accountability for research and for its cost-effective application in management. A logical government response to user-group demands for accountability is to bring them into the management system by strengthening their participation in decisions about research and management.

Even in the absence of cost-recovery policies, governments are increasingly devolving fishery management authority to user groups. As the chapters in this book illustrate, devolution has a range of motivations, including a desire to shift intractable problems to user groups, a lack of government capability to finance management, an assumption that conflict resolution and enforcement will be more effective away from the centre, and sometimes a desire to increase democratization of the political system. Whatever the motivation for devolution in a particular setting, the trend toward devolution is clear.

Within some countries a parallel trend to devolution is the establishment of systems of property rights in fisheries. These property rights may range from community-based to private, and may take the forms of community development quotas, territorial use rights, gear quotas, or ITQs. The goal of establishing systems of property rights is to provide assurance of tenure in the fishery and to end the race for fish. When property rights achieve this goal, they are similar to cost-recovery in their effect of creating expectations among user groups that they have legitimate claims to participate in management decisions. Co-management is one process by which participation in management can strengthen. It is compatible with any property rights system.

A growing international trend is the interest in removing some ocean areas from fishery use. These exclusions are implemented through spatial protections such as marine protected areas. The push to implement nonuse areas is often driven by the perception that conventional fishery management has failed to sustain fisheries. Where implemented, marine protected areas are likely to be areas of ocean of a large enough scale to be problematic for management through co-management processes. However, as Chapter 12 demonstrates, scope for co-management is provided in the design and enforcement aspects of marine protected areas.

4. CO-MANAGEMENT IN THE CONTEXT OF CHANGE

The expansion of co-management processes in fisheries reflects a philosophical change about who should make fishery management decisions as well as a practical response to poor management performance under more centralized authority. Implementing philosophical change while repairing damaged management performance sets up difficult conditions for co-management.

The international trend toward participatory management processes is layered against a background of fishery management ineffectiveness. Co-management processes are often implemented when other forms of management are in trouble: when conflict among user groups increases, regulation effectiveness decreases, and management costs inflate. These conditions become the baseline against which co-management processes are implemented, creating serious challenges to the development of effective new processes.

4.1. The Co-Management Dilemma

Co-management faces a number of dilemmas in reconciling its properties and functions with trends in the larger fishery management context. Chapters 2 and 15, for example, describe how the tendency to address many fishery management problems on large scales has led to a reliance on aggregated scientific information that is difficult to relate to the more location-specific information on which co-management decisions are likely to be based. Difficulties can also arise, as observed in Chapter 14, between co-management and the movement toward administrative decentralization, a set of problems that is nicely illustrated in Chapter 7 by the problems co-management is facing within the European Common Fisheries Policy. The growing influence of the environmental movement on fisheries management is illustrated both as possibility in Chapter 12's discussion of marine protected areas and as problem in Chapter 10's report on the distrust that environmental groups in Australia and New Zealand have toward participatory approaches to management.

How we resolve these and other dilemmas will strongly influence the future path and effectiveness of co-management processes. Tending to the design requirements of effective fishery co-management in the small scale while at the same time remaining adaptive to developments in fishery management in the large scale poses for those developing and implementing co-management a set of difficult problems. The first, and most basic problem, is presented in Chapter 4 as a definitional one: the use of the term 'co-management' to cover a wide variety of arrangements weakens co-management's meaning and thereby the understanding of its functions or the development of design improvements. Other problems identified throughout the chapters relate to institution building, institutional integration, management objectives, representation and participation, knowledge production, behavioural incentives, conflict resolution and operational support.

Institution Building

- How to build new institutional processes while recovering overexploited resources.
- How to develop exclusionary property rights strong enough to conserve fish stocks when fisheries are seen as 'sink' resources to increasing populations of fishermen.

Institutional Integration

- How to integrate co-management with rights and responsibilities embedded in national and international institutions.
- How to accommodate international cooperation on fisheries issues while developing decentralized processes of decisionmaking that strengthen institutions at community levels.
- How to apply co-management processes to large-scale spatial management approaches, like MPAs.

Management Objectives

- How to broaden the definition of community to include non-spatial alignments of interests and reconcile those communities with the spatial basis of ecosystems.
- How to develop management objectives that accommodate divergent views of social equity.
- How to accommodate tradition and accomplish institutional flexibility and adaptation.

Representation and Participation

- How to develop representation and participation on a scale that is administratively effective while also ecologically functional.
- How to build trust, develop management capacity and maintain management commitment across diverse participants.

Knowledge Production

- How to integrate diverse sources of knowledge at the same time that more restrictive regulations and precautionary limits require finer-scale understanding of fish populations and ecological system function.
- How to reconcile different approaches of scientists and fishermen to information gathering.

Behavioural Incentives

- How to contain coordination costs and maintain incentives consistent with stewardship goals at community, regional, national and international scales.
- How to maintain a balance between accountability and independence.
- How to minimize opportunities to undermine or circumvent management decisions.

Conflict Resolution

- How to develop conflict resolution processes that accommodate heterogeneous participants and multidimensional change while maintaining the legitimacy of management.

Operational Support

- How to obtain operational support for co-management processes sufficient to fund basic management tasks in a climate of government reluctance to invest further in fishery management.

4.2. The Properties of Co-management

The dilemmas facing the expansion of co-management processes in fisheries will need to be resolved in ways that reflect the basic requirements of fishery management to coordinate multiple objectives and tasks, generate information, develop decision processes and provide a long-term perspective for fishery sustainability.

The resolution of management dilemmas through co-management processes must also reflect co-management's properties. The chapters collectively emphasize two important general properties of co-management: its decentralized decision-making authority and its dynamic as a process. Co-management is presented as a valuable process not only for the decentralized benefits of participation collaboration, democracy and fairness, but also for its dynamism in addressing technical management problems through learning and adaptation.

The chapters also, in their focus on specific aspects of co-management, highlight important properties of co-management that must be accommodated if co-management is to be reconciled with the larger fishery management context. These include the motivation for co-management implementation, institutional integration, scale, knowledge base, participation and representation, and management performance.

4.2.1 Motivation for Co-Management Implementation

Several of the chapters note that the motivations for developing co-management processes in fisheries often complicate their implementation. That these motivations are often more the negative aspects of *status quo* fishery management then the positive aspects of participatory decision processes is illustrated in contrasting ways by the emergence of co-management in Africa within a legacy of 'participatory' colonial natural resource management (Chapter 5), in Europe as a part of communities' efforts to deal with an emerging mega-state (Chapter 7) and in Canada as a response to devastating stock collapses (Chapter 9). Governments may reach the limits of their ability of implement and enforce regulations that prevent fish stock overexploitation because of their inability to operate effectively at very small scales (Chapter 11). They may respond to increases in user group conflict by transferring responsibility for conflict resolution to stakeholder groups, a common case in Latin America (Chapter 8). They may react to the increasing cost of management coordination and implementation by shifting those costs onto co-management processes (Chapter 3). Asian governments, for example, are taking this approach to address the extra costs of addressing both fisheries and broader marine ecosystem problems (Chapter 6).

Governments may also have conflicting policies of developing co-management processes while at the same time undermining their effectiveness by encouraging the use of fishery resources as relief-valves for underemployment on other sectors (Chapter 5). Redistributing decision making power and access to resources is a difficult political goal when vested interests are at stake Hence, real change may require a considerable outside push, such as new legislation recognizing indigenous resource rights (Chapters 4 and 9), demands for a growth of an important new sector such as recreational fishing (Chapter 13), or pressure from environmental groups (Chapters 8 and 12) and development donors (Chapters 5 and 8).

At the same time, there is growing recognition by governments of co-management's positive potential to provide incentives for stewardship. Governments, and particularly those in developing regions, are increasingly supportive of the co-management model. The support is expressed in both legislative and programmatic initiatives. Co-management's potential is enhanced by participation in the development of management objectives, regulatory design, implementation and enforcement. As described in Chapter 1, experience in the field of community development offers valuable lessons in how to develop effective participatory approaches that address both heterogeneity and conflict.

Motivations for implementing co-management, both negative and positive, determine the initial conditions facing co-management. Co-management may be saddled with the burden of fishery reconstruction even as it attempts to become an effectively functioning process.

4.2.2 Institutional Integration

The importance of ensuring integration of co-management processes in fisheries with higher and lower levels of authority is underscored by many of the examples presented in this book. The Australia and New Zealand cases (Chapter 10) show that co-management is fragile and easily undermined at early stages, creating a clear need for strong government support from higher levels and for trust among participants. The fragility also places a value on transparency and subsidiarity, as emphasized in the Introduction, where rights are visible and consistent across levels. This may, as experience in Asia has shown (Chapter 6), require that new legal, institutional and administrative arrangements be developed to recognize and clarify co-management's role. Signals and feedbacks must be clear and transparent to all participants.

If under co-management the collective rights of the group supercede individual rights, as argued in Chapter 4, institutional integration must also refer to internal integration of co-management participants and include a recognition that co-management may involve multiple horizontal arrangements with other groups. Because co-management processes involve disparate interests, values and objectives each subject to continual change and adaptation, the internal integration is necessarily multidimensional and dynamic and will also involve conflict (Chapter 11). Knowing this, it is particularly important that mechanisms be developed to integrate participants in the co-management process if a collective persona is to be maintained and institutional fragmentation is to be avoided.

4.2.3 Scale

Scale plays a critical role in fishery co-management processes (Chapter 11). Much of co-management's vulnerability in its early stages of implementation appears to relate to questions of scale: the optimal scale of representation (Chapters 1 and 16), and whether decision authority encompasses a scale large enough to relate to ecosystem functions (Chapters 2 and 9) but small enough to be cost-effective in coordinating among participants (Chapter 3). Fragmentation of management authority is common, as illustrated so dramatically by the European Union case (Chapter 7), and misfits between the scale of the ecosystem and its management may be a major contributor to the mixed effectiveness of co-management processes in practice (Chapter 9).

The co-management literature, and the chapters in this book as well, contain widespread reference to the scale concepts 'local' and 'community'. These concepts are slippery to define. What is 'local' can vary according to technology, government structure, demographic patterns and economic linkages. However local is defined, not all fishery management problems can be addressed a local scale, as demonstrated by fishery management in the European Union (Chapter 7), where the need for international cooperation is great. For all fishing nations, international agreements are increasingly important mechanisms in the management of fisheries. The construction of 'community' is similarly changeable. A community can be defined by geo-political boundaries, by business interests, or even by professional training. Each may encompass different scales, and each contributes to the rich foundation of social capital that can serve as a basis for sustainable fishery management (Chapter 17).

Scale also relates to the generation, analysis, legitimacy and interpretability of the knowledge base for decisionmaking. Arguments between fishermen and stock assessment scientists over the appropriate scale of observation are endemic in fishery management, as noted in Chapters 2 and 15. As Chapter 2 illustrates, much of the argument may be traced to the different perspectives on scale brought to management by the two groups, with stock assessment scientists focussed on large-scale dynamics and fishermen attention placed at the vessel operational scale. Co-management processes offer the potential to reconcile these different foci, build trust among participants and enhance the legitimacy of the knowledge base by creating more opportunities for scientists and fishermen to interact and collaborate (Chapter 15). These are among the lessons learned from the field of community development that have direct educational value to researchers and practitioners of fishery co-management (Chapter 1).

4.2.4 Knowledge Base

Co-management processes bring user groups much closer to the knowledge base of fishery management. Co-management participants become sources of knowledge as well as users of that knowledge in management. For many management systems this represents an important change in how knowledge is produced and used (Chapters 2 and 15).

Under more centralized management, scientific knowledge is produced using formal processes in conjunction with standardized management protocols. Knowledge is produced to cover large geographic areas and long time periods. Data collection, analysis and interpretation become the purview of specialists trained in sampling, population dynamics, and risk assessment. As highlighted by the example of fishery science (Chapter 2), the specialization of knowledge often leads to a distancing of user groups from the knowledge base used to develop the regulations that affect their livelihoods, providing scope for user groups to challenge its legitimacy. Distortion of communication among groups through institutional rules is also common (Chapter 15).

The integration of user groups as co-managers introduces different types of information into the knowledge production process. A nontrivial and central question of this integration is how to combine the large-scale scientific knowledge with detailed localized knowledge of fishery participants in ways that enhance the effectiveness of management (Chapter 2). As Chapter 15 demonstrates, the definition of relevant knowledge, how is it generated, how is it analysed, and to whom the results of that analysis are presented are operational questions central to the resulting knowledge base of fishery management. Under co-management processes, the authority to decide these questions broadens from a group of specialists to all participants, raising important questions about the requirements for effective and legitimate representation (Chapter 16).

Participating in the production of knowledge is a way to shape management decisions ultimately based on that knowledge. Yet there is a natural tension between the production of scientific knowledge and participation in management decisions, as discussed in Chapter 15. To overcome this tension the respective contributions of scientific process versus 'local knowledge' must be accepted among participants. As well, the influence of social power and relationships among user groups on knowledge must be acknowledged. When co-managers become co-producers of knowledge, building trust and legitimacy among all participants becomes a not inconsiderable challenge.

4.2.5 Participation and Representation

Co-management is based on the active participation of all fishery interests, which naturally generates questions about scale and representation. Many of these questions are addressed in the chapters of this book. For example, the Introduction points out that co-management can encompass a variety of levels of participation, and Chapter 1 concludes that what is key is to ensure that participation is 'authentic participation', echoed by Chapter 16's discussion of 'effective representation'. Thus the emphasis in many co-management development projects, such as those described for Africa (Chapter 5) and Southeast Asia (Chapter 6), placed on developing community leadership and providing appropriate skill sets to co-managers.

As with any participatory process, the question of representation is key to perceptions of legitimacy. Representation is made necessary by the infeasibility of direct participation of all users in any but the smallest fisheries (Chapter 16). For co-management to be accepted as a valid process of management, all interests must be represented. But fisheries are traditionally composed of highly individualistic enterprises, making the task of defining the full range of interests a problematic one. The identification of an interest is not necessarily straightforward and ensuring the full range of interests is represented is an even more difficult task. The appearance that a category of interests is represented does not necessarily mean that the interests of members of that category are visible in decisions.

For example, Chapter 13 illustrates that recreational fishers are important fishery user groups who are often un-represented by formal organizations. They may lack interest in participation as co-managers, and yet their interests are legitimate and in need of representation. Without formal organizations, how are the categories or recreational interests defined and how are reasonable representatives of those interests selected?

Representation necessarily means selection: Chapter 16 addresses the question of legitimate methods to perform this selection. Who are the stakeholders to be represented? Should co-management involve only those directly participating in fisheries or should new claimants of representation, for example non governmental organizations, be recognized? If representation is broadened beyond traditional participants, will the balance between accountability and independence be altered? How will the scale of the co-management process affect the behaviour of representatives? These questions about representation must all be answered within the context of a particular co-management process.

Questions of representation also relate to the dynamics of conflict and power, the focus of Chapter 11. Conflicts arise within communities, among user groups, among state representatives, and between states, communities and users. Conflict is likely to be present to a significant degree when co-management is implemented, as it often is, at the point of fishery overexploitation. Conflict can motivate participation in co-management, but left unresolved it can also undermine it. As the examples of co-management in Africa (Chapter 5), Southeast Asia (Chapter 6), Latin America (Chapter 8) and North America (Chapter 9) all illustrate, mechanisms of conflict resolution consistent with participation and effective representation are required.

4.2.6 Management Performance

The performance of fishery management is widely acknowledged to depend importantly on the acceptance and support of stakeholders. In fishery management systems based on centralized control, stakeholder support for management is often assessed through the degree of compliance with regulations. Stakeholder support under co-management is more complex. When stakeholders are also co-managers, support is expressed not only in compliance with regulations, but also in the leadership skills applied to the tasks of making contributions to the design of regulations, decisions about their implementation, and evaluation of their effectiveness (Chapter 14).

Assessment of management performance has in the past been limited to a small group of biological indicators of the status of stocks. The implication for assessment of using stakeholders as co-managers is that new performance indicators could be developed to reflect additional criteria important to fishery participants. These would include, as both Chapter 2 and 9 discuss, general indicators of the economic and social performance of fisheries and of the effectiveness of the co-management process. The objective would be to define a set of indicators that would be acceptable and understandable to both user-groups and scientists and that could be produced cost-effectively with available management resources. Because they would be developed internally, the particular mix of indicators would depend on the fishery.

5. SUSTAINING CO-MANAGEMENT

Co-management shares, with other processes of fishery management, examples of failures despite good intentions. The lesson from these failures is clear: sustaining co-management processes will require much more than good intentions and good feelings about participation. It will require integration of its properties with larger fishery trends and contexts.

Sustaining co-management will mean resolving the challenging dilemmas facing it, and the requirements for this resolution are multidimensional. They include clarity of roles and authorities, management capital, fiscal resources and a capacity for learning and adaptation all consistent with the social and ecological context of the fishery. They also, importantly, require multidisciplinary research to develop solutions to these problems. Chapter 17 emphasizes that fishery management begins with and depends on civil society, and that the qualities of civil society that form the basis of community, culture and management should be nourished by governments in the development of co-management.

In the past, processes labelled 'co-management' could include many levels of user-group participation designed to assist government in meeting its resource management objectives. Against this history and its embedded expectations, genuine co-management will require that specific areas of authority be clearly specified to prevent repetition of old patterns (Chapters 1, 4 and 9). This is particularly important when co-management is implemented within a context of existing traditional authorities as in the Africa cases (Chapter 5) or competing new authorities, as in the case of marine protected areas (Chapter 12), the entry of recreational participation (Chapter 13) and other new constituents (Chapter 16). Unless roles and responsibilities are clearly specified, attempts to capture authority will consume management resources and divert management from its central task. It is especially important, as illustrated by Chapter 10, that the design of co-management should, through clarity about roles and responsibilities, anticipate and prevent opportunities to sabotage decision authority.

To be maintained, co-management, like all management, is dependent on its management capital to produce a flow of services. Management capital is the leadership, skills, and capabilities of managers, attributes not necessarily present in a broader group of stakeholders (Chapters 1 and 3). When management capital among co-managers is inadequate to the task, it must be developed if co-management is to be sustained. Chapter 17 emphasizes the need to nurture the fishing cultures and communities that are a reservoir of social capital for sustainable management. Tools in biology, economics and decision processes are another necessary part of management capital. Further, experience in Australia and New Zealand (Chapter 10) demonstrates that education of agency staff as well as stakeholders on the characteristics and properties of co-management is also important to the creation of consistent and sustainable expectations for management.

Finally, it is important to emphasize, as Chapter 10 does, the significance of genuine operational support to sustaining co-management processes. The need for adequate fiscal resources for co-management is particularly important to note in light of the tendency of governments to implement co-management as a cost-saving measure. Additionally, Chapters 1, 8 and 14 show that other resources, such as enabling legislation and government services, are needed to ensure that management functions are maintained. Over the long run, however, movements toward fiscal self-sufficiency of co-management processes will provide incentives for participants to seek cost-effective approaches that contain transaction costs (Chapters 3 and 6).

The analysis of co-management processes in application worldwide has generated a body of general knowledge useful to their design. But sustaining co-management in a particular setting will also depend on a process of learning and adaptation in place. Learning will require that co-management processes be flexible enough to allow experimentation to produce new knowledge, manage conflict (Chapter 11) and to allow adaptation on the basis of what is learned. The Latin American experience (Chapter 8) demonstrates the need for such flexibility on the part of government. However, it should also be noted that flexibility in rulemaking may create costs in lost consistency as well as benefits in adaptability and learning. To sustain co-management, an acceptable balance between these costs and benefits must be found (Chapter 3).

The conditions for sustaining co-management processes must, in the end, reflect the important dimensions of the human-fishery interface. It is at the interface that we read the success and failures of management actions. Co-management, as a human endeavour, will need to be sustained in the important human spheres: social, cultural, economic and political. As an influence over ecosystem health, it must also be sustained in the ecological spheres of community structure and function. This challenging task will require resources, knowledge, and a generous persistence among all participants.

ACKNOWLEDGEMENTS

This chapter was funded by the NOAA Office of Sea Grant and Extramural Programmes, US Department of Commerce, under grant number NA16RG1039 (project number RCF-10), and by appropriations made by the Oregon State legislature. The views expressed herein do not necessarily reflect the views of any of those organizations.

REFERENCES

Food and Agriculture Organization (FAO) (1997) *The State of World Fisheries and Aquaculture 1996*. Rome: Food and Agriculture Organization of the United Nations.

Index